Non-Linear Differential Equations and Dynamical Systems

Mathematics and Physics for Science and Technology

Series Editor: L.M.B.C. Campos
Director of the Center for Aeronautical
and Space Science and Technology
Lisbon University

Volumes in the series:

Topic A – Theory of Functions and Potential Problems

Volume I (Book 1) – Complex Analysis with Applications to Flows and Fields
L.M.B.C. Campos

Volume II (Book 2) – Elementary Transcendentals with Applications to Solids and Fluids
L.M.B.C. Campos

Volume III (Book 3) – Generalized Calculus with Applications to Matter and Forces
L.M.B.C. Campos

Topic B – Boundary and Initial-Value Problems

Volume IV – Ordinary Differential Equations with Applications to Trajectories and Vibrations
L.M.B.C. Campos

Book 4 – Linear Differential Equations and Oscillators
L.M.B.C. Campos

Book 5 – Non-Linear Differential Equations and Dynamical Systems
L.M.B.C. Campos

Book 6 – Higher-Order Differential Equations and Elasticity
L.M.B.C. Campos

Book 7 – Simultaneous Differential Equations and Multi-Dimensional Vibrations
L.M.B.C. Campos

Book 8 – Singular Differential Equations and Special Functions
L.M.B.C. Campos

Book 9 – Classification and Examples of Differential Equations and their Applications
L.M.B.C. Campos

For more information about this series, please visit: https://www.crcpress.com/Mathematics-and-Physics-for-Science-and-Technology/book-series/CRCMATPHYSCI

Mathematics and Physics for Science and Technology

Volume IV

Ordinary Differential Equations with Applications to Trajectories and Vibrations

Book 5

Non-Linear Differential Equations and Dynamical Systems

By

L.M.B.C. Campos

*Director of the Center for Aeronautical
and Space Science and Technology
Lisbon University*

CRC Press
Taylor & Francis Group
Boca Raton London New York

CRC Press is an imprint of the
Taylor & Francis Group, an **informa** business

CRC Press
Taylor & Francis Group
6000 Broken Sound Parkway NW, Suite 300
Boca Raton, FL 33487-2742

First issued in paperback 2023

Library of Congress Cataloging-in-Publication Data

Names: Campos, Luis Manuel Braga da Costa, author.
Title: Non-linear differential equations and dynamical systems/Luis Manuel Braga da Costa Campos.
Description: First edition. | Boca Raton, FL : CRC Press/Taylor & Francis Group, 2018. | Includes bibliographical references and index.
Identifiers: LCCN 2018047806| ISBN 9780367137199 (hardback : acid-free paper) | ISBN 9780429028991 (ebook)
Subjects: LCSH: Differentiable dynamical systems. | Differential equations, Nonlinear.
Classification: LCC QA614.8 .C365 2018 | DDC 515/.355–dc23
LC record available at https://lccn.loc.gov/2018047806

**Visit the Taylor & Francis Web site at
http://www.taylorandfrancis.com**

**and the CRC Press Web site at
http://www.crcpress.com**

ISBN 13: 978-1-032-65372-3 (pbk)
ISBN 13: 978-0-367-13719-9 (hbk)
ISBN 13: 978-0-429-02899-1 (ebk)

DOI: 10.1201/9780429028991

to Leonor Campos

Contents

Notes and Tables

Notes

Tables

Preface

Volume IV (*"Ordinary Differential Equations with Applications to Trajectories and Oscillations"*) is organized like the preceding three volumes of the series *"Mathematics and Physics Applied to Science and Technology"*: (volume III) *"Generalized Calculus with Applications to Matter and Forces"*; (volume II) *"Elementary Transcendentals with Applications to Solids and Fluids"*; (volume I) *"Complex Analysis with Applications to Flows and Fields"*. These three volumes on complex, transcendental, and generalized functions complete Topic A, "Theory of Functions and Potential Fields". Topic B, "Boundary and Initial-Value Problems" starts with volume IV on *"Ordinary Differential Equations with Applications to Trajectories and Oscillations"*.

Volume IV consists of ten chapters: (i) the odd-numbered chapters present mathematical developments; (ii) the even-numbered chapters contain physical and engineering applications; (iii) the last chapter is a set of 20 detailed examples of (i) and (ii). The first book *"Linear Differential Equations and Oscillators"* of volume IV corresponds to the fourth book of the series and consists of chapters 1 and 2 of volume IV. The present second book, *"Non-linear Differential Equations and Dynamical Systems"*, corresponds to the fifth book of the series and consists of chapters 3 and 4 of volume IV.

Chapter 1 described linear differential equations of any order with constant or power coefficients, and chapter 3 focuses on non-linear differential equations of the first-order, including variable coefficients, with extensions to differentials of order higher than the first and in more than two variables, and applications to the representation of vector fields by potentials. Chapter 2 discussed linear oscillators with damping/amplification and forcing with constant coefficients, including ordinary resonance. Chapter 4 considers linear oscillators with variable coefficients leading to parametric resonance, and non-linear oscillators leading to non-linear resonance amplitude jumps and hysteresis. Together with electromechanical dynamos and non-linear damping, these are examples of dynamical systems that may have bifurcations leading to chaotic motions.

Organization of the Book

The chapters are divided into sections and subsections, for example chapter 3, section 3.1, and subsection 3.1.1. The formulas are numbered by chapters in curved brackets; for example, (3.2) is equation 2 of chapter 3. When referring to volume I the symbol I is inserted at the beginning, for example: (i) chapter I.36, section I.36.1, subsection I.36.1.2; (ii) equation (I.36.33a). The final part of each chapter includes: (i) a conclusion referring to the figures as a kind of visual summary; (ii) the note(s), list(s), table(s), diagram(s), and classification(s) as additional support. The latter (ii) apply at the end of each chapter, and are numbered within the chapter (for example note N3.1, table T4.1); if there is more than one they are numbered sequentially (for example, notes N4.1 to N4.13). The chapter starts with an introductory preview, and related topics may be mentioned in the notes at the end. The "Series Preface" and "Mathematical Symbols" in the first book of volume IV are not repeated, and the "Physical Quantities", "References", and "Index" focus on the contents of the present second book of volume IV.

Acknowledgments

The fourth volume of the series justifies renewing some of the acknowledgments also made in the first three volumes, to those who contributed more directly to the final form of the book, namely, Ms. Ana Moura, L. Sousa, and S. Pernadas for help with the manuscripts; Mr. J. Coelho for all the drawings, and at last, but not least, to my wife as my companion in preparing this work.

About the Author

 **L.M.B.C. Campos** was born on March 28, 1950, in Lisbon, Portugal. He graduated in 1972 as a mechanical engineer from the Instituto Superior Tecnico (IST) of Lisbon Technical University. The tutorials as a student (1970) were followed by a career at the same institution (IST) through all levels: assistant (1972), assistant with tenure (1974), assistant professor (1978), associate professor (1982), chair of Applied Mathematics and Mechanics (1985). He has served as the coordinator of undergraduate and postgraduate degrees in Aerospace Engineering since the creation of the programs in 1991. He is the coordinator of the Scientific Area of Applied and Aerospace Mechanics in the Department of Mechanical Engineering. He is also the director and founder of the Center for Aeronautical and Space Science and Technology.

In 1977, Campos received his doctorate on "waves in fluids" from the Engineering Department of Cambridge University, England. Afterwards, he received a Senior Rouse Ball Scholarship to study at Trinity College, while on leave from IST. In 1984, his first sabbatical was as a Senior Visitor at the Department of Applied Mathematics and Theoretical Physics of Cambridge University, England. In 1991, he spent a second sabbatical as an Alexander von Humboldt scholar at the Max-Planck Institut fur Aeronomic in Katlenburg-Lindau, Germany. Further sabbaticals abroad were excluded by major commitments at the home institution. The latter were always compatible with extensive professional travel related to participation in scientific meetings, individual or national representation in international institutions, and collaborative research projects.

Campos received the von Karman medal from the Advisory Group for Aerospace Research and Development (AGARD) and Research and Technology Organization (RTO). Participation in AGARD/RTO included serving as a vice-chairman of the System Concepts and Integration Panel, and chairman of the Flight Mechanics Panel and of the Flight Vehicle Integration Panel. He was also a member of the Flight Test Techniques Working Group. Here he was involved in the creation of an independent flight test capability, active in Portugal during the last 30 years, which has been used in national and international projects, including Eurocontrol and the European Space Agency. The participation in the European Space Agency (ESA) has afforded Campos the opportunity to serve on various program boards at the levels of national representative and Council of Ministers.

His participation in activities sponsored by the European Union (EU) has included: (i) 27 research projects with industry, research, and academic institutions; (ii) membership of various Committees, including Vice-Chairman of the Aeronautical Science and Technology Advisory Committee; (iii) participation on the Space Advisory Panel on the future role of EU in space. Campos has been a member of the Space Science Committee of the European Science Foundation, which works with the Space Science Board of the National Science Foundation of the United States. He has been a member of the Committee for Peaceful Uses of Outer Space (COPUOS) of the United Nations. He has served as a consultant and advisor on behalf of these organizations and other institutions. His participation in professional societies includes member and vice-chairman of the Portuguese Academy of Engineering, fellow of the Royal Aeronautical Society, Astronomical Society and Cambridge Philosophical Society, associate fellow of the American Institute of Aeronautics and Astronautics, and founding and life member of the European Astronomical Society.

Campos has published and worked on numerous books and articles. His publications include 10 books as a single author, one as an editor, and one as a co-editor. He has published 152 papers (82 as the single author, including 12 reviews) in 60 journals, and 254 communications to symposia. He has served as reviewer for 40 different journals, in addition to 23 reviews published in *Mathematics Reviews*. He is or has been member of the editorial boards of several journals, including *Progress in Aerospace Sciences, International Journal of Aeroacoustics, International Journal of Sound and Vibration,* and *Air & Space Europe.*

Campos's areas of research focus on four topics: acoustics, magnetohydrodynamics, special functions, and flight dynamics. His work on acoustics has concerned the generation, propagation, and refraction of sound in flows with mostly aeronautical applications. His work on magnetohydrodynamics has concerned magneto-acoustic-gravity-inertial waves in solar-terrestrial and stellar physics. His developments on special functions have used differintegration operators, generalizing the ordinary derivative and primitive to complex order; they have led to the introduction of new special functions. His work on flight dynamics has concerned aircraft and rockets, including trajectory optimization, performance, stability, control, and atmospheric disturbances.

The range of topics from mathematics to physics and engineering fits with the aims and contents of the present series. Campos's experience in university teaching and scientific and industrial research has enhanced his ability to make the series valuable to students from undergraduate level to research level.

Campos's professional activities on the technical side are balanced by other cultural and humanistic interests. Complementary non-technical interests include classical music (mostly orchestral and choral), plastic arts (painting, sculpture, architecture), social sciences (psychology and biography), history (classical, renaissance and overseas expansion) and technology (automotive, photo, audio). Campos is listed in various biographical publications, including *Who's Who in the World* since 1986, *Who's Who in Science and Technology* since 1994, and *Who's Who in America* since 2011.

Physical Quantities

The location of first appearance is indicated, for example "2.7" means 'section 2.7' "6.8.4" means "subsection 6.8.4", "N8.8" means "note 8.8", and "E10.13.1" means "example 10.13.1".

1 Small Arabic Letters

a — moment arm: 4.7.3

a_n — coefficients of a series: 4.4.10

h — amplitude of excitation of parametric resonance: 4.3.2

q — non-linearity parameter for pendular motion: 4.7.13

2 Capital Arabic Letters

A — anharmonic factor: 4.4.10

$\vec{A}$ — vector potential: 3.9.9

D — drag force: 4.8.13

F — force: 3.9.11

I — moment of inertia: 4.7.3

J — electric current: 4.7.1

L — induction of a coil or self: 4.4.7.1

— lift force: 4.8.13

Q — heat: 3.9.11

R — electrical resistance: 4.7.1

S — entropy: 3.9.11

— dynamo parameter: 4.7.3

T — temperature: 3.9.11

— thrust: 4.8.13

U — internal energy: 3.9.11

W — work: 3-9.11
  — weight: 4.8.13

---

3 Small Greek Letters

α — angle of cylindrical helix: 3.9.5
  — coefficient of the cubic term of the quartic potential: 4.5.1
β — coefficient of the quartic term in the biquadratic potential: 4.4.4
γ — dimensionless non-linearity parameter: 4.4.8
ω_0 — natural frequency: 4.3.2
ω_e — excitation frequency of parametric resonance: 4.3.2
ψ — non-linearity parameter for a quartic potential: 4.6.1

---

4 Capital Greek Letters

Φ — scalar potential: 3.9.9
  — mechanical potential energy: 4.4.2
Θ — Euler potential: 3.9.9
Ω — angular velocity of rotation: 4.7.1
Ψ — Euler potential 3.9.9
Ξ — Clebsch potential: 3.9.9

3

Differentials and First-Order Equations

If a differential equation involves derivatives with regard to one (several) variables, it is an ordinary (partial) differential equation; ordinary differential equations will be considered in the present volume IV, and partial equations in the next volume V. In the dynamical example of the second-order linear system with constant (variable) coefficients (chapter 2(4)), the independent variable is time only, leading to an ordinary differential equation. If the particle or system has one (several) degrees of freedom, for example its position is specified by one (several) coordinates, then the motion is specified by an equation (system of equations); thus the solution of an ordinary differential equation (system of N simultaneous ordinary differential equations), is a function (N functions) of one variable. The derivative of highest order appearing in a differential equation specifies the order of that equation; for example in the dynamics of a particle with one degree of freedom the ordinary differential equation is of order two, because the highest order derivative of position with regard to time is the second, specifying the acceleration. Not all differential equations are readily solvable; this is why it is important to classify them into (i) ordinary or partial, (ii) equations or systems, (iii) order one or higher, and (iv) particular standards or sub-types. The aim is to identify classes of differential equations that have certain properties, making possible specific methods of solution. The starting point was the ordinary linear differential equations of any order N with constant or homogeneous coefficients that have a characteristic polynomial of degree N (chapter 1); by analogy other equations with one characteristic polynomial. Considering (i) linear differential equations with variable coefficients or (ii) non-linear differential equations, it is simpler to start with those of first order (chapter 3).

After the discussion of some general properties of first-order differential equations (section 3.1), methods of solution are presented for eight classes or standards: (i) the separable equation (section 3.2), for which the derivative is the ratio of two separate functions of the independent and dependent variable, which can be solved by quadratures, that is the solution reduces to one (or two) integration(s); (ii) the linear unforced equation, a particular case of a separable equation and hence always solvable (section 3.2); (iii) the solution of the linear forced equation is obtained from its unforced part by the method of variation of parameters (section 3.3), used before (notes 1.2–1.5 and section 2.9); (iv) the Bernoulli equation, which is non-linear but can be transformed to the linear type, and hence is also always solvable (section 3.4); (v) the Riccati equation, which, while there is no known general solution, is a non-linear generalization of the Bernoulli equation (section 3.6), and it can be shown (section 3.5) that if one/two/three particular integrals are known, then the general integral can be obtained via two/one/zero quadratures.

A sixth method is (vi) a change of variable that may render a differential equation solvable; for example the homogeneous first-order differential equation, in which the dependent and independent variable appear only through their ratio, can be transformed into a separable type, and hence integrated in all cases (section 3.7). Proceeding from the first six to the next two classes, any first-order differential equation is equivalent to a first-order differential in two variables (section 3.8) leading to two cases: (vii) if it satisfies a condition of integrability, it is an exact differential that is the differential of a function, that equated to an arbitrary constant supplies the general integral; (viii) if the integrability condition is not satisfied, the first-order differential equation is equivalent to an inexact differential, and it is not the differential of a function, though it becomes so when multiplied by an integrating factor that always exists. This is no longer the case for a first-order differential in three variables (section 3.9), when there are three possibilities: (i/ii) exact (inexact) differential satisfying (not satisfying) an integrability condition and not needing (needing) an integration factor, as for a first-order differential in two variables; (ii/iii) the existence (non-existence) of an integrating factor depends on the satisfaction (non-satisfaction) of a more general integrability condition. The first-order differential may be extended to: (i) more than three variables (notes 3.1–3.15); (ii) homogeneous differentials (notes 3.16–3.20); (iii) higher-order differentials (notes 3.21–3.24).

3.1 General Properties of First-Order Equations

The general properties include a classification of solutions (subsection 3.1.1) and the determination of the arbitrary constant in the general integral (subsection 3.1.2). The differential equation is considered to be solved when it is reduced to an integration (subsection 3.1.3) or quadrature (subsection 3.1.4) that may be elementary or not. The general integral of a first-order differential equation is one family (or several families) of integral curves [subsection 3.1.1 (3.1.5)], and each value of the arbitrary constant corresponds to one curve of the (of each) family.

3.1.1 General Integral and Integral Curves

An ordinary differential equation of the first order is a relation between a function $y(x)$, its variable x, and its first derivative y':

$$y' \equiv \frac{dy}{dx}: \qquad\qquad F(x, y; y') = 0. \qquad\qquad (3.1a, b)$$

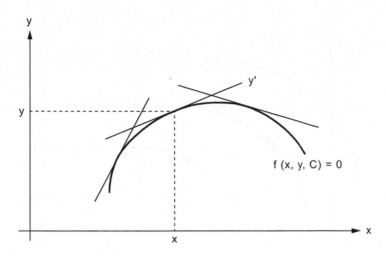

FIGURE 3.1
An explicit first-order differential equation specifies one slope at each point of the plane and its solution is an integral curve with the corresponding tangents.

The function $y(x)$ is the **dependent variable,** and the variable x is the **independent variable.** A solution of the differential equation is a function $y(x)$ that satisfies it (3.1b) when substituted together with its first derivative (3.1a). The solution of the differential equation (3.1a, b), may be given a geometric interpretation (Figure 3.1): the first-order ordinary differential equation (3.1b) specifies at each point of the (x, y)-plane one (or more) slopes y'. A solution must be a curve, whose tangent at the point (x, y), has a slope y' and is called an **integral curve.** Since it specifies one (or more) slopes or tangents at each point, the most general solution is a **family of curves** (Figure 3.2):

$$f\left(x,y,C\right)=0, \tag{3.2}$$

involving a parameter C, and this is the **general integral** of the first-order ordinary differential equation (3.1b). Giving to the parameter C a particular value $C = C_1$, specifies a particular curve, or a **particular integral** of the differential equation. Thus *an ordinary differential equation of first-order (3.1b) relates a family of curves (3.2) to the slopes (3.1a) of its tangents at each point (x, y).* The cases in which the tangent is not unique may lead to more than one family of integral curves or to integral curves with multiple points, the simplest being as double points (Figure 3.3). If the family of integral curves has an envelope this a singular or special integral (section 1.1); the locus of double points or other loci may also lead to special integrals (sections 5.1–5.4).

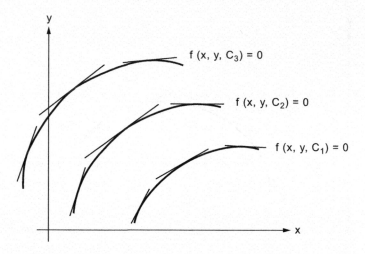

FIGURE 3.2
Each integral curve is a particular integral of the first-order differential equation for a particular value of the constant of integration; the general integral is the family of curves with the value of the arbitrary constant identifying each curve.

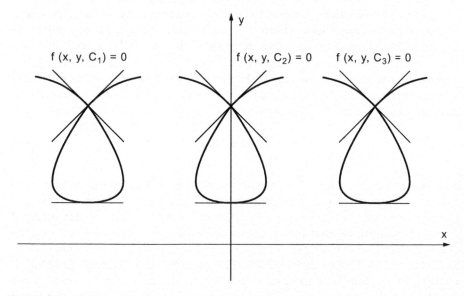

FIGURE 3.3
If there are multiple slopes at some points these are multiple points, for example double points through which the curve crosses itself in two different directions.

3.1.2 Constant of Integration and Boundary Condition

The preceding geometric property may be re-stated in analytical terms: *the general integral (3.2) of the first-order ordinary differential equation (3.1a, b), involves one arbitrary constant of integration C, and includes, for particular values of C, all particular integrals.* A rigorous proof will be given in section 9.1, and here is provided a heuristic explanation; there may also be singular integrals not included in the general integral (chapter 5). If the general integral (3.2) is differentiated with regard to x it yields:

$$\frac{\partial f}{\partial x} + y' \frac{\partial f}{\partial y} = 0, \tag{3.3}$$

where the partial derivatives may depend on the constant of integration C. Taking C as a parameter, and eliminating it between (3.2) and (3.3), leads back to the differential equation of first order (3.1b). If the general integral (3.2) involved more than one constant of integration, say two C, C_o, then after elimination of C with (3.3) the remaining constant C_o would have appeared in (3.1b). If the general integral (3.2) involved no constant of integration, then elimination with (3.3) for (3.1b) would generally be impossible, unless the two equations were redundant, that is the case of singular or special (sections 5.1–5.4) integrals. Excepting the case of special integrals, the general integral (3.2) of a first-order (3.1a) differential equation (3.1b) involves one arbitrary constant of integration that can be determined from a boundary condition:

$$f(x_0, y_0, C) = 0, \tag{3.4}$$

selecting the integral curve passing through a given point, assuming there is one and only one.

3.1.3 Algebra, Analysis, and Differential Equations

A differential equation may be considered a **third-level problem** in the sense that it is considered solved when it reduces to a problem of analysis, like an integration. For example, the simplest ordinary differential equation of first order (3.5b):

$$f \in \mathcal{E}(|R), \qquad y' = f(x), \qquad y = \int^{x} f(\xi)\,d\xi + C, \tag{3.5a–c}$$

where f(x) is an integrable function (3.5a), has general integral (3.5c) obtained by a single integration or **quadrature**, involving one arbitrary constant of integration C. Performing the integration, regardless of whether it is elementary or not, is a **second-level problem of analysis**. The solution of the problem of analysis may lead to a general integral in implicit form (3.2) or in explicit form (3.6) involving the roots of an algebraic equation, such as the characteristic polynomial of a linear differential equations with constant (homogeneous) coefficients [sections (1.3–1.5 (1.6–1.8)]. Finding the roots or obtaining one or more explicit solution(s) in the form:

$$y = y(x; C),$$ (3.6)

is a **first-level problem of algebra.** Thus there is a four-level **hierarchy of problems:** *(i) algebraic at level one; (ii) analysis at level two involving integrations or quadratures; (iii) differential equations at level three, involving relations between functions and derivatives; and (iv) variational problems at level four whose solution lead to differential equations.*

3.1.4 Solution by Quadratures and Indefinite Integrals

A differential equation is solved by **quadratures** when its solution is expressed by an integral, for example (3.5c) is the solution by quadratures of the differential equation (3.5b) where the function f is integrable (3.5a). The general integral (3.5b) specifies a family of integral curves, differing from each other by a translation along the y-axis (Figure 3.3); hence only one curve passes through each point (x_0, y_0):

$$y - y_0 = \int^x f(\xi) d\xi - \int^{x_0} f(\xi) d\xi = \int_{x_0}^x f(\xi) d\xi,$$ (3.7a)

thus the indefinite integral in the general solution (3.5c) is replaced by a definite integral (3.7a) in the particular solution corresponding to the constant of integration:

$$C \equiv y_0 - \int^{x_0} f(\xi) d\xi.$$ (3.7b)

An example of first-order ordinary differential is the case of linear slope (3.8a):

$$y' = 2x, \qquad y = x^2 + C,$$ (3.8a, b)

whose general integral (3.8b) is a family of parabolas (Figure 3.4) cutting the y-axis $x = 0$ at $y = C$; for example the parabola through the origin is the particular integral $C = 0$. The solution of a first-order differential equation may consist of N families of functions if it is of degree N (subsection 5.4.4); next is given an example of degree $N = 2$ (subsection 3.1.5).

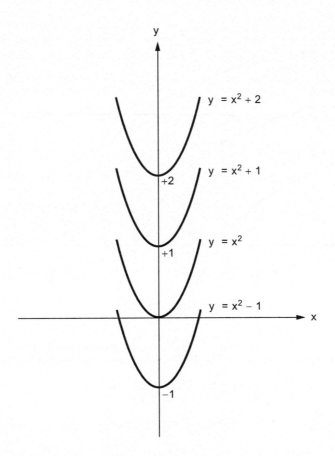

FIGURE 3.4
Through the regular points of a differential equation passes only one integral curve, for example the family of parabolas symmetric relative to the *y*-axis with unit curvature at the apex has particular integrals identified by the intersection with the *y*-axis.

3.1.5 Double Points of Quadratic Differential Equation

A first-order differential equation is **explicit** if it can be solved for the slope (3.9a), and is of **degree M** if is a polynomial of the derivatives (3.9b):

$$y' = f(x,y), \qquad \prod_{m=1}^{M}\left[y' - f_m(x,y)\right] = 0. \qquad (3.9a, b)$$

A simple case of quadratic differential equation is (3.10a):

$$0 = y'^2 - \left[f_+(x) + f_-(x)\right]y' + f_+(x)f_-(x) = \left[y' - f_+(x)\right]\left[y' - f_-(x)\right], \qquad (3.10a)$$

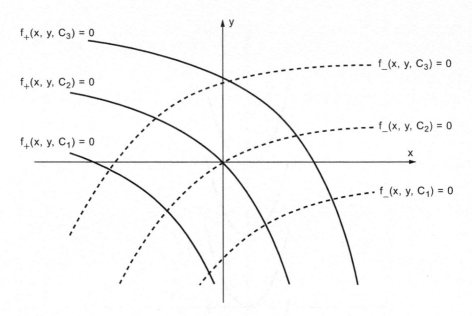

FIGURE 3.5
If there are multiple slopes at all points, then the general integral consists of several families of curves, for example two for a first-order differential equation quadratic on the slope.

whose general integral is two families of curves (3.10b, c):

$$y_\pm(x) = C + \int^x f_\pm(\xi)d\xi. \qquad (3.10b, c)$$

For each value of the constant C there are two curves, and thus the regular points in the plane (Figure 3.5) are double points; unlike those in Figure 3.3 they do not arise from the curve crossing itself. The general integral is (3.10d):

$$0 = \left(y_+(x) - \int^x f_+(\xi)d\xi - C \right)\left(y_-(x) - \int^x f_-(\xi)d\xi - C \right), \qquad (3.10d)$$

because it is satisfied by either of (3.10b, c); there is only one constant of integration in (3.10b, c) $\equiv$ (3.10d) because the ordinary differential equation (3.10a) is of the first-order.

3.2 Integration by Quadratures of a Separable Equation

Integration by quadratures is possible for a separable equation (subsection 3.2.1) for which several examples can be given, including the particular case of the linear unforced equation (subsection 3.2.2).

3.2.1 Explicit and Separable Differential Equation

The first-order explicit differential equation (3.1b) is explicit (3.9b) $\equiv$ (3.11a) if it can be solved for the slope:

$$\frac{dy}{dx} = y' = g(x,y) = -\frac{X(x)}{Y(y)};$$
(3.11a, b)

*if, in addition, the function g(x, y) splits into the ratio of two functions, one of each variable differential equation (3.11b) is **separable** (standard L):*

$$\int^{y} Y(\eta)d\eta = C - \int^{x} X(\xi)d\xi,$$
(3.11c)

and can be integrated by quadratures (3.11c).
  For example, the separable equation (3.12a):

$$\frac{dy}{dx} = ye^{x}: \qquad\qquad y(x) = C \exp(e^{x}),$$
(3.12a, b)

has the general integral (3.12b) because:

$$\log y = \int \frac{dy}{y} = \int e^{x}\,dx + C_1 = \exp(e^{x}) + \log C.$$
(3.13a, b)

In passing from (3.13a) to (3.12b), exp (C_1) was replaced by C because the exponential (or any function) of an arbitrary constant is another arbitrary constant; alternatively, the constant in (3.13b) can be designated $C_1 = \log C$ for consistency (3.13a) $\equiv$ (3.13b). Differentiating (3.12b) with regard to x:

$$y' = e^{x}C\exp(e^{x}) = e^{x}y,$$
(3.14)

provides a check that it is indeed a solution of the differential equation (3.12a). The integral curves (3.12b) do not pass through the origin $y = 0$ for any finite value of x, and are asymptotic to the horizontal line $y \to C$ for $x \to -\infty$. The boundary condition (3.15a):

$$1 = y(0) = Ce: \qquad\qquad y(x) = \frac{\exp(e^{x})}{e} = \exp(e^{x} - 1),$$
(3.15a, b)

specifies the constant and leads to the unique solution (3.15b).

The differential equation of first order (3.16a) is separable (3.16b):

$$\frac{dy}{dx} = \cos x\sqrt{1-y^2}, \qquad \int \frac{dy}{\sqrt{1-y^2}} = \int \cos x \, dx, \qquad \text{(3.16a, b)}$$

and has general integral (3.17a) equivalent to (3.17b):

$$arc\sin y = C + \sin x, \qquad y(x) = \sin(C + \sin x). \qquad \text{(3.17a, b)}$$

The boundary condition (3.18a) leads to an infinity of values of the arbitrary constant (3.18b):

$$1 = y(0) = \sin C, \qquad C = \frac{\pi}{2} + 2n\pi:$$

$$y(x) = \sin\left(2n\pi + \frac{\pi}{2} + \sin x\right) = \sin\left(\frac{\pi}{2} + \sin x\right) = \cos(\sin x), \qquad \text{(3.18a–c)}$$

all corresponding to the same solution (3.18c). Another first-order separable differential equation is considered in the EXAMPLE 10.6.

3.2.2 Linear Unforced Differential Equation

If the function $g(x, y)$ in the explicit differential equation (3.10a) is linear in the independent variable y and unforced with a coefficient $P(x)$ function of the dependent variable (3.19a):

$$\frac{dy}{dy} = P(x)y, \qquad \log y = \int \frac{dy}{y} = \int P(x)dx + \log C, \qquad \text{(3.19a, b)}$$

then the equation is separable (3.11b), and hence solvable by quadratures:

$$X(x) \equiv \exp\left\{\int^x P(\xi)d\xi\right\}: \qquad y(x) = CX(x). \qquad \text{(3.20a, b)}$$

Thus *the **linear, unforced** (standard LI) first-order differential equation (3.20a) has general integral (3.21a, b).*
For example, the differential equation (3.22a):

$$y' = \left(\frac{1}{x} - x\right)y: \qquad y(x) = Cx\exp\left(-\frac{x^2}{2}\right), \qquad \text{(3.21a, b)}$$

has general integral (3.21b)≡(3.22b):

$$P(x) = \frac{1}{x} - x: \quad y(x) = C\exp\int^x\left(\frac{1}{\xi} - \xi\right)d\xi = C\exp\left(\log x - \frac{x^2}{2}\right), \quad (3.22a, b)$$

where in this case was applied (3.20a) with (3.22a). The boundary condition (3.23a) specifies the constant:

$$1 = y(1) = \frac{C}{\sqrt{e}}: \quad y(x) = x\sqrt{e}\exp\left(-\frac{x^2}{2}\right) = x\exp\left(\frac{1-x^2}{2}\right) \quad (3.23a, b)$$

in the unique solution (3.23b).

Another example of linear unforced differential equation of first order is (3.24a) that has general integral (3.24b):

$$y' = y\tan x: \quad\quad\quad y(x) = C\sec x, \quad\quad\quad (3.24a, b)$$

as follows from (3.19a) with (3.25a) leading to (3.25b) ≡ (3.24b):

$$P(x) = \tan x: \quad y(x) = C\exp\left\{\int^x \tan\xi\ d\xi\right\} = C\exp\{-\log(\cos x)\} = \frac{C}{\cos x}.$$

$$(3.25a, b)$$

The boundary condition (3.26a):

$$1 = y(0) = C: \quad\quad\quad y(x) = \sec x, \quad\quad\quad (3.26a, b)$$

leads to the unique solution (3.26b).

3.3 Linear Forced Differential Equation

The **method of variation of parameters** (Bernoulli 1697) can be used to obtain a particular integral of the corresponding forced differential equation from the general integral of a linear unforced differential equation with variable coefficients (Notes 1.2–1.4). For example, for: (i) the linear second-order system with constant coefficients (section 2.9) and (ii) the generalization of (3.19a) as the first-order linear forced differential equation with variable coefficients (standard LII):

$$y' = P(x)y + Q(x), \quad\quad\quad (3.27)$$

so designated because y' is a linear function of y, with coefficients depending on x. It will be solved by the method of variation of constants, starting from the solution (3.20a, b) of the unforced equation (3.19a), and replacing the arbitrary constant C by a function $C(x)$ so that the forced equation (3.27) is satisfied. Thus, the solution of the linear forced equation (3.27) is sought in the form:

$$y(x) = C(x)X(x) \equiv C(x)\exp\left\{\int^x P(\xi)d\xi\right\};\qquad\qquad(3.28)$$

differentiating yields:

$$y' = (C' + CP)\exp\left\{\int^x P(\xi)d\xi\right\} = Py + C'X.\qquad\qquad(3.29)$$

The equation (3.29) coincides with (3.27), provided that the still-undetermined function $C(x)$ satisfies (3.30a):

$$Q = C'X,\qquad C(x) = C + \int^x \frac{Q(\xi)}{X(\xi)}d\xi,\qquad\qquad(3.30a, b)$$

that is, a first-order differential equation whose solution is (3.30b). Substituting (3.30b) in (3.28):

$$Y(x) \equiv \int^x \frac{Q(\xi)}{X(\xi)}d\xi:\qquad y(x) = X(x)[C + Y(x)],\qquad\qquad(3.31a, b)$$

it follows that *the complete integral (3.31b) of the* **linear forced** *(standard LII) first-order differential equation (3.27) consists: (i) of the general integral (3.20b) of the unforced equation (3.19a), that is, the first term on the r.h.s. (3.31b), involving one arbitrary constant besides the function (3.20a); and (ii) the latter also appears in the particular integral of the forced equation (3.27), that is, the second term on the r.h.s. of (3.31b) that involves (3.31a) and no constants of integration.*
For example the linear forced first-order differential equation (3.32a):

$$y' = -\frac{y}{x} + \exp(x^2):\qquad y(x) = \frac{C}{x} + \frac{e^{x^2}}{2x},\qquad\qquad(3.32a, b)$$

has complete integral (3.32b), obtained substituting (3.33a–d)

$$P(x)=-\frac{1}{x}: \quad X(x)=\exp\left(-\int^{x}\frac{d\xi}{\xi}\right)=\exp(-\log x)=\frac{1}{x}, \qquad \text{(3.33a, b)}$$

$$Q(x)=e^{x^2}: \quad Y(x)=\int^{x}\xi e^{\xi^2}\,d\xi=\frac{1}{2}e^{x^2}, \qquad \text{(3.33c, d)}$$

in (3.31b). The boundary condition (3.34a):

$$1=y(1)=C+\frac{e}{2}: \quad y(x)=\left(1-\frac{e}{2}\right)\frac{1}{x}+\frac{e^{x^2}}{2x}=\frac{1}{x}+\frac{e^{x^2}-e}{2x}, \qquad \text{(3.34a, b)}$$

leads to the unique solution (3.34b).
   Another linear first-order forced first-order differential equation with variable coefficients is (3.35a):

$$y' = y\cot x + \csc x: \qquad\qquad y(x)=C\sin x-\cos x, \qquad \text{(3.35a, b)}$$

whose complete integral is (3.35b), as follows substituting (3.36a–d):

$$P(x)=\cot x: \quad X(x)=\exp\left\{\int^{x}\cot\xi\,d\xi\right\}=\exp\left[\log(\sin x)\right]=\sin x, \qquad \text{(3.36a, b)}$$

$$Q(x)=\csc x: \qquad\qquad Y(x)=\int^{x}\csc^2\xi\,d\xi=-\cot x, \qquad \text{(3.36c, d)}$$

in (3.31b). The boundary condition (3.37a):

$$1=y\left(\frac{\pi}{2}\right)=C: \qquad\qquad y(x)=\sin x-\cos x, \qquad \text{(3.37a, b)}$$

leads to the unique solution (3.37b). Another linear forced first-order differential equation is considered in the EXAMPLE 10.6.

3.4 Transformation of Bernoulli (1695) Non-Linear Differential Equation

The Bernoulli (1695) first-order ordinary differential equation (standard LIII):

$$y' = P(x)y + Q(x)y^n, \tag{3.38}$$

reduces: (i) for $n = 0$ to the linear forced (3.27) standard LII whose complete integral is (3.20a; 3.31a, b); (ii) for $n = 1$ to the linear unforced (3.19a) standard LI with $P \to P + Q$ whose general integral (3.20a, b) is (3.39b):

$$n = 1: \qquad y(x) = C \exp\left\{ \int^x \left[P(\xi) + Q(\xi) \right] d\xi \right\}. \tag{3.39a, b}$$

In all other cases $n \neq 0, 1$, the **Bernoulli (1695) equation** (3.38) is a non-linear first-order differential equation (standard LIII) generalizing the linear unforced (forced) standard LII (LI). Dividing it by y^n:

$$n \neq 1: \qquad Q + Py^{1-n} = y^{-n}y' = \frac{\left(y^{1-n} \right)'}{1-n}, \tag{3.40a, b}$$

it becomes a linear forced differential equation (3.27), via the transformations:

$$y, P, Q \quad \to \quad y^{1-n}, (1-n)P, (1-n)Q; \tag{3.41a–c}$$

the solution (3.31a, b) of (3.40b) is thus (3.31a) [(3.31b)] with the substitutions (3.41b) [(3.41a)] leading to (3.42a) [(3.42b)]:

$$X_n(x) \equiv \exp\left\{ (1-n) \int^x P(\xi) d\xi \right\}, \quad Y_n(x) = (1-n) \int^x \frac{Q(\xi)}{X_n(\xi)} d\xi, \tag{3.42a, b}$$

where $X_0 = X$ and $Y_0 = Y$. The general integral is analogous to (3.31b):

$$n \neq 1: \qquad \left[y(x) \right]^{1-n} = X_n(x) \left[C + Y_n(x) \right], \tag{3.43a, b}$$

that: (i) coincides with (3.31b) for $n = 0$; (ii) breaks down (3.43a) for $n = 1$, when it is replaced by (3.39a, b). Thus *the (standard LIII) Bernoulli (1695) non-linear first-order differential equation (3.38) has solution (Leibnitz 1696): (3.39b) in the* *linear unforced case (3.39a); (3.43a, b) in all non-linear cases $n \neq 0,1$, as well as the linear unforced case $n = 0$ in (3.31a, b). A simple non-linear case (standard LIV) the* **particular Riccati equation** *is (3.44a, b):*

$$n = 2: \quad y' = yP(x) + y^2 Q(x), \quad \frac{1}{y(x)} = X_2(x)[C + Y_2(x)], \quad \text{(3.44a–c)}$$

$$X_2(x) = \exp\left\{-\int^x P(\xi)d\xi\right\}, \quad Y_2(x) = -\int^x \frac{Q(\xi)}{X_2(\xi)}d\xi, \quad \text{(3.44d, e)}$$

whose solution is (3.44c–e). Two examples of the solution of Bernoulli equations are given next, before generalizing (3.44b) to the Riccati equation (section 3.5).

An example of the non-linear Bernoulli differential equation of particular Riccati type is (3.45a):

$$xy' = -y + y^2 \log x, \quad \frac{1}{y(x)} = Cx + 1 + \log x, \quad \text{(3.45a, b)}$$

whose complete integral (3.45b) is obtained from (3.44b) with (3.44a–e) using (3.46a–d):

$$n = 2: \quad P(x) = -\frac{1}{x}, \quad X_2(x) = \exp\left(\int^x \frac{d\xi}{\xi}\right) = \exp(\log x) = x$$

$$Q(x) = \frac{\log x}{x}, \quad Y_2(x) = -\int^x \frac{\log \xi}{\xi^2}d\xi = \frac{1 + \log x}{x}. \quad \text{(3.46a–d)}$$

The last integration (3.46d) can be performed by two methods: (i) integration by parts (3.47):

$$Y_2(x) = \int^x \log \xi \, d\left(\frac{1}{\xi}\right) = \frac{\log x}{x} - \int^x \frac{1}{\xi}d(\log \xi) = \frac{\log x}{x} - \int^x \frac{d\xi}{\xi^2} = \frac{1 + \log x}{x}; \quad \text{(3.47)}$$

or (ii) change of variable (3.48a) leading to the same result (3.48b) $\equiv$ (3.47):

$$\eta \equiv \frac{1}{\xi}: Y_2(x) = -\int^x \log\left(\frac{1}{\xi}\right)d\left(\frac{1}{\xi}\right) = -\int^{1/x} \log \eta \, d\eta = -\left[\frac{1}{x}\log\left(\frac{1}{x}\right) - \frac{1}{x}\right] = \frac{1 + \log x}{x}.$$

$$\text{(3.48a, b)}$$

The primitive of the logarithm (3.49) was used in (3.48b):

$$\frac{d}{d\eta}[\eta\log\eta-\eta]=\log\eta.\qquad(3.49)$$

Substituting (3.46b, d) in (3.44c) proves (3.45b). The boundary condition (3.50a):

$$1=y(1)=C+1:\qquad y(x)=\frac{1}{1+\log x},\qquad(3.50a, b)$$

leads to the unique solution (3.50b).

Another example of a non-linear Bernoulli differential equation (3.38) not of particular Riccati type (3.44a, b) is:

$$y'=-xy(1-x^2y^2):\qquad \frac{1}{[y(x)]^2}=C\exp(x^2)+1+x^2,\qquad(3.51a, b)$$

whose general integral (3.51b) follows from (3.42a, b; 3.43a, b) using (3.52a–e):

$$n=3:\quad P(x)=-x,\quad X_3(x)=\exp\left(2\int^x\xi\,d\xi\right)=\exp(x^2),\qquad(3.52a–c)$$

$$Q(x)=x^3:\quad Y_3(x)=-2\int^x\xi^3\exp(-\xi^2)\,d\xi=\int^x\xi^2d\left[\exp(-\xi^2)\right],\qquad(3.52d, e)$$

where the last integral is evaluated by parts:

$$Y_3(x)-x^2\exp(-x^2)=-\int^x\exp(-\xi^2)\,d(\xi^2)=-2\int^x\xi\exp(-\xi^2)\,d\xi=\exp(-x^2).$$
$$(3.53)$$

Substitution of (3.53) together with (3.52c) in (3.43b) with $n=3$ proves (3.51b). The boundary condition (3.54a):

$$1=y(0)=C+1:\qquad y(x)=\frac{1}{\sqrt{1+x^2}},\qquad(3.54a, b)$$

leads to the unique solution (3.54b). Another Bernoulli differential equation is considered in the EXAMPLE 10.6.

3.5 Solution of the Riccati (1724) – D'Alembert (1763) Differential Equation

The Riccati (1724) equation (sections 3.5–3.6) is a joint generalization (D'Alembert 1763) of: (i) the linear forced equation (section 3.3) adding a term quadratic in the dependent variable so that the differential equation becomes non-linear; (ii) the non-linear Bernoulli equation (section 3.4) with quadratic non-linear term adding a forcing term. The original Riccati (1824) equation was actually due to Bernoulli (1691), the generalized Riccati equation considered next (sections 3.4–3.5) is a joint generalization of the linear and Bernoulli equations, and the former provides a better starting point for analogies and differences. The general integral of the linear equation involves two quadratures (section 3.3), and (subsection 3.5.1): (i) if one particular integral is known the general integral can be obtained by one quadrature; (ii) if two distinct particular integrals are known the general integral can be obtained without any quadratures by the constancy of the three-point cross-ratio. In the case of the Riccati equation the general integral is not known, but (iii/iv) if one (two distinct) particular integral(s) is (are) known the general integral can be obtained [subsection 3.5.2 (3.5.3)] via two quadratures (one quadrature); (v) if three distinct particular integrals are known, then the general integral can be obtained without any integrations from the constancy of the four-point cross-ratio (subsection 3.5.5). The properties (i) to (v) are summarized in Table 3.1, and (ii) [(v)] imply the following analogy (subsection 3.5.4): (a) any three (four) solutions of the linear (Riccati) ordinary first-order differential equation have a constant three (four)-point cross-ratio; and (b) choosing one to be the general integral it follows that it is a linear (bilinear) function of the constant of integration.

3.5.1 Linear Equation and Three-Point Cross-Ratio

The general integral of the first-order linear forced differential equation (3.27): (i) is a linear function (3.31b) of the arbitrary constant of integration; and (ii) involves two

TABLE 3.1

General Integral of the Linear and Riccati Differential Equations

Number of quadratures needed for general integral: $q =$	2	1	0
Number of particular integrals known: $p =$			
Linear forced differential equation: $p + q = 2$	0	1	2
Riccati differential equation: $p + q = 3$	1	2	3

Note: Number of quadratures q needed to obtain the general integral of a linear first-order (non-linear Riccati) equation when p linearly independent (distinct but not necessarily linearly independent) particular integrals are known.

quadratures (3.20a; 3.31a). If a particular integral is known (3.55a), the general integral satisfies the linear unforced equation (3.55b) ≡ *(3.19a):*

$$y'_1 = Py_1 + Q: \quad y' - y'_1 = P(y - y_1), \quad y(x) = y_1(x) + C_1 X(x), \tag{3.55a–c}$$

and hence (3.20a, b) can be obtained by one quadrature (3.55c). If a second linearly independent particular integral is known (3.56a), then from (3.55c) it follows (3.56b):

$$y'_2 = Py_2 + Q: \qquad\qquad y(x) - y_2(x) = C_2 X(x), \tag{3.56a, b}$$

*that: (i) three solutions of a linear first-order differential equation (3.27, 3.55a; 3.56a) have a constant **three-point cross-ratio** (3.57a):*

$$\frac{y(x) - y_1(x)}{y(x) - y_2(x)} = \frac{C_1}{C_2} \equiv C \quad \Leftrightarrow \quad y(x) = \frac{y_1(x) - Cy_2(x)}{1 - C}, \tag{3.57a, b}$$

(ii) if two particular integrals of a linear first-order differential equation are known (3.55a; 3.56a) the general integral can be obtained without any quadrature (3.57b). As shown in the Table 3.1 for the linear forced first-order differential equation, the sum of (i) the number of known linearly independent particular integrals p and (ii) the number of quadratures in the general integral q is always $2 = p + q$, *because adding one unit to (i) subtracts one unit to (ii). Although the general integral of the Riccati equation is not known, it can be shown (subsections 3.5.2–3.5.4) that if* $p = 1, 2, 3$ *distinct particular integrals are known, the general integral can be obtained with* $q = 3 - p = 2, 1, 0$ *quadratures so that the sum is* $p + q = 3$ *as indicated in the Table 3.1.*

3.5.2 Solution by Two Quadratures if One Particular Integral Is Known

A famous first-order non-linear differential equation for which no general solution is known is the **Riccati (1724) – D'Alembert (1763) equation:**

$$y' = R(x)y^2 + P(x)y + Q(x). \tag{3.58}$$

For $R = 0$ it reduces to the forced linear differential equation (3.27), whose complete integral is (3.20a; 3.31a, b); for $Q = 0$ it is a particular quadratic case (3.44a, b) of Bernoulli's equation (3.38), hence also solvable (3.44c–e). In the case $P = 0$, the general solution is not known; for example, the equation $y' = y^2 + x^2$ has no elementary solution. The general integral y of the Riccati equation (3.58) can be determined if at least one particular integral y_1 is known (3.59a):

$$Q + Py_1 + Ry_1^2 - y'_1 = 0; \qquad y = y_1 + \frac{1}{v}, \tag{3.59a, b}$$

changing the dependent variable (3.59b) into (3.58) leads to (3.60a):

$$0 = Q + Py_1 + \frac{P}{v} + Ry_1^2 + \frac{2Ry_1}{v} + \frac{R}{v^2} - y_1' + \frac{v'}{v^2} = \frac{P + 2Ry_1}{v} + \frac{R + v'}{v^2}, \quad \text{(3.60a, b)}$$

bearing in mind (3.59a). From (3.60a) it follows that v satisfies a linear first-order differential equation (3.60b) ≡ (3.61):

$$v' = -(P + 2Ry_1)v - R. \tag{3.61}$$

The solution of (3.61) ≡ (3.27) involves the substitutions:

$$y, P, Q \quad \rightarrow \quad v, -P - 2Ry_1, -R. \tag{3.62a–c}$$

From (3.20a; 3.31a, b) it follows that:

$$Z(x) \equiv \frac{1}{X(x)} = \exp\left\{\int^x \left[P(\xi) + 2R(\xi)y_1(\xi)\right]d\xi\right\}, \tag{3.63a, b}$$

$$Y(x) = -\int^x \frac{R(\xi)}{X(\xi)}d\xi = -\int^x R(\xi)Z(\xi)d\xi. \tag{3.63c, d}$$

Substitution of (3.63a) in (3.31b) specifies (3.64a):

$$v(x) = \frac{C + Y(x)}{Z(x)}, \quad y(x) = y_1(x) + \frac{Z(x)}{C + Y(x)}, \tag{3.64a, b}$$

and hence (3.59b) is the general integral (3.64b) of the Riccati equation (3.58). It has been shown that *if a particular integral (3.59a) of the Riccati equation (3.58) is known (standard LV), the general integral (3.64b) is specified by two quadratures (3.63b, d).*

As an example, consider the Riccati differential equation (3.65a):

$$y' = (1 - xy)f(x) - y^2: \qquad y_1(x) = \frac{1}{x}, \tag{3.65a, b}$$

that has a particular integral (3.65b). The non-linear first-order differential equation (3.65a) is of the Riccati type with coefficients:

$$Q(x) = f(x), \quad P(x) = -x f(x), \quad R(x) = -1, \tag{3.66a–c}$$

that lead to (3.63b, d) the quadratures:

$$Z(x) = \exp\left\{-\int^x \left[\xi f(\xi) + \frac{2}{\xi}\right] d\xi\right\} = \exp\left[-\int^x \xi f(\xi) d\xi\right] \exp(-2\log x) = \frac{F(x)}{x^2},$$

(3.67a)

$$F(x) \equiv \exp\left[-\int^x \xi f(\xi) d\xi\right], \qquad\qquad Y(x) = \int^x \frac{F(\xi)}{\xi^2} d\xi.$$

(3.67b, c)

Substitution of (3.65b; 3.67a, c) in (3.64b) leads to:

$$y(x) = \frac{1}{x} + \frac{x^{-2} F(x)}{C + Y(x)},$$

(3.68)

as the general integral (3.68; 3.67b, c) of the Riccati differential equation (3.65a) where $f(x)$ is an arbitrary function.

A particular example (3.69a) of the Riccati differential equation (3.65a) is (3.69b):

$$f(x) = \frac{1}{x^3}: \qquad\qquad y' = \frac{1}{x^3} - \frac{y}{x^2} - y^2.$$

(3.69a, b)

Substitution of (3.69a) evaluates the integrals (3.67b, c):

$$F(x) = \exp\left(-\int^x \frac{d\xi}{\xi^2}\right) = \exp\left(\frac{1}{x}\right),$$

(3.70a)

$$Y(x) = \int^x \exp\left(\frac{1}{\xi}\right) \frac{d\xi}{\xi^2} = -\int^x \exp\left(\frac{1}{\xi}\right) d\left(\frac{1}{\xi}\right) = -\exp\left(\frac{1}{x}\right).$$

(3.70b)

Substitution of (3.70a, b) in (3.68) leads to:

$$y(x) = \frac{1}{x} + \frac{x^{-2} \exp(1/x)}{C - \exp(1/x)} = \frac{1}{x} + \frac{1}{x^2} \frac{1}{C\exp(-1/x) - 1},$$

(3.71)

as the general integral (3.71) of the Riccati differential equation (3.69b). The boundary condition (3.72a) implies (3.72b):

$$0 = y(1) = 1 + \frac{1}{C/e - 1}: \qquad C = 0, \qquad y(x) = \frac{1}{x} - \frac{1}{x^2},$$

(3.72a–c)

and leads to the unique solution (3.72c).

3.5.3 Solution by One Quadrature if Two Particular Integrals Are Known

If a second linearly independent particular integral y_2 of the Riccati differential equation is available (3.73a), one quadrature can be dispensed with (standard LVI). Knowledge of two particular integrals (3.59a; 3.73a) specifies (3.59b) ≡ (3.73b), a function v_1 satisfying (3.61) ≡ (3.73c):

$$Q + Py_2 + Ry_2^2 - y_2' = 0: \qquad y_2 - y_1 = \frac{1}{v_1}, \qquad 0 = R + (P + 2y_1 R) v_1 + v_1'.$$

$$(3.73a\text{--}c)$$

The general integral of the Riccati equation (3.58) is sought in the form (3.74a):

$$y = y_1 + \frac{1}{v_1 u}, \qquad v = v_1 u, \qquad (3.74a, b)$$

that is equivalent by (3.59b) to (3.74b). Substituting (3.74b) into (3.61) and using (3.73c) leads to:

$$0 = R + \{(P + 2y_1 R) v_1 + v_1'\} u + v_1 u' = R(1 - u) + v_1 u'. \qquad (3.74c)$$

The latter, (3.74c), is a separable (3.11b) differential equation (3.75a):

$$\frac{du}{u - 1} = \frac{R}{v_1} dx, \qquad \log[u(x) - 1] = \log C + \int^x \frac{R(\xi)}{v_1(\xi)} d\xi, \qquad (3.75a, b)$$

$$u(x) - 1 = C \exp \left\{ \int^x \frac{R(\xi)}{v_1(\xi)} d\xi \right\} \equiv CS(x); \qquad (3.75c)$$

in (3.75c) the following function (3.73b) appears:

$$S(x) \equiv \exp \left\{ \int^x R(\xi) [y_2(\xi) - y_1(\xi)] d\xi \right\}, \qquad (3.76)$$

using (3.73b). Substitution of (3.75c) and (3.73b) in (3.74a) gives:

$$y(x) = y_1(x) + \frac{y_2(x) - y_1(x)}{1 + CS(x)}, \qquad (3.77)$$

as the general integral of Riccati's equation. Thus, *if two particular integrals (3.59a; 3.73a) of the (standard LVI) Riccati equation (3.58) are known, the general integral can be obtained (3.77) by one quadrature (3.76).*

As an example, consider (3.78b) as the particular case (3.78a) of the Riccati differential equation (3.65a):

$$f(x) = \frac{1+x^2}{1-x^2}: \qquad \left(y' + y^2\right)\left(1 - x^2\right) = \left(1 - xy\right)\left(1 + x^2\right). \tag{3.78a, b}$$

Besides (3.65b) another particular integral is (3.79a), leading by (3.66c; 3.76) to (3.79b):

$$y_2(x) = x: \qquad S(x) = \exp\left\{\int^x \left(\frac{1}{\xi} - \xi\right) d\xi\right\} = x \exp\left(-\frac{x^2}{2}\right). \tag{3.79a, b}$$

Substitution of (3.65b; 3.79a, b) in (3.77) specifies:

$$y(x) = \frac{1}{x} + \frac{x^2 - 1}{x + Cx^2 \exp\left(-x^2/2\right)}, \tag{3.80}$$

as the general integral (3.80) of the Riccati differential equation (3.78b).

3.5.4 Four-Point Cross-Ratio for the Riccati Equation

The preceding results (subsections 3.5.2–3.5.3) suggest that the more particular integrals of the Riccati differential equation are known, the less quadratures are needed to find the general integral (Table 3.1); it might be extrapolated that if three particular integrals of the Riccati differential equation are known, then (standard LVII) the general integral can be found without quadratures. This is proved considering a third linearly independent particular integral (3.81a) of the Riccati equation, besides (3.59a; 3.73a):

$$0 = Q + Py_3 + Ry_3^2 - y_3'; \qquad y_3 - y_1 = \frac{1}{v_2}. \tag{3.81a, b}$$

In addition to the function v_1 defined by (3.73b), the function v_2 defined by (3.81b) also satisfies a linear differential equation (3.61) $\equiv$ (3.73c). Thus, as in (3.57a) the functions (v, v_1, v_2) in (3.59b; 3.73b; 3.81b) have a constant three-point cross-ratio (3.82):

$$C = \frac{v - v_1}{v - v_2} = \frac{\dfrac{1}{y - y_1} - \dfrac{1}{y_2 - y_1}}{\dfrac{1}{y - y_1} - \dfrac{1}{y_3 - y_1}} = \frac{y_2 - y}{\left(y - y_1\right)\left(y_2 - y_1\right)} \frac{\left(y - y_1\right)\left(y_3 - y_1\right)}{y_3 - y}. \tag{3.82}$$

Simplifying (3.82) leads to (3.83a) ≡ (3.83b):

$$\frac{y - y_2}{y - y_3} \frac{y_1 - y_3}{y_1 - y_2} = C \quad \Leftrightarrow \quad y = \frac{y_2(y_1 - y_3) - Cy_3(y_1 - y_2)}{y_1 - y_3 - C(y_1 - y_2)}, \quad \text{(3.83a, b)}$$

proving that: *(i) four solutions of the Riccati equation (3.58; 3.59a; 3.73a; 3.81a) have a constant **four-point cross-ratio** (3.83a); (ii) if (3.59a; 3.93a; 3.81a) are three distinct particular solutions of the Riccati equation (3.58), the general integral is given (standard LVII) by (3.83b) as a **bilinear function** (3.84a) of the constant of integration C:*

$$y(x) = \frac{\gamma(x) + C\varepsilon(x)}{\alpha(x) + C\beta(x)}: \quad \{\alpha, \beta, \gamma, \varepsilon\}(x) = \{y_1 - y_3, y_2 - y_1, y_2\alpha, y_3\beta\}, \quad \text{(3.84a–e)}$$

involving the four functions (3.84b–e). Thus the general integral (3.55c) [(3.83b)] of a linear first-order (3.27) [Riccati (3.58)] differential equation is a linear (bilinear) function [sections I.35.1–4 (I.35.5–9)] of the arbitrary constant of integration, in agreement with the constancy of the three- (3.57a) [four- (3.83a)] point cross-ratio.

3.5.5 Solution without Quadratures if Three Particular Integrals Are Known

It has been shown that as more particular integrals of the Riccati equation are known (subsections 3.5.2–3.5.4) then less quadratures are needed to find the general integral. An example is the Riccati equation:

$$2x^2 y' = (x-1)(y^2 - x^2) + 2xy, \quad \text{(3.85)}$$

that corresponds to (3.58) with coefficients:

$$R = \frac{x-1}{2x^2}, \qquad P = \frac{1}{x}, \qquad Q = \frac{1-x}{2}. \quad \text{(3.86a–c)}$$

Noting that (3.87a) is a particular integral of the Riccati differential equation (3.85), the general integral can be obtained (3.63b, d) by two quadratures (3.87b, c):

$$y_1(x) = x: \quad Z(x) = \exp\left\{ \int^x \left(\frac{1}{\xi} + \frac{\xi - 1}{\xi} \right) d\xi \right\} = \exp\left(\int^x d\xi \right) = e^x, \quad \text{(3.87a, b)}$$

$$Y(x) = -\frac{1}{2} \int^x \left(\frac{1}{\xi} - \frac{1}{\xi^2} \right) e^\xi \, d\xi = -\frac{e^x}{2x}; \quad \text{(3.87c)}$$

substitution of (3.87a–c) in (3.64b) leads to:

$$y(x) = x + \frac{e^x}{C_1 - \dfrac{e^x}{2x}} = x + \frac{2x}{2C_1 x e^{-x} - 1},$$
(3.88a, b)

as the general integral (3.88a, b) of the Riccati differential equation (3.85).

Noting that (3.89a) is a distinct particular integral of the Riccati equation (3.85), even though it is linearly dependent from (3.87a) it can be used to reduce to one (3.76) number of quadratures (3.89b):

$$y_2(x) = -x: \qquad S(x) = \exp\left\{ -\int^x \frac{\xi - 1}{\xi} d\xi \right\} = \exp\left(-\int^x d\xi + \log x \right) = x e^{-x};$$

(3.89a, b)

substitution of (3.87a; 3.89a, b) in (3.77) leads to (3.90b):

$$C_2 = -2C_1: \qquad y(x) = x - \frac{2x}{1 + C_2 x e^{-x}},$$
(3.90a, b)

that coincides with (3.88b) ≡ (3.90b) with the arbitrary constants of integration related by (3.90a).

Choosing the arbitrary constant (3.91a, b), the general integral (3.88b) ≡ (3.90b) of the Riccati differential equation (3.85) is given by (3.91c–e):

$$C = 2C_1 = -C_2: \qquad y(x) = x + \frac{2x}{C x e^{-x} - 1} = x + \frac{2x e^x}{C x - e^x} = \frac{C x^2 + x e^x}{C x - e^x},$$
(3.91a–e)

as a bilinear function (3.91e) ≡ (3.84a) of the constant of integration involving the four functions (3.84b–e) ≡ (3.92a–d):

$$\{\alpha(x), \beta(x), \gamma(x), \varepsilon(x)\} = \{-e^x, x, x e^x, x^2\}.$$
(3.92a–d)

Taking the four values (3.93a–d) of the constant in the general integral (3.91e) leads to four particular integrals (3.93e–h):

$$C = \infty, 0, 1, -1: \qquad \{y_{1,2}; y_{3,4}\} = \left\{ \pm x, x \frac{x \pm e^x}{x \mp e^x} \right\}, \qquad C = \frac{1}{2},$$
(3.93a–i)

of which: (i) the first two are the particular integrals (3.87a; 3.89a); and (ii) the last two are also particular integrals that are less obvious on inspection of (3.85). The four-point cross-ratio (3.83a):

$$C = \frac{y_4 - y_2}{y_4 - y_3} \frac{y_1 - y_3}{y_1 - y_2} = \frac{2x^2}{x + e^x} \frac{x^2 - e^{2x}}{-4x^2 e^x} \frac{e^x}{e^x - x} = \frac{1}{2}, \tag{3.94}$$

takes the constant value (3.94) ≡ (3.93i).

3.6 Linear Second-Order Differential Equation with Variable Coefficients

The Riccati equation is a non-linear first-order differential equation (section 3.5) that can be transformed via a change of variable (subsection 3.6.1) to a linear second-order differential equation (section 3.6). The general integral of the latter is not known, and indeed a wide variety of special functions (sections 9.4–9.9 and notes 9.1–9.47) are solutions of second-order linear differential equations with simple forms of the variable coefficients (subsection 3.6.2); this leaves scant hope of finding a general integral of the Riccati equation, although several specific cases of solution are known, for example: (i) using known particular integrals (subsections 3.5.2–3.5.5); (ii) using the relation with linear second-order equations, the simplest being constant (homogeneous) coefficients [subsection 3.6.3 (3.6.4)]; and (iii) direct solution for less general sub-cases of the Riccati equation (subsection 3.6.5).

3.6.1 Transformation between the Riccati and a Linear O.D.E.

The change of variable (3.95a) leads to (3.95b, c):

$$y = -\frac{u'}{Ru}: \quad y' = -\frac{u''}{Ru} + \frac{u'^2}{Ru^2} + \frac{u'R'}{R^2 u}, \quad Ry^2 = \frac{u'^2}{Ru^2}, \tag{3.95a–c}$$

where the non-linear term involving u'^2 is the same. Thus the non-linear terms cancel when substituting (3.95a–c) in the Riccati equation (3.58):

$$u'' - \left(P + \frac{R'}{R} \right) u' + QRu = 0, \tag{3.96}$$

leading to a linear second-order differential equation (3.96). Thus *the change of variable (3.95a) ≡ (3.97a, b):*

$$y(x) = -\frac{1}{R(x)}\frac{d}{dx}\{\log[u(x)]\} \quad \Leftrightarrow \quad u(x) = C\exp\left\{-\int^x R(\xi)y(\xi)d\xi\right\},$$

$$(3.97a, b)$$

transforms (standard LVIII) the Riccati equation (3.58) into (3.96) ≡ (3.98a–c), a linear second-order differential equation (3.98a) with coefficients (3.98b, c):

$$u'' + Au' + Bu = 0: \qquad A = -P - \frac{R'}{R}, \qquad B = QR. \qquad (3.98a\text{–}c)$$

The general integral of the linear second-order differential equation (3.98a) is a linear combination of two linearly independent particular integrals (3.99b) involving two arbitrary constants:

$$C \equiv \frac{C_2}{C_1}: \quad u(x) = C_1 u_1(x) + C_2 u_2(x); \quad y(x) = -\frac{1}{R(x)}\frac{u_1'(x) + C u_2'(x)}{u_1(x) + C u_2(x)},$$

$$(3.99a\text{–}c)$$

it follows (3.95a) ≡ (3.97a) that the solution of the Riccati equation (3.58) is a bilinear function (3.99c) of a single constant of integration (3.99a).

3.6.2 Relation with the Theory of Special Functions

The transformations (3.97a, b) hold in both directions: (i) if y is a solution of the Riccati equation (3.58), then (3.97b) is a solution of the linear second-order differential equation (3.98a–c); (ii) if a solution u of the linear differential equation (3.98a–c) is known, then (3.97a) specifies a solution of the Riccati equation (3.100):

$$y' = \frac{B}{R} - \left(A + \frac{R'}{R}\right)y + Ry^2. \qquad (3.100)$$

The solution of linear second-order differential equations specifies a wide variety of special functions, for example Legendre (hyperspherical Legendre) polynomials [section III.8.3 (III.9.6)] and hypergeometric functions (notes I.23.2-I.23.3). These special functions generally have infinite representations, like the power series for the Gaussian (sections I.29.9 and II.3.9) confluent (section III.3.9) and generalized (example I.30.20 and note II.3.3) hypergeometric series and also hypogeometric series (section II.1.9). The theory of linear differential equations with variable coefficients and related special functions is considered subsequently (sections 9.4–9.9 and notes 9.1–9.47). Since there is no known general integral of (i) a linear second-order differential equation with variable coefficients (3.98a), a general integral of the (ii) Riccati equation (3.58) is also an unlikely prospect. The relation between the two may be used, for example considering the solution of the Riccati equations related [subsection 3.6.3 (3.6.4)] to the simplest linear second-order differential equations, namely with constant (homogeneous) coefficients [sections 1.3 (1.6)].

3.6.3 Riccati Equation Corresponding to Constant Coefficients

The simplest linear second-order differential equation has constant coefficients (section 1.3), and the roots (a, b) of the characteristic polynomial correspond to:

$$D \equiv \frac{d}{dx}: \quad 0 = \{P_2(D)\}u(x) = \{(D-a)(D-b)\}u(x) = u'' - (a+b)u' + abu.$$

$$(3.101a, b)$$

If the roots are distinct (3.102a), the particular integrals (3.102b, c) are linearly independent, and the general integral is (3.102d):

$$a \neq b: \quad u_1(x) = e^{ax}, \quad u_2(x) = e^{bx}, \quad u(x) = C_1 e^{ax} + C_2 e^{bx}; \qquad (3.102a\text{–}d)$$

the differential equation (3.101b) $\equiv$ (3.98a) corresponds (3.103a, b) to (standard LIX) the Riccati (3.100) equation (3.103c):

$$\{A, B\} = \{-a-b, ab\}: \quad y' = \frac{ab}{R} + \left(a+b-\frac{R'}{R}\right)y + Ry^2, \qquad (3.103a\text{–}c)$$

whose solution (3.99c) involves (3.104b) one constant of integration (3.104a):

$$C \equiv \frac{C_2}{C_1}: \quad -R(x)y(x) = \frac{C_1 a e^{ax} + C_2 b e^{bx}}{C_1 a^{ax} + C_2 e^{bx}} = \frac{a + Cb\exp[(b-a)x]}{1 + C\exp[(b-a)x]}. \qquad (3.104a, b)$$

The preceding results do not apply if the characteristic polynomial has a double root:

$$0 = \{P_2(D)\}u(x) = \{(D-a)^2\}u(x) = u'' - 2au' + a^2u, \qquad (3.105)$$

in which case the linearly independent particular integrals (3.106a, b) lead to the general integral (3.106c):

$$u_1(x) = e^{ax}, u_2(x) = xe^{ax}: \quad u(x) = e^{ax}(C_1 + C_2 x). \qquad (3.106a\text{–}c)$$

The differential equation (3.105) ≡ (3.98a) corresponding (3.107a, b) to (standard LX) the Riccati (3.100) equation (3.107c):

$$\{A,B\}=\left\{-2a,a^2\right\}: \qquad y'=\frac{a^2}{R}+\left(2a-\frac{R'}{R}\right)y+Ry^2, \qquad (3.107\text{a–c})$$

has (3.106a, b; 3.99c) general integral (3.108a, b):

$$C_0\equiv\frac{C_1}{C_2}: \qquad -R(x)y(x)=\frac{C_2+C_1a+C_2ax}{C_1+C_2x}=\frac{1+C_0+ax}{C_0+x}. \qquad (3.108\text{a, b})$$

It has been shown that the *linear second-order differential equation with constant coefficients (3.101a) whose characteristic polynomial has two distinct roots (3.101b; 3.102a) [a double root (3.105)] has: (i) linearly independent particular integrals (3.102b, c) [(3.106a, b)] and general integral (3.102d) [(3.106c)] involving two arbitrary constants of integration; and (ii) the corresponding Riccati equation (3.103c) [(3.107c)] has [standard LIX (LX)] general integral (3.104b) [(3.108b)] involving one arbitrary constant (3.104a) [≡(3.108a)].*

3.6.4 Riccati Equation Corresponding to Homogeneous Coefficients

Replacing ordinary (3.101a) by homogeneous (3.109a) derivatives leads from the linear second-order differential equation with constant (3.101b) to (section 1.6) homogeneous (3.109b) coefficients:

$$\vartheta\equiv x\frac{d}{dx}: \quad 0=\left\{P_2(\vartheta)\right\}u(x)=\left\{(\vartheta-a)(\vartheta-b)\right\}u(x)=x^2u''-(a+b-1)xu'+abu.$$

$$(3.109\text{a, b})$$

If the roots are distinct (3.110a), the particular integrals (3.110b, c) are linearly independent and the general integral (3.110d) is a linear combination:

$$a\neq b: \quad u_1(x)=x^a, \quad u_2(x)=x^b, \quad u(x)=C_1x^a+C_2x^b. \qquad (3.110\text{a–d})$$

The differential equation (3.109b) ≡ (3.98a) corresponds (3.111a, b) (standard LXI) to the Riccati (3.100) equation (3.111c):

$$\{A,B\}=\left\{\frac{1-a-b}{x},\frac{ab}{x^2}\right\}: \quad y'=\frac{ab}{x^2R}+\left[\frac{a+b-1}{x}-\frac{R'}{R}\right]y+Ry^2, \qquad (3.111\text{a–c})$$

whose general integral (3.99c) is (3.112a, b):

$$a \neq b: \qquad -Ry(x) = \frac{C_1 a x^{a-1} + C_2 b x^{b-1}}{C_1 x^a + C_2 x^b} = \frac{a + bCx^{b-a}}{x + Cx^{b-a+1}}, \qquad (3.112a, b)$$

using (3.104a).

The preceding solution (3.112b) does not (3.112a) hold if the characteristic polynomial (3.109b) has a double root:

$$0 = \{P_2(\vartheta)\} u(x) = \{(\vartheta - a)^2\} u(x) = x^2 u'' - (2a-1)xu' + a^2 u; \qquad (3.113)$$

in this case (3.114a) the linearly independent particular integrals (3.114b, c) lead (standard LXII) to the general integral (3.114d):

$$a = b: \quad u_1(x) = x^a, \quad u_2(x) = x^a \log x; \quad u(x) = x^a (C_1 + C_2 \log x). \qquad (3.114a\text{--}d)$$

The equation (3.113) ≡ (3.98a) has (3.115a, b) corresponding (3.100) Riccati equation (3.115c):

$$\{A, B\} = \left\{ \frac{1-2a}{x}, \frac{a^2}{x^2} \right\}: \quad y' = \frac{a^2}{x^2 R} + \left[\frac{2a-1}{x} - \frac{R'}{R} \right] y + Ry^2, \qquad (3.115a\text{--}c)$$

whose general integral is:

$$-R(x)y(x) = \frac{C_1 a + C_2 + C_2 a \log x}{x(C_1 + C_2 \log x)} = \frac{1 + C_0 a + a \log x}{x(C_0 + \log x)}, \qquad (3.116a, b)$$

using (3.108a).

It has been shown that *the linear second-order differential equation with homogeneous coefficients with characteristic polynomial with distinct roots (3.109b; 3.110a) [a double root (3.113)] has: (i) linearly independent particular integrals (3.110b, c) [(3.114b, c)] and general integral (3.110d) [(3.114d)] involving two arbitrary constants of integration; and (ii) the corresponding [standard LXI (LXII)] Riccati equation (3.111c) [(3.115c)] has general integral (3.112b) [(3.116b)] involving one arbitrary constant (3.104a) [(3.104a)].* Besides the solution of the generalized Riccati equation (3.58) using (i) particular integrals (subsections 3.5.2–3.5.5) or (ii) transformation to a linear second-order equation (subsections 3.6.3–3.6.4), (iii) direct solution is also possible, for example for the original Riccati equation with restricted constant coefficients (subsection 3.6.5).

3.6.5 Direct Solution of the Original Riccati Equation

The generalized Riccati equation (3.58) is an extension of the original Riccati equation (3.117d) corresponding to the coefficients (3.117a–c):

$$\{R(x),Q(x),P(x)\}=\left\{b,\frac{\lambda}{x},ax^n\right\}:\qquad y'=ax^n+\frac{\lambda}{x}y+by^2. \qquad\qquad (3.117\text{a–d})$$

The solution is sought in the form (3.118a) leading to (3.118b):

$$y(x)=x^\lambda\,u(x):\qquad\qquad x^\lambda u'=ax^n+bu^2x^{2\lambda}. \qquad\qquad (3.118\text{a, b})$$

In the particular case (3.119a) the differential equation (3.118a) simplifies to (3.119b) that is separable (3.119c):

$$n=2\lambda:\qquad x^{-\lambda}u'=a+bu^2,\qquad \int\frac{du}{a+bu^2}-C=\int x^\lambda\,dx=\frac{x^{1+\lambda}}{1+\lambda}. \qquad (3.119\text{a–c})$$

The integration on the r.h.s. of (3.119c) is immediate and that on the l.h.s. is distinct (3.120b) [(3.121b)] for (a, b) with the same (3.120a) [opposite (3.121a)] signs:

$$k\equiv\frac{b}{a}>0:\quad \int\frac{du}{a+bu^2}=\frac{1}{a}\int\frac{du}{1+ku^2}=\frac{1}{a\sqrt{k}}\int^{u\sqrt{k}}\frac{dv}{1+v^2}=\frac{1}{\sqrt{ab}}arc\tan\left(u\sqrt{k}\right)$$

$$=\frac{1}{\sqrt{ab}}arc\tan\left(x^{-\lambda}y\sqrt{\frac{b}{a}}\right),$$

$$(3.120\text{a, b})$$

$$k\equiv-\frac{b}{a}>0:\quad \int\frac{du}{a+bu^2}=\frac{1}{a}\int\frac{du}{1-ku^2}=\frac{1}{a\sqrt{k}}\int^{u\sqrt{k}}\frac{dv}{1-v^2}=\frac{1}{\sqrt{-ab}}arc\tanh\left(u\sqrt{k}\right)$$

$$=\frac{1}{\sqrt{-ab}}arg\tanh\left(x^{-\lambda}y\sqrt{-\frac{b}{a}}\right),$$

$$(3.121\text{a, b})$$

where: (i) was used the change of variable (3.122a, b) in both cases:

$$v=u\sqrt{k},\qquad du=\frac{dv}{\sqrt{k}}:\qquad\qquad u(x)=x^{-\lambda}y(x), \qquad\qquad (3.122\text{a–c})$$

(ii) used the circular (hyperbolic) tangent (II.7.124a) [(II.7.124b)]; and (iii) substituted (3.118a) ≡ (3.122c). Solving (3.120b) [(3.121b)] for y and substituting in (3.119c) it follows that:

$$y(x) = x^\lambda \times \begin{cases} \sqrt{\dfrac{a}{b}} \tan\left\{ \sqrt{ab}\left[w(x) + C \right] \right\} & \text{if} \quad ab > 0, \qquad (3.123a) \\[4mm] \sqrt{-\dfrac{a}{b}} \tanh\left\{ \sqrt{-ab}\left[w(x) + C \right] \right\} & \text{if} \quad ab < 0, \qquad (3.123b) \end{cases}$$

the solution of the (standard LXIII) Riccati equation (3.117d; 3.119a) ≡ (3.124a) is (3.123a) [(3.123b)] for (a, b) with the same (opposite) signs:

$$y' = ax^{2\lambda} + \frac{\lambda}{x} y + by^2: \qquad\qquad w(x) = \frac{x^{1+\lambda}}{1+\lambda}. \qquad (3.124a, b)$$

The function (3.124b) appears in both cases (3.123a) [(3.123b)] in the general integral (3.125b) [(3.126b)]:

$$C_+ \equiv \tan\left(C\sqrt{ab} \right): \qquad x^{-\lambda} y(x) \sqrt{\frac{b}{a}} = \tan\left[\sqrt{ab}\,(w + C) \right] = \frac{\tan\left(w\sqrt{ab} \right) + C_+}{1 - C_+ \tan\left(w\sqrt{ab} \right)},$$

$$(3.125a\text{–}c)$$

$$C_2 = \tanh\left(C\sqrt{-ab} \right): \qquad x^{-\lambda} y \sqrt{\frac{b}{a}} = \tanh\left[\sqrt{-ab}\,(w + C) \right] = \frac{\tanh\left(w\sqrt{-ab} \right) + C_-}{1 + C_- \tanh\left(w\sqrt{-ab} \right)},$$

$$(3.126a\text{–}c)$$

that is, a bilinear function (3.125c) [(3.126c)] of the arbitrary constant (3.125a) [(3.126a)]. The passage from (3.125b) [(3.126b)] to (3.125c) [(3.126c)] used the formula of addition of arguments for the circular (hyperbolic) tangent (II.5.48a) [(II.5.50a)]. Another Riccati equation is considered in the EXAMPLE 10.6.

3.7 Non-Linear First-Order Homogeneous Differential Equation

The designation homogeneous or inhomogeneous differential equation is used with three distinct meanings, of which two are (one is not) related. The first meaning (i) is a differential equation (1.1a, b) is homogeneous (inhomogeneous) if it has (does not have) the trivial solution $y = 0$; for example the linear differential equation is homogeneous (inhomogeneous) if unforced (1.54) [forced (1.33)]. The designation forced (unforced) linear differential equation is preferable, because it is unrelated to homogeneous (inhomogeneous) differential equation in the other two meanings. The second meaning (ii) is the linear differential equation with homogeneous coefficients, that can be unforced (1.291a) [forced (1.285)] and of any order; it involves homogeneous (1.300b) rather than ordinary (1.53a) derivatives, and is unchanged by a rescaling (1.300a, b) of the independent variable. In the case of a linear first-order differential equation with homogeneous coefficients the derivative y' is a linear function of y/x. This can be generalized to the third meaning (iii) of a homogeneous first-order differential equation (section 3.7) for which the derivative y' is any function, generally non-linear, of the ratio of the dependent and independent variables y/x. The linear differential equation with homogeneous coefficients (section 1.6) can be solved for any order using the roots of the corresponding characteristic polynomial. The non-linear case can be solved by quadratures for the first-order differential equation (subsection 3.7.2). The non-linear homogeneous first-order differential equation is unaffected by the same re-scaling of both the independent and dependent variables (subsections 3.7.1); in the case of the linear homogeneous differential equation of any order (section 1.6), the invariance also applies to re-scaling of the independent variable (subsection 1.6.1). Thus a homogeneous differential equation can be solved: (i) for any order in the linear case (sections 1.6–1.8); (ii) in the non-linear case for the first-order (subsections 3.7.1–3.7.3).

3.7.1 Re-Scaling of Independent and Dependent Variables

A **first-order homogeneous differential equation** (standard LXIV) expresses the derivative as a function of the ratio of the dependent and independent variables:

$$\frac{dy}{dx} = f\left(\frac{y}{x}\right). \tag{3.127}$$

If the function (standard LXV) is linear (3.128a), the differential equation (3.127) becomes (3.128b):

$$f(v) = av + b: \quad xy' = ay + bx, \quad \{A_1, A_0\} = \{1, -a\}, \quad B(x) = bx, \quad \text{(3.128a–e)}$$

that is equivalent to the linear differential equation with homogeneous coefficients (1.285) of order one $N = 1$, with coefficients (3.128c, d) and forcing term (3.128e). *The non-linear homogeneous first-order differential equation (3.127)* is **invariant** (3.129d) with regard to re-scaling both the dependent and independent variables (3.121b, c) by the same constant factor (3.129a):

$$\lambda = const: \qquad \{\bar{y}, \bar{x}\} = \lambda \{y, x\}: \qquad \frac{d\bar{y}}{d\bar{x}} = f\left(\frac{\bar{y}}{\bar{x}}\right). \qquad (3.129a\text{–}d)$$

In the linear case the invariance of the homogeneous differential equation of first-order (3.128b) [any order (1.285)] also applies to re-scaling only the independent variable:

$$b = 0; \qquad \lambda = const: \qquad \bar{x} = \lambda x, \qquad \bar{x} y' = ay, \qquad (3.130a\text{–}d)$$

in the absence of forcing (3.130a–d) [(1.300a–c)].

3.7.2 Linear and Non-Linear Homogeneous Differential Equations

The form of the homogeneous first-order differential equation (3.127) suggests the change of variable (3.131a):

$$v = \frac{y}{x}: \qquad y' = (vx)' = v'x + v, \qquad (3.131a, b)$$

that transforms (3.131b) to a differential equation (3.132a) that is separable (3.132b):

$$f(v) - v = x\frac{dv}{dx}, \qquad \int \frac{dv}{f(v) - v} + \log C = \int \frac{dx}{x} = \log x; \qquad (3.132a, b)$$

the solution follows by integration or quadrature:

$$x = C \exp\left\{\int^{y/x} \frac{dv}{f(v) - v}\right\}. \qquad (3.133)$$

Thus, *the (standard LXIV) homogeneous non-linear first-order differential equation (3.127) has general integral (3.133).*
   In the linear case (3.128a) use (3.134a) of (3.133) leads to (3.134b):

$$f(v) = av + b: \quad x = C \exp\left\{\int^{y/x} \frac{dv}{(a-1)v + b}\right\} = C \exp\left\{\frac{\log\left[(a-1)y/x + b\right]}{a-1}\right\},$$

$$(3.134a, b)$$

as the general integral (3.135b, c):

$$a \neq 1: \qquad x = C\left[(a-1)\frac{y}{x}+b\right]^{1/(a-1)}, \qquad y(x) = \frac{x}{a-1}\left[\left(\frac{x}{C}\right)^{a-1}-b\right], \qquad (3.135a\text{–}c)$$

of the linear homogeneous first-order differential equation (3.128a, b), in the case (3.135a); the remaining case (3.136a) leads (3.136b) by (3.133) to (3.136c):

$$a = 1: \quad f(x) = v+b, \quad x = C_0 \exp\left\{\frac{1}{b}\int^{y/x} dv\right\} = C_0 \exp\left(\frac{y}{bx}\right), \qquad (3.136a\text{–}c)$$

showing that (3.136e):

$$xy' = y+b, \qquad y(x) = bx\log\left(\frac{x}{C_0}\right), \qquad (3.136d,\ e)$$

is the complete integral of (3.136d).

The linear homogeneous differential equation (3.128b) can be also be solved using (1.292a–c) the characteristic polynomial (3.137b):

$$a \neq 1: \qquad P_1(\delta) = \delta - a: \quad y(x) = C_1 x^a + \frac{bx}{P_1(1)} = C_1 x^a + \frac{bx}{1-a}, \qquad (3.137a\text{–}c)$$

leading to the complete integral (3.137c) if (3.137a) is met; otherwise (3.138a) the general integral is (3.138b):

$$a = 1: \qquad\qquad y(x) = C_2 x + \frac{bx\log x}{P_1'(1)} = bx\log x + C_2 x. \qquad (3.138a,\ b)$$

Thus *the (standard LXV) first-order linear homogeneous differential equation (3.128b) ≡ (3.127; 3.128a) [(3.136d)] can be solved by the method of change of variable (characteristic polynomial) leading to the same general integral, namely for $a \neq 1 (a = 1)$ in (3.135a, c) ≡ (3.137a, c) [(3.136a–e) ≡ (3.138a, b)] with the constants of integration related by (3.139a) [(3.139b)]:*

$$\frac{1}{C_1} = (a-1)C^{a-1}, \qquad\qquad C_2 = -b\log C_0. \qquad (3.139a,\ b)$$

In the case of the non-linear first-order homogeneous differential equation, only the first method of change of variable can be expected to lead to a solution (subsection 3.7.3).

3.7.3 Change of Variable and Integration by Quadratures

As an example, the solution of the homogeneous non-linear differential equation (3.140a):

$$y' = \frac{y+x}{y-x}, \quad C = y^2 - 2xy - x^2, \quad y = x \pm \sqrt{2x^2 + C}, \quad \text{(3.140a–c)}$$

is (3.140b) ≡ (3.140c), because (3.140a) corresponds to (3.127) with $f(v)$ specified by (3.141a):

$$f(v) = \frac{v+1}{v-1}: \quad \frac{x}{\sqrt{C}} = \exp\left\{\int^{y/x}\left(\frac{v+1}{v-1} - v\right)^{-1} dv\right\} = \exp\left\{-\int^{y/x}\frac{v-1}{v^2 - 2v - 1} dv\right\}$$

$$= \exp\left\{-\frac{1}{2}\log\left(\frac{y^2}{x^2} - 2\frac{y}{x} - 1\right)\right\} = \left|\frac{y^2}{x^2} - 2\frac{y}{x} - 1\right|^{-1/2},$$

$$\text{(3.141a, b)}$$

so that (3.133) becomes (3.141b); the solution (3.141b) is an implicit relation (3.140b) between x and y that may be rendered explicit (3.140c) by solving it as a quadratic equation in y.

Another example is the non-linear first-order homogeneous differential equation (3.142a) corresponding (3.127) to (3.142b):

$$y' = \frac{x^2 + y^2}{xy - x^2}: \quad f(v) = \frac{v^2 + 1}{v - 1}, \quad f(v) - v = \frac{v+1}{v-1}; \quad \text{(3.142a–c)}$$

substituting (3.142c) in the quadrature (3.133) leads to (3.143):

$$\log\left(\frac{x}{C}\right) = \int^{y/x}\frac{v-1}{v+1} dv = \int^{y/x}\left(1 - \frac{2}{v+1}\right) dv = \frac{y}{x} - 2\log\left(1 + \frac{y}{x}\right), \quad \text{(3.143)}$$

corresponding to (3.144a):

$$\frac{x}{C} = \left(1 + \frac{y}{x}\right)^{-2}\exp\left(\frac{y}{x}\right), \quad (x+y)^2 = Cx\exp\left(\frac{y}{x}\right), \quad \text{(3.144a, b)}$$

that is, the general integral (3.144a) ≡ (3.144b) of (3.142a).

3.8 Exact and Inexact First-Order Differentials

A first-order differential equation is equivalent to a differential in two variables, leading to two cases depending on whether the vector of coefficients has zero (non-zero) curl; whether the differential is an exact (inexact) differential [subsections 3.8.1–3.8.2 (3.8.3–3.8.11)], that is, whether it coincides (does not coincide) with the differential of a function. Thus, the integrability condition for a differential is that the vector of coefficients has zero curl (subsection 3.8.1); in this case it is the differential of a function, that, equated to an arbitrary constant, specifies the general integral. The general integral may be found by inspection (subsection 3.8.1) or by formal integration (subsection 3.8.2).

If the differential in two variables is inexact, it can be made exact (subsection 3.8.3) and hence, solved by multiplying by an integrating factor (subsection 3.8.4); the integrating factor satisfies a partial differential equation (subsection 3.8.5) and any solution will do, the simpler the better. These are infinitely many integrating factors (subsection 3.8.6) and the ratio of two of them is a solution of the inexact differential. To obtain the solution of the inexact differential it is sufficient to find one integration factor, for example: (i) a function of one variable (subsection 3.8.7); (ii) a symmetric function of two variables, such as [subsection 3.8.8 (3.8.9)] the sum (product); (iii) the ratio of two variables (subsection 3.8.10); or (iv) a non-symmetric non-homogeneous function of two variables (subsection 3.8.11).

3.8.1 Vector of Coefficients with Zero Curl

An explicit (3.9a) first-order differential equation can be written as a **differential** in two variables (3.145b):

$$\bar{X} = \{X, Y\}: \qquad 0 = X(x,y)dx + Y(x,y)dy = \bar{X}.d\bar{x}, \qquad (3.145\text{a, b})$$

that is, all the differential equations considered before (sections 3.1–3.7) are instances of the type (3.145b) that specifies (Figure 3.6) a family of plane curves orthogonal to the vector field (3.145a). This is an **exact differential** (standard LXVI) if it coincides with the differential (3.146b) of a differentiable function (3.146a):

$$\Phi \in \mathcal{D}\left(|R^2\right): \qquad 0 = d\Phi = \frac{\partial\Phi}{\partial x}dx + \frac{\partial\Phi}{\partial y}dy; \qquad \Phi(x,y) = C. \qquad (3.146\text{a–c})$$

In this case the general integral is obtained by equating the function to an arbitrary constant (3.146c). A necessary condition that (3.145b) is an exact

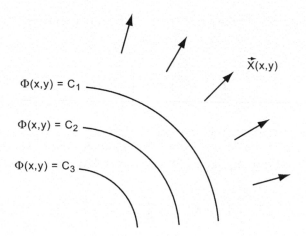

FIGURE 3.6
A family of plane curves in the plane may be specified by its tangents (normals) as a solution [Figure 3.2 (3.6)] of the differential equation (associated differential in two-variables) of first-order.

differential is that a differentiable function Φ exists such that (3.145b) $\equiv$ (3.146b), implying that the vector field equals its gradient (3.147a) $\equiv$ (3.147b, c):

$$\vec{X} = \nabla\Phi: \qquad X = \frac{\partial\Phi}{\partial x}, \qquad Y = \frac{\partial\Phi}{\partial y}. \qquad (3.147a\text{–}c)$$

If the function has continuous second-order derivatives (3.148a), then the cross-derivatives are equal (3.148b), leading to the **integrability condition** (3.148c) for the exact differential (3.145b):

$$\Phi \in C^2\left(|R^2\right): \qquad \frac{\partial^2\Phi}{\partial x\,\partial y} - \frac{\partial^2\Phi}{\partial y\,\partial x} = 0 \quad \Leftrightarrow \quad \Omega \equiv \frac{\partial Y}{\partial x} - \frac{\partial X}{\partial y} = 0. \qquad (3.148a\text{–}c)$$

The conditions (3.148a–c) are equivalent to:

$$\vec{X} \in C^1\left(|R^2\right): \qquad 0 = \nabla \wedge \vec{X} = \begin{vmatrix} \dfrac{\partial}{\partial x} & \dfrac{\partial}{\partial y} \\ X & Y \end{vmatrix} = \left(\frac{\partial Y}{\partial x} - \frac{\partial X}{\partial y}\right)\vec{e}_z = \vec{e}_z\Omega, \qquad (3.149a,\ b)$$

the statement that the vector of coefficients (3.145a) is continuously differentiable (3.149a) and has zero curl (3.149b).

The integrability condition (3.148a–c) $\equiv$ (3.149a, b) is not only necessary but also sufficient, that is, if (3.149b) is met, then (3.145b) is an exact differential.

The proof uses the **Stokes (1875) or curl loop theorem** (III.5.210a, b) ≡ (3.150a, b):

$$\vec{X} \in C^1(|D): \qquad\qquad \int_D (\nabla \wedge \vec{X}).d\vec{S} = \int_{\partial D} \vec{X}.d\vec{x}, \qquad\qquad (3.150a, b)$$

that applies to a continuously differentiable vector (3.150a) in a plane domain bounded by a closed regular curve (3.150b). The integrability condition of zero curl (3.149b) ≡ (3.151a) implies (3.151b) by the Stokes theorem (3.150a, b):

$$\nabla \wedge \vec{X} = 0: \qquad 0 = \int_D (\nabla \wedge \vec{X}).d\vec{S} = \int_{\partial D} \vec{X}.d\vec{x} = \int_{\partial D} d\Phi; \qquad (3.151a, b)$$

from (3.151b) it follows that the integrand is an exact differential (3.152) ≡ (3.145b):

$$d\Phi = \vec{X}.d\vec{x} = X\,dx + Y\,dy = 0, \qquad\qquad (3.152)$$

and (3.152) is satisfied by (3.146c). It has been shown that *a necessary and sufficient condition that the differential (3.145b) ≡ (3.153b) with continuously differentiable vector of coefficients (3.153a) is exact (standard LXVI) is that the curl is zero (3.149b) ≡ (3.153c)*:

$$X, Y \in C^1(|R^2): \quad 0 = X\,dx + Y\,dy = d\Phi \quad \Leftrightarrow \quad \Omega \equiv \frac{\partial Y}{\partial x} - \frac{\partial X}{\partial y} = 0 \Leftrightarrow \Phi(x, y) = C;$$

$$(3.153a\text{--}d)$$

in that case the general integral is specified equating to an arbitrary constant the function (3.153d), that that is twice continuously differentiable (3.148a).

For example, the equation (3.154a):

$$0 = 2y^3 x\,dx + 3x^2 y^2\,dy = d(x^2 y^3), \qquad\qquad x^2 y^3 = C, \qquad (3.154a, b)$$

has general integral (3.154b). It can be checked that for the differential (3.154a) ≡ (3.145b) with coefficients (3.155a, b):

$$X = 2xy^3, \qquad Y = 3x^2 y^2: \qquad \frac{\partial Y}{\partial x} = 6xy^2 = \frac{\partial X}{\partial y}, \qquad\qquad (3.155a\text{--}c)$$

that the integrability condition (3.148c) ≡ (3.155c) is met.

3.8.2 Formal Integration of an Exact Differential

In the case of inability to identify the integral (3.146b, c) on inspection of the (standard LXVI) exact equation (3.145b), it can be obtained by formally integrating (3.147b, c). The integration of (3.147b) [(3.147c)] leads to (3.156a) [(3.156b)]:

$$\Phi(x,y) = \int X(x,y)dx + u(y) = \int Y(x,y)dy + v(x), \qquad (3.156a, b)$$

involving the arbitrary function $u(y)$ [($v(x)$]) that is obtained (3.157a) [(3.157b)] by substitution in (3.147c) [(3.147b)]:

$$\frac{du}{dy} = \frac{\partial \Phi}{\partial y} - \int \frac{\partial X}{\partial y}dx = Y - \int \frac{\partial X}{\partial y}dx, \qquad \frac{dv}{dx} = \frac{\partial \Phi}{\partial x} - \int \frac{\partial Y}{\partial x}dy = X - \int \frac{\partial Y}{\partial x}dy.$$

$$(3.157a, b)$$

Thus, *if the differential (3.145b) is exact (3.147b, c), its general integral (3.146c) can be obtained from the quadrature (3.156a) [(3.156b)] involving an arbitrary function $u(y)$ [$v(x)$] satisfying (3.157a) [(3.157b)].*

For example, the first-order differential equation or differential in two variables:

$$y \log ye^x (1+x)dx + xe^x (1+\log y)dy = 0, \qquad (3.158)$$

has (3.145b) coefficients:

$$X = y \log ye^x (1+x), \qquad Y = xe^x (1+\log y), \qquad (3.159a, b)$$

showing that it satisfies the integrability condition (3.148c):

$$\frac{\partial Y}{\partial x} = e^x (1+x)(1+\log y) = \frac{\partial X}{\partial y}, \qquad (3.160)$$

and hence is an exact differential. The integral (3.156a) is given by:

$$\Phi(x,y) = y \log y \int e^x (1+x)dx + u(y) = y \log ye^x x + u(y), \qquad (3.161)$$

where the function $u(y)$ satisfies (3.157a) leading to (3.162a):

$$\frac{du}{dy} = (1+\log y)\left\{ xe^x - \int e^x (1+x)dx \right\} = 0; \quad \rightarrow \quad u = C = 0, \qquad (3.162a\text{--}c)$$

and hence is a constant $u = C$; any value may be chosen for the constant, such as zero (3.162c), so that the general integral (3.161) of (3.158) is by (3.163):

$$\Phi(x,y) = xe^x y \log y = C; \tag{3.163}$$

it can be checked that the differential of (3.163) is (3.158). Three other exact differentials are considered in the EXAMPLE 10.8.

3.8.3 Transformation between Exact and Inexact Differentials

The solution of a differential (3.145b) is unchanged when multiplying by an arbitrary function $\lambda(x,y)$; this multiplication may change an exact differential into an inexact differential, since if (X, Y) meet the integrability condition (3.148c), then generally $(\lambda X, \lambda Y)$ do not; the reverse is always possible, that is, if (X, Y) do not meet the integrability condition (3.148c), a function λ can be found so that $(\lambda X, \lambda Y)$ do. This function is called an **integrating factor** and before addressing its general properties (subsections 3.8.4–3.8.5) and giving examples (subsections 3.8.6–3.8.11), the two preceding exact differentials (3.154a, b) and (3.158; 3.163) are reconsidered. The exact differential (3.154a) [(3.158)] in two variables is equivalent to the first-order differential equation (3.164a) [(3.165a)]:

$$y' = -\frac{2xy^3}{3x^2y^2} = -\frac{2y}{3x}, \tag{3.164a, b}$$

$$y' = -\frac{e^x(1+x)y \log y}{xe^x(1+\log y)} = -\left(1+\frac{1}{x}\right)\frac{y \log y}{1+\log y}, \tag{3.165a, b}$$

that may be simplified to (3.164b) [(3.165b)]. The latter lead to the differentials (3.166a) [(3.167a)]:

$$2y\,dx + 3x\,dy = 0: \qquad \frac{\partial Y}{\partial x} = 3 \neq 2 = \frac{\partial X}{\partial y}, \tag{3.166a, b}$$

$$\left(1+\frac{1}{x}\right)y \log y\,dx + (1+\log y)\,dy = 0: \qquad \frac{\partial Y}{\partial x} = 0 \neq \left(1+\frac{1}{x}\right)(1+\log y) = \frac{\partial X}{\partial y}, \tag{3.167a, b}$$

that are no longer exact (3.166b) [(3.167b)] because they do not satisfy the integrability condition (3.148c); thus, simplifying an exact differential form by dividing out by a common factor may render it inexact.

Conversely, multiplying an inexact differential by a suitable integrating factor may render it exact, and hence lead to the general integral. For example the differential:

$$0 = d\Phi = \tan y \, dx + \tan x \, dy,$$ (3.168)

is inexact since the coefficients (3.169a, b) do not satisfy (3.169a) the integrability condition (3.148c):

$$\{X,Y\} = \{\tan y, \tan x\}: \qquad \frac{\partial Y}{\partial x} = \sec^2 x \neq \sec^2 y = \frac{\partial X}{\partial y}.$$ (3.169a–c)

When multiplied by the integrating a factor (3.169a):

$$\lambda = \cos x \cos y: \qquad 0 = \lambda \, d\Phi = \cos x \sin y \, dx + \cos y \sin x \, dx,$$ (3.169a, b)

the inexact differential form (3.168) becomes (3.169b), which is an exact differential (3.170a):

$$0 = \lambda \, d\Phi = d(\sin x \sin y), \qquad \sin x \sin y = C,$$ (3.170a, b)

and leads to the general integral (3.170b) of (3.168). This raises the question of the existence of an integrating factor for an inexact differential in two variables that is proved next (subsection 3.8.4).

3.8.4 Existence of Integrating Factor for an Inexact Differential

Any explicit differential equation (3.145b), even if it is inexact, can be put in the form:

$$\frac{dy}{dx} = -\frac{X(x,y)}{Y(x,y)}.$$ (3.171)

It has a solution (3.146c) implying (3.146b) and hence:

$$\frac{dy}{dx} = -\frac{\partial \Phi / \partial x}{\partial \Phi / \partial y}.$$ (3.172)

Equating (3.171) $\equiv$ (3.172) leads to (3.173a):

$$\frac{X}{Y} = \frac{\partial \Phi / \partial x}{\partial \Phi / \partial y}, \qquad \frac{1}{X}\frac{\partial \Phi}{\partial x} = \frac{1}{Y}\frac{\partial \Phi}{\partial y} \equiv \lambda,$$ (3.173a–c)

that is equivalent to (3.173b); the common ratio (3.173c) is called the **integrating factor** because it leads from the inexact (3.174a) to the exact (3.174b) differential:

$$\frac{\partial Y}{\partial x} \neq \frac{\partial X}{\partial y} : \qquad \lambda\left(X\,dx + Y\,dy\right) = \frac{\partial \Phi}{\partial x}dx + \frac{\partial \Phi}{\partial y}dy = d\Phi. \qquad (3.174a, b)$$

It follows that *the inexact (3.174a) differential (3.145b), has an integrating factor* λ *satisfying (3.173b, c) that leads by multiplication to an exact differential (3.174b), specifying the general integral (3.146c).* It could be expected that since the differential (3.145b) involves two functions (X, Y): (i) a single function Φ cannot lead to an exact differential, unless the integrability condition (3.148c) is satisfied; and (ii) two functions (λ, Φ) could lead to an exact differential (3.174b), as has just been proved, and in addition there are infinitely many choices of λ, as shown next (subsections 3.8.5–3.8.6).

3.8.5 Partial Differential Equation for the Integrating Factor

In order to determine the equation satisfied by the **integrating factor,** (3.173b, c) is substituted in (3.148a, b) $\equiv$ (3.175a, b) leading to (3.175c):

$$\Phi \in C^2\left(|R^2\right): \qquad 0 = \frac{\partial^2 \Phi}{\partial x\,\partial y} - \frac{\partial^2 \Phi}{\partial y\,\partial x} = \frac{\partial}{\partial x}(\lambda Y) - \frac{\partial}{\partial y}(\lambda X), \qquad (3.175a\text{–}c)$$

that is equivalent to (3.175c) $\equiv$ (3.176b):

$$\lambda \in C^1\left(|R^2\right): \qquad \Omega \equiv \frac{\partial Y}{\partial x} - \frac{\partial X}{\partial y} = \frac{1}{\lambda}\left(X\frac{\partial \lambda}{\partial y} - Y\frac{\partial \lambda}{\partial x}\right) = \left(X\frac{\partial}{\partial y} - Y\frac{\partial}{\partial x}\right)\log \lambda,$$

$$(3.176a, b)$$

where the integrating factor must be continuously differentiable (3.176a). Thus *an inexact (3.177c) differential (3.177a, b):*

$$X, Y \in C^1\left(|R^2\right): \qquad 0 = X(x,y)dx + Y(x,y)dy, \qquad \Omega = \frac{\partial Y}{\partial x} - \frac{\partial X}{\partial y} \neq 0, \qquad (3.177a\text{–}c)$$

has a differentiable (3.176a) integrating factor (3.174a, b) that satisfies the partial differential equation (3.176b). The partial differential equation (3.176b) is generally more complicated than the inexact differential (3.145b); however, only a particular solution of (3.176b), the simpler the better, is needed to obtain the general integral of (3.145b). If (3.145b) is an exact differential (3.153c), the integrating factor (3.176b) is any constant; it follows that the integrating factor is not unique as shown next.

3.8.6 Infinitely Many Integrating Factors

The integrating factor λ for a general inexact differential (3.145b) is a function $\lambda(x,y)$ that is not unique. For example, substituting (3.178a) in (3.176b) leads to (3.178b):

$$\lambda = \mu\Phi: \qquad \frac{\partial Y}{\partial x} - \frac{\partial X}{\partial y} + \left(Y\frac{\partial}{\partial x} - X\frac{\partial}{\partial y}\right)\log\mu = \left(X\frac{\partial}{\partial y} - Y\frac{\partial}{\partial x}\right)\log\Phi; \quad \text{(3.178a, b)}$$

from (3.173a) ≡ (3.179a) if follows that the r.h.s. of (3.178b) vanishes, leading to (3.179b):

$$Y\frac{\partial\Phi}{\partial x} = X\frac{\partial\Phi}{\partial y}: \qquad \frac{\partial Y}{\partial x} - \frac{\partial X}{\partial y} + \left(Y\frac{\partial}{\partial x} - X\frac{\partial}{\partial y}\right)\log\mu = 0, \qquad \text{(3.179a, b)}$$

that proves that μ is also an integrating factor because it satisfies (3.176b). This result can be stated in two equivalent ways: *(a) if λ, μ are integrating factors (3.176b; 3.179b) of the (standard LXVII) inexact differential (3.145b), their ratio $\Phi = \lambda / \mu$ in (3.178a) is a solution (3.146c); or (b) if Φ is a solution (3.146c) of the inexact differential (3.145b) and μ an integrating factor (3.179b), then (3.178a) and also (3.180b) are also integrating factors (3.176b), where (3.180a) is any differentiable function of the solution:*

$$G \in \mathcal{D}(|R): \qquad\qquad \lambda = \mu G(\Phi). \qquad\qquad \text{(3.180a, b)}$$

The latter result follows from either of the statements: (i) *if (3.148c) is a solution of the inexact differential (3.145b), then (3.181a) is also a solution:*

$$G\big(\Phi(x,y)\big) = G(C) \equiv C_1; \qquad \lambda G(\Phi)\big(X\,dx + Y\,dy\big) = G(\Phi)d\Phi, \quad \text{(3.181a, b)}$$

(ii) *if the inexact differential (3.145b; 3.174a) is transformed into an exact differential (3.174b) by multiplying by the integrating factor λ, then (3.181b) is also an exact differential proving that λG is also an integrating factor.*

3.8.7 Integrating Factor Depending Only on One Variable

Since the integrating factor is not unique it may be sought in several forms that exist only if they satisfy (3.176a, b). The simplest integrating factor that is a function of only the variable $x(y)$ is given (3.176b) by (3.182a) [(3.182b)]:

$$\Omega \equiv \frac{\partial Y}{\partial x} - \frac{\partial X}{\partial y} = -Y\frac{d}{dx}\big\{\log[\lambda(x)]\big\}, \qquad X\frac{d}{dy}\big\{\log[\lambda(y)]\big\}, \quad \text{(3.182a, b)}$$

implying that *the inexact (3.177c) differential (3.177a, b) whose coefficients satisfy (3.183a) [(3.184a)] has [standard LXVIII (LXVIX)] an integrating factor (3.174) that depends (3.183b) [(3.184b)] only on x(y)*:

$$\frac{1}{Y}\left(\frac{\partial X}{\partial y}-\frac{\partial Y}{\partial x}\right)=-\frac{\Omega}{Y}\equiv f(x):\qquad \lambda(x)=\exp\left\{\int^{x}f(\xi)d\xi\right\},\qquad\text{(3.183a, b)}$$

$$\frac{1}{X}\left(\frac{\partial Y}{\partial x}-\frac{\partial X}{\partial y}\right)=\frac{\Omega}{X}\equiv g(y):\qquad \lambda(y)=\exp\left\{\int^{y}g(\eta)d\eta\right\}.\qquad\text{(3.184a, b)}$$

As an example, consider (standard LI) the linear unforced first-order differential equation (3.19a) ≡ (3.185a) written as two distinct differentials (3.185b, c):

$$y'=Py:\qquad\qquad Py\,dx-dy=0\quad\Leftrightarrow\quad P\,dx-\frac{dy}{y}=0.\qquad\text{(3.185a–c)}$$

One differential (3.185b) ≡ (3.186a) [(3.185c) ≡ (3.187a)] is inexact (3.186b, c) [exact (3.187b, c)] with non-zero (zero) curl (3.186d) [(3.187d)]:

$$Py\,dx-dy=0:\qquad\{\,X,Y\,\}=\{\,Py,-1\,\},\qquad\Omega=-P,\qquad\text{(3.186a–d)}$$

$$P\,dx-\frac{dy}{y}=0:\qquad\{\,X,Y\,\}=\left\{P,-\frac{1}{y}\right\},\qquad\Omega=0.\qquad\text{(3.187a–d)}$$

The exact (3.187d) differential (3.185c) is separable and leads (3.19b) ≡ (3.188a) to the general integral (3.20a, b) ≡ (3.188b):

$$\int P\,dx-\log y=-\log C:\qquad y(x)=C\exp\left\{\int^{x}P(\xi)d\xi\right\}.\qquad\text{(3.188a, b)}$$

The inexact differential (3.186a–d) leads (3.183a) [(3.184a)] to integrating factor (3.183b) [(3.184b)] depending only on x(y), that is (3.189a, b) [(3.190a, b)]:

$$-\frac{\Omega}{Y}=-P=\frac{d}{dx}(\log\lambda):\qquad \lambda(x)=\exp\left\{-\int P\,dx\right\},\qquad\text{(3.189a, b)}$$

$$\frac{\Omega}{X}=-\frac{1}{y}=\frac{d}{dy}(\log\mu):\qquad \mu(y)=\frac{1}{y}.\qquad\text{(3.190a, b)}$$

Multiplying the inexact differential (3.185b) ≡ (3.186a) by the integrating factors (3.189b) [(3.190b)] leads to the exact differentials (3.191) [(3.92)]:

$$0 = \exp\left(-\int P\,dx\right)(Py\,dx - dy) = -d\left[y\exp\left(-\int P\,dx\right)\right], \qquad (3.191)$$

$$0 = \frac{1}{y}(Py\,dx - dy) = P\,dx - \frac{dy}{y} = d\left[\int P\,dx - \log y\right]. \qquad (3.192)$$

From (3.191) [(3.192)] follow the general integrals (3.193a) [(3.193b)] that coincide with (3.188b) with suitable identification of the constants of integration:

$$y\exp\left\{-\int^{x} P(\xi)\,d\xi\right\} = C, \qquad \int^{x} P(\xi)\,d\xi - \log y = -\log C; \qquad (3.193a, b)$$

the two integrating factors are independent because one depends (3.189b) on x and the other (3.190b) on y and their ratio:

$$C = \frac{\lambda(x)}{\mu(y)} = y\exp\left\{-\int^{x} P(\xi)\,d\xi\right\}, \qquad (3.194)$$

is the general integral (3.188b) ≡ (3.193a) ≡ (3.193b) ≡ (3.194) in agreement with (3.180b). The four methods of solution of the linear unforced first-order differential equation (3.19a) have been used to illustrate the properties of integrating factors. The (standard LII) linear forced first-order differential equation (3.27) can also be solved (EXAMPLE 10.7) using an integrating factor depending only on x. Between the simplest integrating factors depending (subsection 3.8.7) only on one variable (x or y) and the most general (subsection 3.8.11) depending arbitrarily on both variables (x and y), are the integrating factors depending: (i/ii) symmetrically on the two variables through [subsection 3.8.8 (3.8.9)] their sum (product); and (iii) on the ratio of variables (subsection 3.8.10) as for an homogeneous equation (section 3.7).

3.8.8 Integrating Factor Depending Symmetrically on Two Variables

An example of a symmetric integrating factor is a function of the sum of the variables (3.195a) implying (3.195b, c):

$$z = x + y: \qquad \frac{\partial}{\partial x} = \frac{\partial z}{\partial x}\frac{d}{dz} = \frac{d}{dz} = \frac{\partial z}{\partial y}\frac{d}{dz} = \frac{\partial}{\partial y}; \qquad \Omega = (X - Y)\frac{d}{dz}\left\{\log[\lambda(z)]\right\},$$

$$(3.195a\text{--}d)$$

substituting (3.195b, c) in (3.176b) follows that *if (standard LXX) the coefficients of the inexact (3.177c) differential (3.177a, b) satisfy (3.196a):*

$$\frac{1}{X-Y}\left(\frac{\partial Y}{\partial x}-\frac{\partial X}{\partial y}\right)=\frac{\Omega}{X-Y}=h(x+y): \qquad \lambda(x+y)=\exp\left\{\int^{x+y}h(z)dz\right\},$$

(3.196a, b)

there is an integrating factor (3.174b) depending only on the sum of variables (3.196b). For example, the differential (3.197a) has coefficients (3.197b, c) leading to a non-zero curl (3.197d) proving that it is inexact:

$$0=2y\,dx+(x+3y)dy: \qquad \{X,Y\}\ =\ \{2y,x+3y\}, \qquad \Omega=-1. \qquad (3.197\text{a–d})$$

The condition (3.196a) is met (3.198a):

$$\frac{\Omega}{X-Y}=\frac{1}{x+y}=h(x+y): \qquad \lambda(x+y)=\exp\left\{\int^{x+y}\frac{dz}{z}\right\}=x+y, \qquad (3.198\text{a, b})$$

leading (3.196b) to the integrating factor (3.198b). Multiplying the inexact differential (3.197a) by the integrating factor (3.198b) leads to the exact differential (3.199a):

$$0=(x+y)\left[2y\,dx+(x+3y)dy\right]=(x+y)\left[2y(dx+dy)+(x+y)dy\right]$$

$$=yd\left[(x+y)^2\right]+(x+y)^2\,dy=d\left[y(x+y^2)\right],$$

(3.199a)

and hence to:

$$C=y(x+y)^2=x^2y+2xy^2+y^3,$$

(3.199b)

as the general integral (3.199a, b) of the inexact differential (3.197a).

3.8.9 Integrating Factor Depending on the Sum or Product of Variables

An integrating factor involving symmetrically the two variables $(x,\ y)$ depends either on the sum (3.195a–d) [or on the product (3.200a–c)]:

$$u=xy: \qquad \left\{\frac{\partial}{\partial x},\frac{\partial}{\partial y}\right\}=\left\{\frac{\partial u}{\partial x},\frac{\partial u}{\partial y}\right\}\frac{d}{du}=\{y,x\}\frac{d}{du}; \qquad \Omega=(Xx-Yy)\frac{d}{du}\{\log[\lambda(u)]\};$$

(3.200a–d)

substitution of (3.200b, c) in (3.176b) leads to (3.200d), implying:

$$\frac{\Omega}{Xx - Yy} = \frac{\partial Y/\partial x - \partial X/\partial y}{Xx - Yy} = j(xy): \qquad \lambda(xy) = \exp\left\{\int^{xy} j(u)\,du\right\}. \qquad \text{(3.201a, b)}$$

It has been shown that *if (standard LXXI) the inexact (3.177c) differential (3.177a, b) has coefficients satisfying (3.201a), then there exists an integrating factor (3.174b) depending only on the product of variables (3.201b)*. As an example, the differential (3.202a) has coefficients (3.202b, c) and hence is inexact (3.202d):

$$y(1 + xy)\,dx + x\,dy = 0: \qquad \{X, Y\} = \{y + xy^2, x\}, \qquad \Omega = -2xy; \qquad \text{(3.202a–d)}$$

since (3.201a) is satisfied by (3.203a) there is (3.201b) an integrating factor (3.203b):

$$\frac{\Omega}{Xx - Yy} = -\frac{2}{xy} = j(xy): \qquad \lambda(xy) = \exp\left\{-2\int^{xy}\frac{du}{u}\right\} = \frac{1}{x^2 y^2}. \qquad \text{(3.203a, b)}$$

Multiplying the inexact differential (3.202a) by the integrating factor (3.203b) leads to the exact differential (3.204a):

$$0 = \frac{y(1 + xy)\,dx + xd\,y}{x^2 y^2} = \frac{dx}{x} + \frac{dx}{x^2 y} + \frac{dy}{xy^2} = d\left(\log x - \frac{1}{xy}\right); \qquad C = \log x - \frac{1}{x\,y}, \qquad \text{(3.204a, b)}$$

thus (3.204b) is the general integral of the inexact differential (3.202a).

3.8.10 Integrating Factor a Homogeneous Function

Analogous with the first-order homogeneous differential equation, (3.127) is considered an integrating factor depending on the ratio of variables (3.205a) leading to (3.205b, c):

$$v = \frac{y}{x}: \qquad \left\{\frac{\partial}{\partial x}, \frac{\partial}{\partial y}\right\} = \left\{\frac{\partial v}{\partial x}, \frac{\partial v}{\partial y}\right\}\frac{d}{dv} = \left\{-\frac{y}{x^2}, \frac{1}{x}\right\}\frac{d}{dv};$$

$$\text{(3.205a–d)}$$

$$\Omega = \left(\frac{X}{x} + \frac{Yy}{x^2}\right)\frac{d}{dv}\{\log[\lambda(v)]\};$$

substituting (3.205b, c) in (3.176b) leads to (3.205d); it follows that *if (standard LXXII) the inexact (3.177c) differential (3.177a, b) has coefficients satisfying (3.206a):*

$$\frac{x^2\Omega}{Xx+Yy} = \frac{x^2}{Xx+Yy}\left(\frac{\partial Y}{\partial x} - \frac{\partial X}{\partial y}\right) = k\left(\frac{y}{x}\right): \qquad \lambda\left(\frac{y}{x}\right) = \exp\left\{\int^{y/x} k(v)dv\right\},$$

(3.206a, b)

there is an integrating factor (3.174) depending only on the ratio of variables (3.206b). As an example, the differential (3.207a) with coefficients (3.207b, c) is inexact (3.207d):

$$0 = (2x+y)\,dx + (2y-x)\,dy: \qquad \{X,Y\} = \{2x+y, 2y-x\}, \qquad \Omega = -2;$$

(3.207a–d)

the condition (3.206a) is satisfied (3.208a):

$$\frac{\Omega x^2}{Xx+Yy} = -\frac{x^2}{x^2+y^2} = -\frac{1}{1+y^2/x^2} = k\left(\frac{y}{x}\right),$$

(3.208a)

$$\lambda\left(\frac{y}{x}\right) = \exp\left\{-\int^{y/x} \frac{dv}{1+v^2}\right\} = \exp\left[-arc\tan\left(\frac{y}{x}\right)\right],$$

(3.208b)

leading to the integrating factor (3.208b), whose differential is:

$$d\lambda = -\lambda d\left[arc\tan\left(\frac{y}{x}\right)\right] = -\lambda\frac{d(y/x)}{1+y^2/x^2} = \lambda\frac{y\,dx - x\,dy}{x^2+y^2}.$$

(3.208c)

   Multiplying the inexact differential (3.207a) by the integrating factor (3.208b) leads to an exact differential:

$$0 = \lambda\left[(2x+y)dx - (x-2y)dy\right] = \lambda(y\,dx - x\,dy) + 2\lambda(x\,dx + y\,dy)$$
$$= (x^2+y^2)d\lambda + \lambda d(x^2+y^2) = d\left[\lambda(x^2+y^2)\right],$$

(3.209)

where (3.208c) was used. Equating the term in square brackets to an arbitrary constant in (3.209) and using (3.208b) it follows that:

$$x^2+y^2 = \frac{C}{\lambda} = C\exp\left[arc\tan\left(\frac{y}{x}\right)\right],$$

(3.210)

is the general integral (3.210) of the inexact differential (3.207a).

3.8.11 Unsymmetric Non-Homogeneous Integrating Factor

As an example of an integrating factor that depends on both variables (unlike subsection 3.8.6) and is neither symmetric (subsections 3.8.8–3.8.9) nor homogeneous (subsection 3.8.10), consider the differential (3.211a) that is inexact for $a \neq b$ and has an integrating factor (3.211b):

$$0 = ay\,dx + bx\,dy: \qquad \lambda = x^{a-1}y^{b-1}, \qquad (3.211a, b)$$

as follows from their product (3.212a):

$$0 = \lambda(ay\,dx + bx\,dy) = ax^{a-1}y^b dx + bx^a y^{b-1}dy = d(x^a y^b): \quad x^a y^b = C, \qquad (3.212a, b)$$

this proves that (3.212b) is the general integral of the inexact differential (3.211a). Replacing (a, b) by (α, β), it follows that the inexact differential (3.213a) has integrating factor (3.213b) leading to the general integral (3.213c):

$$0 = \alpha y\,dx + \beta x\,dy, \qquad \mu = x^{\alpha-1}y^{\beta-1}, \qquad x^\alpha y^\beta = C. \qquad (3.213a\text{–}c)$$

The ratio of the integrating factors (3.213b; 3.211b) is (3.214a):

$$\frac{\mu}{\lambda} = x^{\alpha-a}y^{\beta-b}: \qquad 0 = (ay\,dx + bx\,dy) + x^{\alpha-a}y^{\beta-b}(\alpha y\,dx + \beta x\,dy), \qquad (3.214a, b)$$

implying that the inexact differential (3.214b) has integrating factor (3.211b) because their product is an exact differential:

$$0 = \lambda\big[(ay\,dx + bx\,dy) + x^{\alpha-a}y^{\beta-b}(\alpha y\,dx + \beta x\,dy)\big]$$
$$= x^{a-1}y^{b-1}(ay\,dx + bx\,dy) + x^{\alpha-1}y^{\beta-1}(\alpha y\,dx + \beta x\,dy) = d(x^a y^b + x^\alpha y^\beta). \qquad (3.215)$$

It has been proved that the inexact differential (3.214b) has integrating factor (3.211b) leading (standard LXXIII) to the general integral:

$$C = x^a y^b + x^\varepsilon y^\beta. \qquad (3.216)$$

The integrating factor (3.211b) is: (i) symmetric if $a = b$; (ii) homogeneous if $a - 1 = 1 - b$ or $a + b = 2$; and (iii) both (i) and (ii) if $a = 1 = b$ in which case the integrating factor is unity, and is not needed because the differential is exact. In the case (i), that includes (iii), the condition (3.217a) leads in (3.211a) to an exact differential (3.217b), whose solution (3.217c):

$$a = b: \qquad 0 = a(y\,dx + x\,dy) = d(axy), \qquad xy = C, \qquad (3.217a\text{–}c)$$

and does not require an integrating factor. In the case (3.217a) the integrating factor (3.211b) simplifies to (3.218a), leading to (3.218b):

$$\lambda = (xy)^a: \qquad \lambda(y\,dx + x\,dy) = (xy)^a\,d(xy) = d\left[\frac{(xy)^{a+1}}{a+1}\right], \qquad (3.218a,\,b)$$

that specifies the same general integral as before (3.217c). Five other cases of integrating factors for inexact differentials are considered in the EXAMPLE 10.8.

3.9 Existence or Non-Existence of Integrating Factor

In addition to the two cases (i, ii) for a first-order differential in two variables (section 3.8) there is a third case (iii) for three variables, that is non-existence of an integrating factor leading to the **Pfaff problems**. For a first-order differential in two variables an integrating factor always exists, implying that there are plane curves orthogonal to a continuous vector field specified by the coefficients. For a first-order differential in three variables there are three cases (i–iii). The case (i) is an exact differential if the curl of the vector of coefficients is zero (subsection 3.9.1); in this case the integration is immediate and specifies a family of surfaces orthogonal to the vector field. A family of surfaces orthogonal to the vector field also exists (subsection 3.9.4) in the case (ii), when the differential is inexact and has an integrating factor (subsection 3.9.2); the necessary and sufficient condition of integrability is the Beltrami condition that the vector field of coefficients is orthogonal to its curl (subsection 3.9.3), implying that the tangent curves have no torsion or have zero helicity (subsection 3.9.5).

The simplest Pfaff problem arises in the case (iii) of an inexact differential in three variables for which the vector of coefficients does not satisfy the Beltrami condition (subsection 3.9.6): there is no family of surfaces orthogonal to the vector field, but on any regular surface there is a family of curves orthogonal to the vector field (subsection 3.9.7). The general inexact first-order differential in three variables leads to a general representation of a vector field by three scalar Clebsch potentials (subsection 3.9.8). Two alternative representations exist (subsection 3.9.9): (i) by a different set of three scalar potentials; and (ii) replacing two of the scalar potentials by a vector potential. The work of a three-dimensional force is, in general, (a) an inexact first-order differential in three variables without an integrating factor (subsection 3.9.10); it becomes an exact differential in the particular case (b) of a conservative force field that is the gradient of a single scalar potential. The general case (a) leads to the first principle of theormodynamics (subsection 3.9.11); that is, to the existence of entropy, temperature, and internal energy.

3.9.1 Exact Differential in Three Variables

The differential with two variables (3.145b) has a vector of coefficients with two components (3.145a), implying that: (i) in order to be an exact differential specified by a single function (section 3.8) an integrability condition must be met; and (ii) for an inexact differential (section 3.9) an integrating factor always exists because it supplies a second function. As the number of variable increases there are more vector components to be satisfied and hence more constraints on (i) exact differentials and (ii) integrating factors. For example, the first-order differential in three variables (3.219):

$$0 = X(x,y,z)dx + Y(x,y,z)dy + Z(x,y,z)dz, \qquad (3.219)$$

is exact if it coincides with the differential of a function of three variables (3.220a), implying that three conditions must be met (3.220b):

$$d\Phi = X\,dx + Y\,dy + Z\,dx: \qquad \{X,Y,Z\} = \left\{ \frac{\partial\Phi}{\partial x}, \frac{\partial\Phi}{\partial y}, \frac{\partial\Phi}{\partial z} \right\}. \qquad (3.220a, b)$$

If the integral function has continuous second-order derivatives (3.221a), the cross second-order derivatives are equal (3.221b–d):

$$\Phi \in C^2(|R^3): \qquad \frac{\partial^2\Phi}{\partial y\,\partial x} = \frac{\partial^2\Phi}{\partial x\,\partial y}, \quad \frac{\partial^2\Phi}{\partial z\,\partial x} = \frac{\partial^2\Phi}{\partial x\,\partial z}, \quad \frac{\partial^2\Phi}{\partial z\,\partial y} = \frac{\partial^2\Phi}{\partial y\,\partial z}.$$

$$(3.221a–d)$$

This implies that the vector of coefficients has continuous first-order derivatives (3.222a) and hence follows (3.222b–d):

$$X,Y,Z \in C^1(|R^3): \qquad \frac{\partial X}{\partial y} = \frac{\partial Y}{\partial x}, \frac{\partial X}{\partial z} = \frac{\partial Z}{\partial x}, \frac{\partial Y}{\partial z} = \frac{\partial Z}{\partial y}. \qquad (3.222a–d)$$

The **curl** of a three-dimensional vector (3.222a) is given in Cartesian coordinates by (III.6.41a) ≡ (3.223a, b):

$$\vec{\Omega} \equiv \nabla \wedge \vec{X} = \begin{vmatrix} \vec{e}_x & \vec{e}_y & \vec{e}_z \\ \dfrac{\partial}{\partial x} & \dfrac{\partial}{\partial y} & \dfrac{\partial}{\partial z} \\ X & Y & Z \end{vmatrix} = \vec{e}_x \left(\frac{\partial Z}{\partial y} - \frac{\partial Y}{\partial z} \right) + \vec{e}_y \left(\frac{\partial X}{\partial z} - \frac{\partial Z}{\partial x} \right) + \vec{e}_z \left(\frac{\partial Y}{\partial x} - \frac{\partial X}{\partial y} \right),$$

$$(3.223a, b)$$

and the conditions (3.222b–d) mean that it vanishes. The integrability conditions (3.222b–d) are necessary; they are also sufficient as follows from the Stokes theorem (3.150a, b; 3.151a, b). It has been shown that the *first-order differential in three-variables (3.219) is (standard LXXIV) **exact** (3.220a) ≡ (3.224b) and has general integral (3.224e):*

$$\vec{X} \in C^1(|R): \quad 0 = \vec{X}.d\vec{x} = d\Phi \quad \Leftrightarrow \quad \vec{\Omega} \equiv \nabla \wedge \vec{X} = 0 \quad \Leftrightarrow \quad \vec{X} = \nabla\Phi \quad \Leftrightarrow \quad \Phi(\vec{x}) = C,$$

$$(3.224a\text{–}e)$$

specifying a family of twice continuously differentiable (3.221e) surfaces orthogonal (3.234b) to the continuously differentiable (3.222a) ≡ (3.224a) vector of coefficients if the latter has zero curl (3.222b-d) ≡ (3.223a, b; 3.224c). For example (3.225a) is an exact differential:

$$0 = \alpha x^{\alpha-1} y^\beta z^\gamma dx + \beta x^\alpha y^{\beta-1} z^\gamma dy + \gamma x^\alpha y^\beta z^{\gamma-1} dz = d\left(x^\alpha y^\beta z^\gamma\right), \quad x^\alpha y^\beta z^\gamma = C,$$

$$(3.225a, b)$$

with general integral (3.225b); it can be checked that the vector of coefficients has zero curl (3.222b–d).

3.9.2 Integrating Factor and Integrability Condition

If the integrability conditions (3.222b–d) are not met, then the differential in three variables (3.219) is inexact. A surface orthogonal to the vector of coefficients will still exist (Figure 3.7) if there is an integrating factor that multiplied by the inexact differential (3.219) renders it exact (3.226a), implying (3.226b):

$$\lambda\left(X\,dx + Y\,dy + Z\,dz\right) = d\Phi: \quad \left\{\frac{\partial\Phi}{\partial x}, \frac{\partial\Phi}{\partial y}, \frac{\partial\Phi}{\partial z}\right\} = \lambda(X, Y, Z). \qquad (3.226a, b)$$

There are: (i) three conditions to be satisfied, namely the three components $\{X, Y, Z\}$ of the vector of coefficients (3.226b); (ii) only two free functions, namely the integral Φ and the integrating factor λ. Thus one constraint or integrability condition must be imposed on the vector of coefficients so that an integrating factor exists. This condition is obtained by substituting (3.226b) in (3.222b–d), assuming that the integral has continuous second-order derivatives:

$$\frac{\partial}{\partial y}(\lambda X) = \frac{\partial}{\partial x}(\lambda Y), \qquad \frac{\partial}{\partial z}(\lambda X) = \frac{\partial}{\partial x}(\lambda Z), \qquad \frac{\partial}{\partial z}(\lambda Y) = \frac{\partial}{\partial y}(\lambda Z); \quad (3.227a\text{–}c)$$

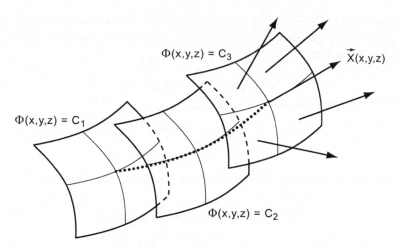

FIGURE 3.7
A two (three)-dimensional vector field in the plane (space) has a family [Figure 3.6 (3.7)] of orthogonal curves (surfaces) always (only if it has zero helicity). A plane vector field always has zero helicity, so a family of plane orthogonal curves always exists (Figure 3.6). A vector field in space may have non-zero helicity, and in that case there is no family of orthogonal curves (Figure 3.8).

the conditions (3.227a–c) are equivalent to:

$$\frac{\partial Y}{\partial x} - \frac{\partial X}{\partial y} = \frac{1}{\lambda}\left(X\frac{\partial\lambda}{\partial y} - Y\frac{\partial\lambda}{\partial x}\right), \quad \frac{\partial X}{\partial z} - \frac{\partial Z}{\partial x} = \frac{1}{\lambda}\left(Z\frac{\partial\lambda}{\partial x} - X\frac{\partial\lambda}{\partial z}\right),$$

$$\frac{\partial Z}{\partial y} - \frac{\partial Y}{\partial z} = \frac{1}{\lambda}\left(Y\frac{\partial\lambda}{\partial z} - Z\frac{\partial\lambda}{\partial y}\right),$$

(3.228a–c)

and specify the integrating factor λ, if it exists.

The preceding results can be put in vector form noting that (3.226a) is equivalent to (3.229a):

$$\nabla\Phi = \lambda\,\vec{X}: \quad 0 = \nabla\wedge(\nabla\Phi) = \nabla\wedge(\lambda\vec{X}) = \lambda(\nabla\wedge\vec{X}) + \nabla\lambda\wedge\vec{X},\quad (3.229a–c)$$

implying (3.229b) ≡ (3.227a–c); that is, equivalent to (3.229c) ≡ (3.228a–c) bearing in mind that:

$$\nabla\lambda\wedge\vec{X} = \begin{vmatrix} \vec{e}_x & \vec{e}_y & \vec{e}_z \\ \dfrac{\partial\lambda}{\partial x} & \dfrac{\partial\lambda}{\partial y} & \dfrac{\partial\lambda}{\partial z} \\ X & Y & Z \end{vmatrix} \begin{aligned} &= \vec{e}_x\left(Z\frac{\partial\lambda}{\partial y} - Y\frac{\partial\lambda}{\partial z}\right) \\ &+ \vec{e}_y\left(X\frac{\partial\lambda}{\partial z} - Z\frac{\partial\lambda}{\partial x}\right) \\ &+ \vec{e}_z\left(Y\frac{\partial\lambda}{\partial x} - X\frac{\partial\lambda}{\partial y}\right); \end{aligned}$$

(3.230a, b)

substitution of (3.223b) and (3.230b) in (3.229c) yields (3.228a–c). The property of the curl of the gradient (3.231) [curl of the product of a scalar and a vector (3.232)] was used in (3.229b) [(3.229c)]:

$$\left[\nabla \wedge (\nabla \Phi)\right]_i \equiv e_{ijk}\, \partial_j \partial_k \Phi = 0,$$ (3.231)

$$\left[\nabla \wedge (\lambda \bar{X})\right]_i = e_{ijk}\, \partial_j (\lambda X_k) = e_{ijk} (\partial_j \lambda) X_k + e_{ijk} \lambda \partial_j X_k = (\nabla \lambda \wedge \bar{X})_i + \lambda (\nabla \wedge \bar{X})_i,$$ (3.232)

that can be proved using the permutation symbol in three-dimensions:

$$
e_{ijk} =
\begin{cases}
+1 & \text{if } (i, j, k) \text{ is an even permutation,} & (3.233a) \\
-1 & \text{if } (i, j, k) \text{ is an odd permutatics of } (1, 2, 3), & (3.233b) \\
0 & \text{otherwise: there are repeated indices.} & (3.233c)
\end{cases}
$$

From (3.229c) follows (3.234b):

$$\lambda \neq 0: \quad 0 = \bar{X}.(\nabla \lambda \wedge \bar{X}) = -\lambda \bar{X}.(\nabla \wedge \bar{X}) \quad \Rightarrow \quad H \equiv \bar{X}.\bar{\Omega} = \bar{X}.(\nabla \wedge \bar{X}) = 0,$$ (3.234a–c)

implying for non-zero integrating factor (3.234a) that the **helicity** is zero (3.234c).

3.9.3 Helicity and Beltrami Vectors

The necessary condition for the existence of an integrating factor (3.234c) ≡ (3.235):

$$0 = H = \bar{X}.(\nabla \wedge \bar{X}) = X\left(\frac{\partial Z}{\partial y} - \frac{\partial Y}{\partial z}\right) + Y\left(\frac{\partial X}{\partial z} - \frac{\partial Z}{\partial x}\right) + Z\left(\frac{\partial Y}{\partial x} - \frac{\partial X}{\partial y}\right),$$ (3.235)

follows alternatively from (3.228a–c) multiplying the l.h.s. respectively by (Z, Y, X) and adding so that the l.h.s. vanishes. The helicity is given equivalently by (3.235) or (3.236a, b):

$$
H =
\begin{vmatrix}
X & Y & E \\
\dfrac{\partial}{\partial x} & \dfrac{\partial}{\partial y} & \dfrac{\partial}{\partial z} \\
X & Y & Z
\end{vmatrix}
= X\left(\frac{\partial Z}{\partial y} - \frac{\partial Y}{\partial z}\right) + Y\left(\frac{\partial X}{\partial z} - \frac{\partial Z}{\partial x}\right) + Z\left(\frac{\partial Y}{\partial x} - \frac{\partial X}{\partial y}\right),
$$ (3.236a, b)

as follows from the inner product (3.234c) of (3.222a) ≡ (3.224a) and (3.223a, b). A vector with zero helicity (3.235) is called a **Beltrami vector**, and (3.234c) ≡ (3.235) is not only a necessary but also a sufficient condition for the existence of an integrating factor (Notes 3.9–3.11).

It has been shown that *a differential in three variables (3.219) ≡ (3.237b) with (standard LXXV) continuously differentiable vector of coefficients (3.222a) ≡(3.237a): (i) is inexact if the curl is not zero (3.237c):*

$$\vec{X} \in C^1\left(|R^3\right): \qquad \vec{X}.d\vec{x} = 0, \qquad \nabla \wedge \vec{X} \neq 0; \qquad (3.237a\text{–}c)$$

(ii) if the helicity is zero (3.234c) ≡ (3.235) ≡ (3.238a); that is, the vector of coefficients is a Beltrami vector orthogonal to its curl there is an integrating factor (3.238b) specified by any solution of (3.229c) ≡ (3.228a–c):

$$0 \equiv H = \vec{X}.\left(\nabla \wedge \vec{X}\right) \quad \Leftrightarrow \quad 0 = \lambda\left(\vec{X}.d\vec{x}\right) = d\Phi \quad \Leftrightarrow \quad \Phi(\vec{x}) = C. \quad (3.238a\text{–}c)$$

and (iii) multiplication of the differential by the integrating factor leads to the general integral (3.238c).

3.9.4 Inexact, Integrable Differential in Three Variables

For example, the differential (3.239a) is inexact because the vector of coefficients (3.239b) has non-zero curl (3.239c):

$$0 = y\,dx + x\,dy + xy\,dz: \qquad \vec{X} = \{\, y, x, xy \,\}, \qquad \nabla \wedge \vec{X} = \{\, x, -y, 0 \,\}, \quad (3.239a\text{–}c)$$

as follows from (3.240a):

$$\nabla \wedge \vec{X} = \begin{vmatrix} \vec{e}_x & \vec{e}_y & \vec{e}_z \\ \dfrac{\partial}{\partial x} & \dfrac{\partial}{\partial y} & \dfrac{\partial}{\partial z} \\ y & x & xy \end{vmatrix} = \vec{e}_x\,x - \vec{e}_y\,y; \qquad \left(\nabla \wedge \vec{X}\right).\vec{X} = 0,$$

$$(3.240a, b)$$

the three-dimensional vector field (3.239b) is a Beltrami vector with zero helicity (3.240b) and thus the inexact differential has an integrating factor, and in fact, infinitely many. The inexact (3.239c) integrable (3.240b) differential (3.239a) can be put into the form (3.241a):

$$0 = d(xy) + xy\,dz, \qquad \lambda(x, y) = \frac{1}{xy}, \qquad (3.241a, b)$$

suggesting that is has the integrating factor (3.241b); it can be confirmed that multiplying factor (3.241b) renders the differential (3.239a) exact (3.242a):

$$\frac{y\,dx + x\,dy + xy\,dz}{xy} = \frac{dx}{x} + \frac{dy}{y} + dz = d\big[z + \log(xy)\big], \quad z + \log(xy) = C,$$

$$(3.242a, b)$$

and specifies the general integral (3.242b). The first-order differential in three variables (3.239a) is inexact (3.239b, c) but has (3.240a, b) an integrating factor (3.241b) leading to the general integral (3.242b); the latter specifies a family of surfaces (Figure 3.7) orthogonal to the vector field (3.239b). If the vector of coefficients has non-zero helicity, an orthogonal family of surfaces does not exist (subsections 3.9.5–3.9.6).

3.9.5 Plane Curves and Curves with Torsion

A plane vector (3.243a) has curl (3.223a, b) orthogonal to the plane (3.243b) and hence the helicity that is the inner product is zero (3.243c):

$$\vec{X} = \vec{e}_x A + \vec{e}_y B : \quad \nabla \wedge \vec{X} = \vec{e}_z\left(\frac{\partial B}{\partial x} - \frac{\partial A}{\partial y}\right), \quad H \equiv \vec{X}.\left(\nabla \wedge \vec{X}\right) = 0. \quad (3.243a\text{–}c)$$

It has been shown *that a plane vector field (3.243a) has zero helicity (3.243c) and hence a differential in two variables always has an integrating factor,* in agreement with subsection 3.9.1; the geometric interpretation (Figure 3.6) is that a continuously differentiable vector field in the plane always has an orthogonal family of curves. The helicity is a measure of the deviation from a plane curve and corresponds to a curve with **torsion** in space. For example, a cylindrical helix with (Figure 3.8) constant radius a and inclination α has cylindrical coordinates (3.244a):

$$\vec{X} = \vec{e}_r a + \vec{e}_z a\phi\sin\alpha : \quad \nabla \wedge \vec{X} = \vec{e}_r\frac{1}{a}\partial_\phi X_z = \vec{e}_r\sin\alpha, \quad H \equiv \vec{X}.\left(\nabla \wedge \vec{X}\right) = a\sin\alpha,$$

$$(3.244a\text{–}c)$$

corresponding to: (i) the curl (3.244b) in cylindrical (III.6.41b) coordinates (r, ϕ, z); and (ii) constant helicity (3.244c). The situation in three-dimensions: (i) is similar for a vector field with zero curl (helicity) for which the differential is exact (is inexact but has an integrating factor), so that an orthogonal family of surfaces exists (Figure 3.7); and (ii) different if the vector of coefficients has non-zero helicity, because an orthogonal family of surfaces does not exist. In the latter case the integrability condition of zero helicity is replaced by an arbitrary function Ψ, that together with the integrating factor λ and general integral Φ satisfies the three components of the vector of coefficients of the differential in three variables (subsection 3.9.6).

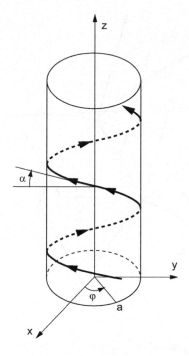

FIGURE 3.8
A circular cylindrical helix with radius a and constant slope α has constant helicity $a \sin \alpha$.

3.9.6 Inexact Differential without an Integrating Factor

Consider *a first-order differential in three variables (3.219) ≡ (3.245b) with (standard LXXVI) continuously differentiable vector of coefficients (3.232a) ≡ (3.245a) with non-zero helicity (3.245d) and hence non-zero curl (3.245c):*

$$\vec{X} \in C^1\left(|R^3\right): \qquad \vec{X}.d\vec{x} = 0, \qquad \nabla \wedge \vec{X} \neq 0 \neq \vec{X}.\left(\nabla \wedge \vec{X}\right) \equiv H. \tag{3.245a–d}$$

Then (i) an arbitrary differentiable function (3.246a) can be imposed (3.246b) specifying the differential of one variable in terms of the others (3.246c):

$$\Psi(x,y,z) \in C^1\left(|R^3\right): \qquad 0 = d\Psi = \frac{\partial \Psi}{\partial x}dx + \frac{\partial \Psi}{\partial y}dy + \frac{\partial \Psi}{\partial z}dz,$$

$$dz = -\frac{\partial \Psi/\partial x}{\partial \Psi/\partial z}dx - \frac{\partial \Psi/\partial y}{\partial \Psi/\partial z}dy; \tag{3.246a–c}$$

(ii) substitution of (3.246c) in the inexact non-integrable differential in three variables (3.219) leads to a differential in two variables (3.247a):

$$0 = \left(X - Z\frac{\partial \Psi/\partial x}{\partial \Psi/\partial z}\right)dx + \left(Y - Z\frac{\partial \Psi/\partial y}{\partial \Psi/\partial z}\right)dy = \lambda^{-1}d\Phi, \qquad \Phi(x,y,z) = C, \tag{3.247a–c}$$

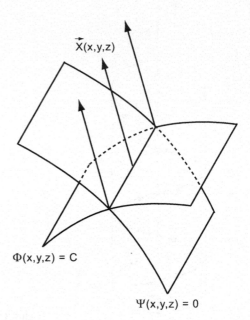

FIGURE 3.9
A vector field in space with non-zero helicity does not have a family of orthogonal surfaces, but given any surface there is on it a family of orthogonal curves. In the plane case (Figure 3.6) there is only one choice of plane and the helicity is always zero.

that has an integrating factor (3.247b) and leads to the general integral (3.247c). The geometrical interpretation (Figure 3.9) is that: (i) on any regular (3.248a) surface (3.248b) chosen 'a priori' there is a curve orthogonal to the vector field (3.247b) ≡ (3.248b):

$$\Psi \in C^1(|R): \qquad \Psi(\bar{x}) = 0, \qquad 0 = \lambda(\bar{X}.d\bar{x}) = d\Phi; \qquad \Phi(\bar{x}) = C, \qquad \text{(3.248a–d)}$$

and (ii) the curve is specified by the intersection of the surface (3.248b) with the general integral (3.247c) ≡ (3.248d).

3.9.7 Curves on a Surface Orthogonal to a Vector Field

For example the first-order differential in three variables (3.249a) has vector of coefficients (3.249b) with non-zero curl (3.249c) and non-zero helicity (3.249d):

$$0 = y\,dx + x\,dy + x\,dz: \quad \bar{X} = \{y, x, x\}, \quad \nabla \wedge \bar{X} = -\vec{e}_y, \quad \bar{X}.(\nabla \wedge \bar{X}) = -x, \qquad \text{(3.249a–d)}$$

as follows from (3.250b) using the notation (3.250a):

$$\{\partial_x, \partial_y, \partial_z\} \equiv \left\{ \frac{\partial}{\partial x}, \frac{\partial}{\partial y}, \frac{\partial}{\partial z} \right\}: \quad \nabla \wedge \bar{X} = \begin{vmatrix} \vec{e}_x & \vec{e}_y & \vec{e}_z \\ \partial_x & \partial_y & \partial_z \\ y & x & x \end{vmatrix} = -\vec{e}_y. \qquad \text{(3.250a, b)}$$

Hence the differential (3.249a) is inexact and has no integrating factor; that is, the vector field (3.249b) has no orthogonal family of surfaces. Choosing the condition (3.251a) transforms (3.251b):

$$0 = z - xy \equiv \Psi(x, y, z): \qquad dz = x\,dy + y\,dx, \qquad\qquad (3.251a, b)$$

the differential from three (3.249a) to two (3.252) variables:

$$0 = y\,dx + x\,dy + xd(xy) = (1 + x)(y\,dx + x\,dy) = (1 + x)d(xy) = d\Phi; \quad (3.252)$$

the integrating factor (3.253a) multiplied by the inexact differential (3.252) leads to the exact differential (3.253b) and the general integral (3.253c):

$$\lambda = \frac{1}{1 + x}: \qquad 0 = \lambda\,d\Phi = \lambda(1 + x)d(xy) = d(xy), \qquad xy = C. \qquad (3.253a\text{--}c)$$

Thus the curves specified by the intersection of the surfaces (3.251a) and (3.253c) are orthogonal to the vector field (3.249b).

Choosing a condition (2.254a) different from (3.251a) leads to (3.254b):

$$0 = z - x^2 y \equiv \Theta(x, y, z): \qquad dz = 2xy\,dx + x^2\,dy, \qquad\qquad (3.254a, b)$$

corresponding to a different surface. Substitution of (3.254a, b) leads from the differential in three (3.249a) to two (3.255) variables:

$$0 = y\,dx + x\,dy + xd(x^2 y) = y(1 + 2x^2)dx + x(1 + x^2)dy. \qquad (3.255)$$

The latter (3.255) $\equiv$ (3.256a) is a separable first-order differential equation in two variables:

$$0 = \frac{1 + 2x^2}{1 + x^2}\frac{dx}{x} + \frac{dy}{y} = \left(\frac{1}{x} + \frac{x}{1 + x^2}\right)dx + \frac{dy}{y} = d\left[\log x\, y\sqrt{1 + x^2}\right]: \quad xy\sqrt{1 + x^2} = C,$$

$$\qquad\qquad\qquad\qquad\qquad\qquad\qquad\qquad\qquad\qquad\qquad\qquad (3.256a, b)$$

that specifies the general integral (3.256b). Thus the first-order differential in three variables (3.249a) is inexact (3.249c) and has no integrating factor (3.249d), leading to the following two equivalent statements: (i) subject to the constraint (3.251a) [(3.254a)] it has general integral (3.253c) [(3.256b)]; and (ii) on the surface (3.251a) [(3.254a)] there exist curves orthogonal to the vector field (3.249b), that is obtained by intersection with the surfaces (3.253c) [(3.256b)].

3.9.8 Representation of a Vector Field by Potentials

The first-order differential in three variables (3.219) in the most general case (3.247a), when it is inexact (3.245c) and has no integrating factor (3.245d), can be written (Notes 3.9–3.11) in the form (3.257a) ≡ (3.314b):

$$0 = \vec{X}.d\vec{x} = d\Phi + \Theta d\Psi = (\nabla\Phi + \Theta\nabla\Psi).d\vec{x}: \qquad \vec{X} = \nabla\Phi + \Theta\nabla\Psi, \qquad (3.257a, b)$$

and implies (3.257b) that the vector of coefficients is the sum of the gradients of two **Euler potentials** (Φ, Ψ), with the second multiplied by a third potential Θ. Another representation of a vector field (III.5.266a–c) ≡ (3.258a) is the sum of an **irrotational (solenoidal) part** that is the gradient (curl) of a **scalar (vector) potential**:

$$\vec{X} = \nabla\Phi + \nabla \wedge \vec{A}; \qquad \nabla \wedge \vec{X} = \nabla \wedge (\nabla\Phi) + \nabla \wedge (\Theta\nabla\Psi) = \nabla\Theta \wedge \nabla\Psi, \quad (3.258a, b)$$

from (3.257b) follows (3.258b) using (3.229a–c) ≡ (3.231; 3.232; 3.233a–c). It follows that the second term on the r.h.s. of (3.258a) has a representation (3.258b) ≡ (3.258c):

$$\nabla \wedge \vec{A} = \nabla\Xi \wedge \nabla\Psi; \qquad \vec{X} = \nabla\Phi + \nabla\Xi \wedge \nabla\Psi, \qquad (3.258c,d)$$

substitution of (3.258c) in (3.258a) leads to (3.258d) representation of a vector field by three **Clebsch potentials**, as the sum of a irrotational (solenoidal) part that is the gradient of a potential (the outer product of the gradients of two potentials).

3.9.9 Choice of Scalar and Vector Potentials

It has been shown that *a continuously differentiable vector field in three dimensions (3.259a) has two particular and three general representations. The two particular representations are as the gradient (3.259c) [curl (3.259e)] of a scalar (vector) potential and hold if [standard LXXVII (LXXVIII)] the vector field is irrotational (3.259b) [solenoidal (3.259e)].*

$$\vec{X} \in C^1\left(|R^3\right): \quad \nabla \wedge \vec{X} = 0 \quad \Leftrightarrow \quad \vec{X} = \nabla\Phi, \quad \nabla.\vec{X} = 0 \quad \Leftrightarrow \quad \vec{X} = \nabla \wedge \vec{A}. \quad (3.259a–e)$$

The three general representations (3.260a):

$$\vec{X} \in C^1\left(|R^3\right): \qquad \Phi, \vec{A} \in C^2\left(|R^3\right) \quad \Rightarrow \quad \vec{X} = \nabla\Phi + \nabla \wedge \vec{A}, \qquad (3.260a–c)$$

$$\Phi, \Psi, \Xi \in C^2\left(|R^3\right) \quad \Rightarrow \quad \vec{X} = \nabla\Phi + \nabla\Xi \wedge \nabla\Psi, \qquad (3.261a, b)$$

$$\Phi, \Psi, \in C^2\left(|R^3\right), \Theta \in C^1\left(|R^3\right) \quad \Rightarrow \quad \vec{X} = \nabla\Phi + \Theta\nabla\Psi, \qquad (3.262a–c)$$

are (i) as (standard LXXIX) the sum of the gradient (curl) of a scalar (vector) potential (3.260c) that have continuous second-order derivatives (3.260b); (ii) as (standard
LXXX) the sum (3.261a) of the gradient of a potential (outer product of the gradient of two potentials) that have continuous second-order derivatives (3.261b); and
(iii) as (standard LXXXI) the sum (3.262c) of the gradients of two potentials with
continuous second-order derivatives (3.262a) with the second term multiplied by a
third potential with continuous first-order derivatives (3.262b). In all three representations (i) to (iii): (a) the irrotational part is the first term on the r.h.s., that is, the
gradient of a scalar potential; (b) the solenoidal part is the second term on the r.h.s.,
that is, the curl of a vector potential (3.263a) that has two alternative representations
(3.263b) ≡(3.263c) in terms of scalar potentials:

$$\vec{X} - \nabla\Phi = \nabla \wedge \vec{A} = \nabla\Xi \wedge \nabla\Psi = \Theta\nabla\Psi. \qquad (3.263a\text{–}c)$$

*The differentiability requirements may be relaxed using generalized functions
(chapter III.5).*

3.9.10 Conservative Forces and Potential and Total Energies

The **work** dW exerted by a **forced** $\vec{F}$ in an infinitesimal **displacement** $d\vec{x}$ is
generally a first-order differential in three variables:

$$dW = \vec{F}.d\vec{x} = F_x\,dx + F_y\,dy + F_z\,dz. \qquad (3.264)$$

If the force field is **conservative***, that is, if it has zero curl (3.265a), it is minus the
gradient of a potential (3.265b), whose sum with the work is conserved (3.265c) and
specifies the total energy (3.265d):*

$$\nabla \wedge \vec{F} = 0: \qquad \vec{F} = -\nabla\Phi, \qquad dW = -\nabla\Phi.d\vec{x} = -d\Phi, \qquad W + \Phi = E. \qquad (3.265a\text{–}d)$$

*A conservative force field (3.265a) is orthogonal (3.265b) to the equipotential sur-
faces; for a displacement on the equipotential surfaces the work is zero. If the force
field is not conservative (3.266a), for example a friction force, then there is no conservation of the total energy, and an integrating factor exists (3.266c) if it is a Beltrami
field (3.266b):*

$$\nabla \wedge \vec{F} \neq 0 = \vec{F}.\left(\nabla \wedge \vec{F}\right): \qquad \lambda\,dW = \lambda\left(\vec{F}.d\vec{x}\right) = -d\Phi; \qquad (3.266a\text{–}c)$$

*in this case there is an integrating factor λ that, multiplied by the force $\lambda\vec{F}$, leads to
the existence of equipotentials, and for a displacement on the equipotential the work
is zero (3.266c). In general, the force is not a Beltrami vector (3.267a) and there is
no equipotential surface where the work is zero (3.266c). In this case it is possible to*

choose an arbitrary surface (3.267b) and on it there is a curve such that, for a displacement along this curve (3.267c), the work is zero (3.267d):

$$\vec{F}.\left(\nabla \wedge \vec{F}\right) \neq 0; \quad \Psi(\vec{x}) = 0: \quad d\vec{x}_* = \nabla\Phi \wedge \nabla\Psi: \quad dW_* = \vec{F}.d\vec{x}_* = 0, \qquad (3.267a\text{–}d)$$

because the force is orthogonal. The last general case (3.267a–d) of a non-conservative force field (3.267a) that is not a Beltrami vector (3.267b) leads (3.314b) to the first principle of thermodynamics (subsection 3.9.11).

3.9.11 First Principle of Thermodynamics

The first-order inexact differential in three variables without an integrating factor (3.257a) leads to the **first principle of thermodynamics**: *the work (3.264) of the force $\vec{F}$ in an infinitesimal displacement $d\vec{x}$ is generally a first-order differential in three variables that may be inexact without an integrating factor but always has (3.257a) the representation (3.268):*

$$\vec{F}.d\vec{x} = dW = dU - TdS, \qquad (3.268)$$

that can be interpreted as follows: (i) the **heat** *(3.269a) is an inexact differential because added to the work it leads to the* **internal energy** *(3.269b) that is an exact differential or* **function of state** *U:*

$$dQ \equiv T\,dS: \qquad\qquad dW + dQ = dU; \qquad\qquad (3.269a, b)$$

(ii) the heat exchanged (3.269a) equals the **temperature** *T multiplied by the variation of the* **entropy** *S, that is a function of state; and (iii) the temperature is the inverse of an integrating factor because dividing the heat, that is an inexact differential, by the temperature leads to an exact differential of a function of state, namely the entropy (3.270a):*

$$dS = \frac{dQ}{T}; \qquad dU = dW + TdS. \qquad (3.270a, b)$$

Substituting (3.269a) in (3.269b) specifies the internal energy (3.270b) as a function of state. Other functions of state can be defined using dual differentials and Legendre transforms (section 5.5).

NOTE 3.1: Multivariable and Higher-Order Differentials

The first-order differential in two (three) variables [section 3.8 (3.9)] can be extended to any number of variables (note 3.2). In the case of more than three variables (notes 3.3–3.15) there are the same three cases as for (section 3.9) three variables: (i) exact differential whose integration is immediate (note 3.3) if the bivector curl of the vector of coefficients is zero; (ii) inexact differential with an integrating factor if the bivector curl is not zero but the helicity trivector is zero (notes 3.4–3.5); and (iii) inexact differential without integrating factor if the helicity trivector is not zero (notes 3.6–3.8). The proof that the Beltrami condition of zero helicity trivector is sufficient for the existence of an integrating factor (notes 3.9–3.11): (i) was used for first-order differentials in three variables (subsection 3.9.3) and leads to (ii) representations of a vector field by potentials (subsections 3.9.8–3.9.9) and to (iii) the first principle of thermodynamics (subsections 3.9.10–3.9.11); (iv) applies to first-order differentials of any order N, and allows transformation into a first-order differential in $(N-1)$ variables so that successive application up to a first-order differential in two variables provides a general method of integration (note 3.12). The first-order differential equations (differentials) include [sections 3.1–3.7 (sections 3.8–3.9 and notes 3.13–3.16)] homogeneous cases [section 3.7 (notes 3.17–3.20)]. The higher-order differential equations (differentials) [sections 5.6–5.9 (notes 3.21–3.24)] may be solved by methods that include factorization into lower orders [subsection 5.8.1 (notes 3.23–3.24)].

NOTE 3.2: Differentials, Vectors, and Hypersurfaces

The first-order differential (3.145b) in two variables has for general integral (3.146c) a family of plane curves (Figure 3.6) orthogonal to the vector field of coefficients (3.145a). A first-order differential in three variables (3.219) ≡ (3.271):

$$0 = X_1(x_1,x_2,x_3)dx_1 + X_2(x_1,x_2,x_3)dx_2 + X_3(x_1,x_2,x_3)dx_3 = \vec{X}.d\vec{x}, \quad (3.271)$$

has for general integral (3.272a), that, if it exists:

$$\Phi(x_1,x_2,x_3) = C: \quad 0 = d\Phi = \frac{\partial\Phi}{\partial x_1}dx_1 + \frac{\partial\Phi}{\partial x_2}dx_2 + \frac{\partial\Phi}{\partial x_3}dx_3 = \nabla\Phi.d\vec{x}, \quad (3.272a\text{–}c)$$

represents (3.272b, c) a family of regular surfaces (3.273a) whose normals (3.273b) are parallel (Figure 3.7) to the vector field (3.273c) of coefficients:

$$\Phi \in C^1(|R^3): \quad \nabla\Phi = \lambda\vec{X}, \quad \vec{X} = \{X_1,X_2,X_3\}. \quad (3.273a\text{–}c)$$

Similarly, for a differential (3.274b) in N variables (3.274a), its general integral (3.274c), if it exists, is a family of hypersurfaces with normals parallel to the vector field of the coefficients (3.274b):

$$n = 1,...,N: \quad 0 = X_n dx_n = \lambda d\Phi, \quad \Phi(x_n) = C, \quad \frac{\partial\Phi}{\partial x_n} = \lambda X_n. \quad (3.274a\text{–}d)$$

The Einstein **summation convention** (3.275b), that a repeated index implies summation over its range of values (3.275a), was used in (3.274b):

$$d\Phi = \sum_{n=1}^{N} \frac{\partial \Phi}{\partial x_n} dx_n \equiv \frac{\partial \Phi}{\partial x_n} dx_n = \lambda X_n \, dx_n. \tag{3.275a, b}$$

All of the preceding are **first-order differentials** respectively in two (3.145a, b; 3.146c), three (3.271; 3.272a–c) and N variables (3.274a–d). Higher-order differentials are considered (notes 3.21–3.24) after completing the analysis of first-order differentials (notes 3.3–3.20).

NOTE 3.3: Exact Differential in N-Variables

A first-order differential in N-variables (3.274a, b) $\equiv$ (3.276a, b) is (standard LXXXII) exact (3.276c) if it coincides with the differential of a function that specifies the general integral (3.276d):

$$n = 1, \ldots, N: \qquad 0 = X_n \, dx_n = d\Phi, \qquad \Phi(x_n) = C; \tag{3.276a–d}$$

this implies that the vector field $\bar{X}$ is orthogonal (3.276b) to the hypersurfaces (3.276d). If the function (3.276d) has continuous second-order derivatives (3.277a), the equality of the cross-derivatives (3.277b) leads to the integrability condition (3.277c):

$$\Phi \in C^2\left(|R^N\right): \qquad \frac{\partial^2 \Phi}{\partial x_m \partial x_n} = \frac{\partial^2 \Phi}{\partial x_m \partial x_n} \quad \Leftrightarrow \quad \frac{\partial X_n}{\partial x_m} = \frac{\partial X_m}{\partial x_n}. \tag{3.277a–c}$$

The conditions (3.277c) correspond to the vanishing of the curl of the vector of coefficients; in N-dimensions the curl of a vector is a bivector, that is, a skew-symmetric matrix (note 3.8). Thus, *a first-order differential in* N *variables* (3.278b) *with (standard LXXXII) continuously differentiable vector of coefficients* (3.278a) *is exact* (3.278c) *if the curl is zero* (3.278d):

$$X_n \in C^1\left(|R^N\right): \quad 0 = X_n \, dx_n = d\Phi \quad \Leftrightarrow \quad \Omega_{mn} \equiv \frac{\partial X_n}{\partial x_m} - \frac{\partial X_m}{\partial x_n} = 0 \quad \Leftrightarrow \quad \Phi(x_n) = C;$$

$$\tag{3.278a–e}$$

in this case the general integral (3.278e) *specifies a family of hypersurfaces orthogonal to the vector field* (3.278a). In two (three) dimensions $N = 2(N = 3)$ then (3.276a–d) simplifies to (3.153a–d) [(3.224a–e)]. For example, the first-order differential in four variables (3.279a):

$$0 = x_2 \, dx_1 + x_1 \, dx_2 + \frac{dx_3}{x_4} - \frac{x_3}{\left(x_4\right)^2} dx_4 = d\left(x_1 x_2 + \frac{x_3}{x_4}\right): x_1 x_2 + \frac{x_3}{x_4} = C, \tag{3.279a, b}$$

is exact, and is the sum of two exact differentials in two variables leading to the general integral (3.279b). It can be checked that the vector of coefficients (3.280a, b) has zero curl (3.280c):

$$n = 1, 2, 3, 4: \qquad X_n = \left\{ x_2\ ,\ x_1\ ,\ \frac{1}{x_4}\ ,\ -\frac{x_3}{(x_4)^2} \right\}, \qquad \frac{\partial X_n}{\partial x_m} = \frac{\partial X_m}{\partial x_n}, \qquad (3.280a\text{–}c)$$

that is, it meets the $\begin{pmatrix} 4 \\ 2 \end{pmatrix} = 4! / (2!\,2!) = 3! = 6$ conditions (3.280c).

NOTE 3.4: Integrating Factor for a Differential in N Variables

The first-order differential in N variables (3.274a, b) will have a family of integral hypersurfaces (3.274c) orthogonal to the vector field of coefficients if an integrating factor (3.274d) exists. If the function has continuous second-order derivatives (3.277a), substitution of (3.274d) in (3.277b) leads to (3.281a):

$$\frac{\partial}{\partial x_n}(\lambda X_m) = \frac{\partial}{\partial x_m}(\lambda X_n): \qquad \frac{\partial X_m}{\partial x_n} - \frac{\partial X_n}{\partial x_m} = \frac{1}{\lambda}\left(X_n \frac{\partial \lambda}{\partial x_m} - X_m \frac{\partial \lambda}{\partial x_n} \right), \qquad (3.281a, b)$$

that is, (3.281a) $\equiv$ (3.281b), the set of equations specifying the integrating factor if it exists. The integrating factor may be eliminated from (3.281b), leading to the integrability condition:

$$0 = X_s\left(\frac{\partial X_m}{\partial x_m} - \frac{\partial X_n}{\partial x_n} \right) + X_m\left(\frac{\partial X_n}{\partial x_s} - \frac{\partial X_s}{\partial x_n} \right) + X_n\left(\frac{\partial X_s}{\partial x_m} - \frac{\partial X_m}{\partial x_s} \right) \equiv H_{mns}. \qquad (3.282)$$

The conditions (3.282) are skew-symmetric in (m, n, s) and simplify to zero helicity (3.234c) $\equiv$ (3.235) in the three-dimensional case $N = 3$. In N-dimensions the helicity is a trivector, that is, tensor with three indices that is skew-symmetric relative to any pair (note 3.8). It has been shown that *the first-order differential in N variables (3.278b), with (standard LXXXIII) continuously differentiable vector of coefficients (3.278a)* $\equiv$ *(3.283a) is inexact if the curl is not zero (3.283b)*:

$$X_m \in C^1\left(|R^N\right): \qquad \Omega_{mn} \equiv \frac{\partial X_n}{\partial x_m} - \frac{\partial X_m}{\partial x_n} \neq 0, \qquad (3.283a, b)$$

but has an integrating factor (3.274d) $\equiv$ *(3.284b) specified by (3.281b), if the integrability conditions (3.282)* $\equiv$ *(3.284a) are met stating that the helicity is zero:*

$$H_{mns} = \sum_{m,n,s}^{cyl} \Omega_{mn} X_s = 0 \quad \Leftrightarrow \quad 0 = \lambda X_n\, dx_n = d\Phi \quad \Leftrightarrow \quad \Phi(x_n) = C; \qquad (3.284a\text{–}c)$$

in this case the general integral is (3.284c), involving a twice continuously differentiable function (3.277a) that specifies a family of hypersurfaces orthogonal to the vector field (3.283a).

NOTE 3.5: Integrable Differential in N Variables

As an example, consider the first-order differential in four variables (3.285):

$$0 = x_2 \, dx_1 + x_1 \, dx_2 + \frac{x_1 x_2}{x_4} \, dx_3 - \frac{x_1 x_2 x_3}{(x_4)^2} \, dx_4, \tag{3.285}$$

with vector coefficients (3.286a) that has (3.278c) non-zero curl (3.286b):

$$X_m = \left\{ x_2, x_1, \frac{x_1 x_2}{x_4}, -\frac{x_1 x_2 x_3}{(x_4)^2} \right\}:$$

$$\{ \Omega_{12}, \Omega_{13}, \Omega_{14}, \Omega_{23}, \Omega_{24}, \Omega_{34} \} = \left\{ 0, \frac{x_2}{x_4}, -\frac{x_2 x_3}{(x_4)^2}, \frac{x_1}{x_4}, -\frac{x_1 x_3}{(x_4)^2}, 0 \right\}.$$

$$\tag{3.286a, b}$$

Since the curl (3.274b) is a bivector, that is, a skew-symmetric 4×4 matrix, it has six independent components (3.286b); the curl (3.286b) is not zero, proving that the differential (3.285) is inexact. Since the integrability conditions are skew-symmetric in three indices (m, n, s) = (1,2,3,4) taking four values, there are four independent conditions, one for each "missing" index in the helicity:

$$\Omega_{12} X_3 + \Omega_{23} X_1 + \Omega_{31} X_2 = 0 = \Omega_{12} X_4 + \Omega_{24} X_1 + \Omega_{41} X_2,$$

$$\tag{3.287a–d}$$

$$\Omega_{13} X_4 + \Omega_{34} X_1 + \Omega_{41} X_3 = 0 = \Omega_{23} X_4 + \Omega_{34} X_2 + \Omega_{42} X_3;$$

since all four integrability conditions (3.287a–d) are met, the differential (3.285) has an integrating factor. The integrating factor may be found noting that: (i) the first two terms of the differential (3.285) are exact (3.288a):

$$x_2 \, dx_1 + x_1 \, dx_2 = d(x_1 x_2); \quad \frac{x_1 x_2}{x_4} \, dx_3 - \frac{x_1 x_2 x_3}{(x_4)^2} \, dx_4 = x_1 x_2 d\left(\frac{x_3}{x_4} \right), \tag{3.288a, b}$$

(ii) the last two terms (3.288b) of the differential (3.285) have an integrating factor (3.289a), since (3.288b) becomes exact multiplying by (3.289a); and (iii) although the first two terms (3.288a) are exact, they remain exact when multiplied by (3.289a). Thus (3.289a) is an integrating factor for the complete differential in four variables (3.285) as follows from (3.289b):

$$\lambda = \frac{1}{x_1 x_2} : \quad \frac{x_2 dx_1 + x_1 dx_2 + (x_1 x_2 / x_4) dx_3 - \left[x_1 x_2 x_3 / (x_4)^2 \right] dx_4}{x_1 x_2}$$

$$= \frac{dx_1}{x_1} + \frac{dx_2}{x_2} + \frac{dx_3}{x_4} - \frac{x_3\, dx_3}{(x_4)^2} = d\left[\log(x_1 x_2) + \frac{x_3}{x_4} \right]; \qquad (3.289a, b)$$

the exact differential (3.289b) specifies the general integral:

$$x_1 x_2 \exp\left(\frac{x_3}{x_4} \right) = C. \qquad (3.290)$$

Thus the first-order differential in four variables (3.285) with vector of coefficients (3.286a) is inexact (3.286b) but has (3.287a–d) an integrating factor (3.289a) leading to the general integral (3.290); the latter specifies a family of hypersurfaces in the four-dimensional space orthogonal to the vector field (3.286a).

NOTE 3.6: Curves on a Surface Orthogonal to a Vector Field

If the first-order differential in N*-variables (3.274a, b) $\equiv$ (3.291b) with (standard LXXXIV) continuously differentiable vector of coefficients (3.283a) $\equiv$ (3.291a) does not meet the **Beltrami conditions** of zero **helicity** (3.282) $\equiv$ (3.291d), then it is inexact; that is, it has non-zero curl (3.291c) and has no integrating factor:*

$$X_n \in C^1\left(|R^N \right): \quad X_n\, dx_n = 0, \quad \Omega_{mn} \equiv \partial_m X_n - \partial_n X_m \neq 0 \neq \sum_{m,n,s}^{cycl} X_s \Omega_{mn} \equiv H_{mns};$$

$$(3.291a–d)$$

*this leads to the **Pffaf problem**: (3.292a) independent (3.292c) conditions (3.292b) can be imposed:*

$$r = 1, ..., N-2: \quad \Psi_r(x_1, ..., x_N) = 0, \quad Ra\left(\frac{\partial \Psi_r}{\partial x_n} \right) = N-2, \qquad (3.292a–c)$$

where the functions (3.293b) are continuously differentiable (3.293a):

$$\Psi_r(x_n) \in C^1\left(|R^N \right): \quad 0 = d\Psi_r = \sum_{n=1}^{N} \frac{\partial \Psi_r}{\partial x_n} dx_n. \qquad (3.293a, b)$$

From (3.292c) it follows that (3.293b) can be solved (3.294a–c) for the differentials $(dx_3, \dots dx_N)$ *in terms of* (dx_1, dx_2):

$$n = 1, \dots, N; \quad s = 3, \dots, N: \qquad dx_s = A_s\, dx_1 + B_s\, dx_2. \qquad (3.294\text{a–c})$$

Substitution of (3.294c) in (3.291b) leads to a differential in two variables (3.295a, b):

$$0 = \sum_{n=1}^{N} X_n dx_n = \left(X_1 + \sum_{n=3}^{N} A_s \right) dx_1 + \left(X_2 + \sum_{s=3}^{N} B_s \right) dx_2 \equiv Y_1\, dx_1 + Y_2\, dx_2. \quad (3.295\text{a, b})$$

The latter (3.295b) has an integrating factor (3.296a) leading to the general integral (3.296b):

$$\lambda\left(Y_1\, dx_1 + Y_2\, dx_2 \right) = d\Phi, \qquad \Phi(x_n) = C. \qquad (3.296\text{a, b})$$

The geometric interpretation is: (i) the N-2 regular (3.293a) hypersurfaces (3.292a, b) intersect (3.292c) in a surface or subspace of dimension M = 2; and (ii) on this surface there are curves, namely the intersection with the general integral (3.296b), that are orthogonal to the vector field (3.291a) of coefficients.

NOTE 3.7: Inexact Differential in Four Variables without Integrating Factor

The first-order differential in four variables:

$$0 = x_2\, dx_1 + x_1\, dx_2 + x_4\, dx_3 - x_3\, dx_4, \qquad (3.297)$$

has no integrating factor, because the: (i) first two terms are an exact differential (3.298):

$$x_2\, dx_1 + x_1\, dx_2 = d(x_1 x_2); \qquad (3.298)$$

(ii) the last two terms (3.299b) have an integrating factor (3.299a):

$$\lambda = \frac{1}{x_3 x_4} : \qquad \frac{x_4\, dx_3 - x_3\, dx_4}{x_3 x_4} = \frac{dx_3}{x_3} - \frac{dx_4}{x_4} = d\left[\log\left(\frac{x_3}{x_4} \right) \right]; \qquad (3.299\text{a, b})$$

and (iii) there is no common overall integrating factor because the exact differential (3.297) multiplied by (3.299a) becomes inexact. It can be checked that the vector of coefficients (3.300a):

$$\bar{X} = \{ x_2, x_1, x_4, -x_3 \} : \qquad \{ \Omega_{12}, \Omega_{13}, \Omega_{14}\Omega_{23}, \Omega_{24}, \Omega_{34} \} = \{ 0, 0, 0, 0, 0, -2 \},$$

$$(3.300\text{a, b})$$

has curl (3.300b) with one non-zero component, arising from the last two terms in (3.297). The Beltrami conditions (3.282) are not met because of the same term; that is, (3.287a, b) hold but not (3.287c, d), that are replaced by:

$$\Omega_{13}X_4 + \Omega_{34}X_1 + \Omega_{41}X_3 = -2x_2, \qquad \Omega_{23}X_4 + \Omega_{34}X_2 + \Omega_{42}X_3 = -2x_1;$$

$$(3.301a, b)$$

this confirms that the helicity trivector is not zero, and hence an integrating factor does not exist. Imposing two conditions (3.302a, b):

$$0 = \Psi_1(x_1, x_2, x_3) = x_3 - x_1 - x_2, \quad 0 = \Psi_2(x_1, x_2, x_4) = x_4 - x_1 + x_2,$$

$$(3.302a, b)$$

reduces the differential from four (3.297) to two (3.303) variables:

$$0 = x_2\, dx_1 + x_1\, dx_2 + (x_1 - x_2)d(x_1 + x_2) - (x_1 + x_2)d(x_1 - x_2) = -x_2\, dx_1 + 3x_1\, dx_2;$$

$$(3.303)$$

the differential equation (3.303) is separable (3.304a):

$$0 = 3\frac{dx_2}{x_2} - \frac{dx_1}{x_1} = d\left(3\log x_2 - \log x_1\right), \qquad x_1 = C(x_2)^3, \quad (3.304a, b)$$

leading to the integral (3.304b). Thus the first-order differential in four variables (3.297) is inexact (3.300b) and has no integrating factor (3.287a, b; 3.301a, b); imposing two conditions (3.302a, b) the general integral is (3.304b). Thus the family of curves (3.304b) on the surface that is the intersection of the two hypersurfaces (3.302a, b) is orthogonal to the vector field (3.300a).

NOTE 3.8: Independent Components of the Curl and Helicity

Given a first-order differential (3.276a, b) in N variables associated (note III.9.46) with the vector of coefficients (3.283a) may be considered: (i) the **curl** (3.283b); that is, a skew symmetric covariant tensor or **bivector**, with N^2 components (3.305a) of which are independent (3.305b), as for the triangle below the diagonal of an $N \times N$ matrix:

$$\#\Omega_{mn} = N^2, \qquad \#^* \Omega_{mn} = \frac{N^2}{2} - \frac{N}{2} = \frac{N(N-1)}{2} = \binom{N}{2}; \qquad (3.305a, b)$$

and (ii) the **helicity** (3.282); that is, a covariant skew-symmetric tensor with three indices or **trivector** that has (3.306a) components of which (3.306b) are independent:

$$\# H_{mns} = N^3, \quad \#^* H_{mns} = \binom{N}{3} = \frac{N(N-1)(N-2)}{6}. \qquad \text{(3.306a, b)}$$

In both cases of the bivector (3.305a, b) [trivector (3.306a, b)]: (i) repeated indices lead to zero, by skew-symmetry; and (ii) the same set of indices in any order gives the same values with plus (minus) sign for even (odd) permutations. Thus for a bivector (3.305a) the first index takes N values, the second index n takes $N-1$ values, and the product is divided by permutations of 2, leading to (3.305b). For a trivector (3.306a) the third index takes $N-2$ values and the product is divided by permutations of $3! = 6$ leading to (3.306b). Thus *an N-dimensional differentiable vector field has a curl bivector (3.291c) [helicity trivector (3.291d)] with (3.305a) [(3.306a)] components of which (3.305b) [(3.306b)] are independent, that are indicated in the Table 3.2 for $N = 1,....10$ together with the number s of hypersurfaces (3.292a, b) with linearly independent normal vectors (3.292c) that appear in the Pfaff problem (standard LXXXIV).*

TABLE 3.2
Number of Independent Components of the Curl and Helicity

Space	Field	Curl	Helicity	Constraints
Coordinates	Vector	Bivector	Trivector	Hypersurfaces
x_m	X_m	$\Omega_{mn} = \partial_m X_n - \partial_n X_m$	$H_{mnr} = \sum_{cycl} X_s(\partial_m X_n - \partial_n X_m)$	$\Psi_e(x_n) = 0$
N	N	$N(N-1)/2$	$N(N-1)(N-2)/6$	$N-2$
$N=1$	1	0	0	–
$N=2$	1	1	0	–
$N=3$	3	3	1	1
$N=4$	4	6	4	2
$N=5$	5	10	10	3
$N=6$	6	15	20	4
$N=7$	7	21	35	5
$N=8$	8	28	56	6
$N=9$	9	36	84	7
$N=10$	10	45	120	8

Note: Total number of components and number of independent components of the bivector curl and trivector helicity of a vector field in N dimensions.

NOTE 3.9: Necessary and Sufficient Condition of Integrability

With respect to a first-order differential with any number of variables (3.278a, b), the necessary condition of immediate integrability (3.278c) [existence of an integrating factor (3.282)] have already been proven. Next are proven [note(s) 3.9 (3.10–3.11)] the sufficient condition for immediate integrability (existence of an integrating factor). The latter proof is based on the reduction of the number of variables of a differential of first order (note 3.10), which may also be used as a method of integration (note 3.11). Next will be proven that the conditions of integrability that were shown to be necessary are also sufficient and apply to differentials of first order with any number of variables; the **conditions of immediate integrability** (3.278c) [**integrability** (3.282)] *requiring the vanishing of the curl bivector (helicity trivector) are both necessary and sufficient for the differential first-order differential in N variables to be an exact differential (3.276b)* ≡ *(3.307a) [to have an integrating factor* λ *in (3.281b), so that upon multiplication by it the form becomes an exact differential (3.284b)* ≡ *(3.307b)]:*

$$\sum_{n=1}^{N} X_n \, dx_n = \sum_{n=1}^{N} \frac{\partial \Phi}{\partial x_n} dx_n = d\Phi, \lambda^{-1} d\Phi. \qquad (3.307a, b)$$

With respect to immediate integrability (3.278c; 3.307a), the necessary condition is proved by the sequence (3.276a–d; 3.277a–c); for the sufficient condition, the starting point is the Stokes theorem (III.9.399a–d) for a surface (3.308a):

$$\int_{D_2} \sum_{n,m=1}^{N} (\partial X_n / \partial x_m - \partial X_m / \partial x_n) dS_{nm} = \int_{D_1} \sum_{n=1}^{N} X_n \, dx_n, \qquad (3.308a)$$

where D_2 is (Figure 3.10) a regular surface with the area element specified by the bivector dS_{nm} supported on the closed regular curve that is its boundary $D_1 \equiv \partial D_2$ along which the displacement is dx_n. Substituting (3.278c) in (3.308a), it follows that the first-order differential, integrated along any closed regular loop, is zero (3.308b):

$$0 = \int_{D_1} \sum_{n=1}^{N} X_n \, dx_n, \qquad d\Phi = \sum_{n=1}^{N} X_n \, dx_n, \qquad (3.308b, c)$$

and hence it must be the differential of a function (3.308c).

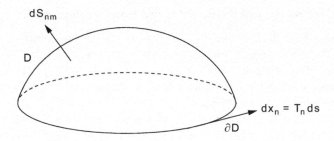

FIGURE 3.10
The proof of the sufficiency condition for a first-order differential to be exact uses the Stokes theorem stating that the flux of the curl of a continuously differentiable vector across a regular surface D with area element dS_{nm} supported on a closed regular curve or loop ∂D equals the circulation of the vector along the boundary with arc length ds and unit tangent vector T_n and displacement dx_n.

NOTE 3.10: Sufficient Condition for the Existence of an Integrating Factor

With respect to the condition of integrability (3.282), its necessity was proved by (3.281a, b); for the sufficiency, it will shown that if the conditions of integrability are satisfied it is possible to reduce a first-order differential with N variables to one with $N-1$ variables. Appling the procedure $N-2$ times leads to a first-order differential in two variables (3.145a, b) that always has an integrating factor. The procedure will be demonstrated by reducing the first-order differential in three variables (3.271) to one in two variables (3.145b). The proof will need the following result: *if the vector (3.309a) satisfies the condition of integrability (3.309b) of zero helicity (3.282), so does (3.309e) any parallel vector field 3.309d), where μ is any continuously differentiable function (3.309c):*

$$X_n \in C^1\left(|R^N\right): \quad \overset{cyl}{\underset{m \rho n,s}{\sum}} X_s\left(\partial_n X_m - \partial_m X_n\right) = 0 \quad \Rightarrow \qquad (3.309\text{a, b})$$

$$\mu \in C^1\left(R^N\right), \quad Y_n = \mu X_n: \quad \overset{cyl}{\underset{m,n,s}{\sum}} Y_s\left(\partial_n Y_m - \partial_m Y_n\right) = 0. \qquad (3.309\text{c–e})$$

This result can be proved directly by substituting (3.309d) in (3.282). Alternatively, note that if the first-order differential (3.274b) has an integrating factor λ, then first-order differential using (3.310a):

$$X_n = \mu Y_n, \qquad df = \lambda \sum_{n=1}^{N} X_n dx_n = \lambda \mu \sum_{n=1}^{N} Y_n dx_n, \qquad (3.310\text{a, b})$$

has (3.310b) the integrating factor $\lambda\mu$. A third proof follows by noting that (3.310a) replaces λ in (3.274b) by $\lambda\mu$, but since this does not appear in the integrability condition (3.282), the latter is unchanged.

Since the reduction of a first-order differential in N variables to $(N - 1)$ variables is similar for all N, the method is demonstrated for a first-order differential in three variables (3.219), where z is taken for the moment as a parameter, so that it reduces to a first-order differential in two variables (3.311) with integrating factor λ:

$$\lambda\left(X\,dx + Y\,dy\right) = du\left(x, y, z\right). \tag{3.311}$$

From (3.311) it follows the partial derivatives of u with regard to x, y are given by (3.312a, b):

$$\frac{\partial u}{\partial x} = \lambda X, \quad \frac{\partial u}{\partial y} = \lambda Y; \quad \frac{\partial u}{\partial z} = \lambda Z - v, \tag{3.312a–c}$$

if z is taken as variable again, $\partial u / \partial z$ need not coincide with λZ, that is, it differs (3.312c) by an undetermined function v. Since the vector of coefficients satisfy the integrability condition (3.282), so does vector of coefficients multiplied by the integrating factor (3.313a) leading by (3.312a–c) to (3.313b) $\equiv$ (3.313c):

$$\vec{Y} = \lambda\vec{X} = \left\{\frac{\partial u}{\partial x}, \frac{\partial u}{\partial y}, \frac{\partial u}{\partial z} - v\right\} = \nabla u - \vec{e}_z v. \tag{3.313a–c}$$

Even if the integrability condition is not satisfied (3.314a), it follows from (3.313c) that:

$$\vec{X}.\left(\nabla \wedge \vec{X}\right) \neq 0: \qquad \vec{Y}.d\vec{x} = du - v\,dz, \tag{3.314a, b}$$

a first-order differential with three variables can be written in the form (3.314b) that: (i) leads to (3.268), the first principle of thermodynamics (subsections 3.9.10–3.9.11); and (ii) proves (3.257a, b) the representation of a continuously differentiable vector field in terms of three scalar potentials (subsections 3.9.8–3.9.9). By using the integrability condition (3.234c), the first-order differential with three variables can be reduced to two variables, as shown next (Note 3.11).

NOTE 3.11: Reduction of the Number of Variables of a Differential

The vector (3.313c) may be substituted in the integrability condition (3.234c) in two steps (3.315a–c):

$$\nabla \wedge \vec{Y} = -\nabla \wedge (\vec{e}_z v) = \left\{ -\frac{\partial v}{\partial y}, \frac{\partial v}{\partial x}, 0 \right\}, \tag{3.315a}$$

$$0 = \vec{Y}.\left(\nabla \wedge \vec{Y}\right) = -\frac{\partial u}{\partial x}\frac{\partial v}{\partial y} + \frac{\partial u}{\partial y}\frac{\partial v}{\partial x} = - \begin{vmatrix} \partial u/\partial x & \partial u/\partial y \\ \partial v/\partial x & \partial v/\partial y \end{vmatrix}. \tag{3.315b, c}$$

The first line of the determinant (3.315c) does not vanish, so the determinant can be zero only if the second line is proportional to the first:

$$\frac{\partial v}{\partial x} = \mu \frac{\partial u}{\partial x}, \qquad \frac{\partial v}{\partial y} = \mu \frac{\partial u}{\partial y}, \tag{3.316a, b}$$

implying (3.316c):

$$dv = \frac{\partial v}{\partial x}dx + \frac{\partial v}{\partial y}dy = \mu\, du, \qquad \mu = \frac{dv}{du}, \tag{3.316c, d}$$

that the function v depends only on u in (3.316db), and possibly on the parameter z. It follows that:

$$\lambda\left(X\, dx + Y\, dy + Z\, dz\right) = du - v dz, \tag{3.317}$$

where (3.312a–c) was used, is a first-order differential with two variables u, z; hence it has (3.318a) an integrating factor η:

$$dw = \eta\left(du - v dz\right) = \lambda\eta\left(X\, dx + Y\, dy + Z dz\right), \quad w(x, y, z) = C, \tag{3.318a–c}$$

implying that the first-order differential with three variables (3.219) has an integrating factor $\lambda\eta$ in (3.318b) and hence its solution is (3.318c). QED.

NOTE 3.12: Integration of a First-Order Differential with Four Variables

The preceding proof of sufficiency of the integrability conditions also provides a method (standard LXXXV) of integration of a first-order differential with N variables by reducing it to two variables in $N - 2$ steps. This is illustrated next by $N-2=2$ steps for an integrable first-order differential in $N=4$ variables. Consider the first-order differential in four variables:

$$\left(2x_1 + x_2^2 + 2x_1x_4 - x_3\right)dx_1 + 2x_1x_2\,dx_2 - x_1\,dx_3 + x_1^2\,dx_4 = 0, \qquad (3.319)$$

whose vector of coefficients is:

$$\{X_1, X_2, X_3, X_4\} = \left\{2x_1 + x_2^2 + 2x_1x_4 - x_3,\; 2x_1x_2,\; -x_1,\; x_1^2\right\}; \qquad (3.320)$$

the curl bivector (3.278c) of the vector of coefficients (3.320) is zero, and thus (3.319) is an exact first-order differential in four variables. Taking x_1, x_2 as parameters (3.321a, b), the first-order differential (3.319) reduces (3.321c) to two variables x_3, x_4:

$$x_1, x_2 = \text{const}: \qquad 0 = -x_1 dx_3 + x_1^2 dx_4 = du, \qquad u \equiv -x_1 x_3 + x_1^2 x_4, \qquad (3.321a\text{–}d)$$

and has integral (3.321d). Next consider x_1 to be a variable, leaving x_2 constant (3.322a) and leading to (3.322b):

$$x_2 = \text{const}: \qquad du = -x_1\,dx_3 + x_1^2\,dx_4 + \left(2x_1x_4 - x_3\right)dx_1, \qquad (3.322a, b)$$

where the first two terms coincide with two terms of the first-order differential (3.319) in three variables x_1, x_3, x_4:

$$x_2 = \text{const}: \qquad 0 = \left(2x_1 + x_2^2 + 2x_1x_4 - x_3\right)dx_1 - x_1 dx_3 + x_1^2 dx_4; \qquad (3.323a, b)$$

substitution of (3.322b) in (3.323b) leads to (3.324a), whose primitive is (3.324b):

$$0 = \left(2x_1 + x_2^2\right)dx_1 + du = dv, \qquad v \equiv u + x_1^2 + x_1 x_2^2. \qquad (3.324a, b)$$

Taking all x_1, x_2, x_3, x_4 in (3.324b) as variables leads to (3.325a):

$$dv = d\left(u + x_1^2 + x_1 x_2^2\right) = du + \left(2x_1 + x_2^2\right)dx_1 + 2x_1x_2\,dx_2 = 0, \qquad (3.325a, b)$$

that coincides with (3.319) by (3.322b), implying (3.325b). From (3.324b) and (3.321d) follows (3.326):

$$C = v = u + x_1^2 + x_1 x_2^2 = x_1^2 + x_1 x_2^2 - x_1 x_3 + x_1^2 x_4, \qquad (3.326)$$

that is the solution (3.326) of the first-order differential (3.319). It can be checked that differentiation of (3.326) leads back to (3.319). The main results on first-order differentials are summarized next for two/three/more variables (notes 3.13/3.14/3.15).

NOTE 3.13: Exact Differentials and Integrating Factor

A first-order differential in one variable (3.327a) is always exact for an integrable coefficient (3.327a), and its solution reduces to an integration (3.227c):

$$X \in E(|R): \quad X(x)dx = d\Phi \quad \Leftrightarrow \quad \Phi(x) = \int^{x} X(\xi)d\xi; \quad\quad (3.327a\text{–}c)$$

in this case there is no curl or helicity. A first-order differential in two variables (3.145b) ≡ (3.328a) may have a vector of coefficients with non-zero curl (3.243b) ≡ (3.328b) and has zero helicity (3.243c) ≡ (3.328c):

$$X_1\,dx_1 + X_2\,dx_2 = 0: \quad \Omega = \frac{\partial X_2}{\partial x_1} - \frac{\partial X_1}{\partial x_2} \neq 0, \quad H = 0. \quad\quad (3.328a\text{–}c)$$

Thus there are two cases: (i) an exact differential (3.329a) if the curl is zero; and (ii) an integrating factor for an inexact differential (3.329b) if the curl is not zero:

$$X_1 dx_1 + X_2 dx_2 = \begin{cases} d\Phi & \text{if} \quad \Omega = 0, & (3.329a) \\ \lambda^{-1} d\Phi & \text{if} \quad \Omega \neq 0. & (3.329b) \end{cases}$$

In both cases (Figure 3.6) the plane vector field has a family of orthogonal curves.

NOTE 3.14: Inexact Differentials without an Integrating Factor

A differential in three variables (3.330a) can have a vector of coefficients with non-zero curl (3.330b) and non-zero helicity (3.330c):

$$0 = X_1 dx_1 + X_2 dx_2 + X_3 dx_3: \quad \vec{\Omega} \equiv \nabla \wedge \vec{X} \neq 0, \quad H = \vec{X}.\left(\nabla \wedge \vec{X}\right) \neq 0. \quad\quad (3.330a\text{–}c)$$

Thus there are three cases: (i) an exact differential (3.331a) for zero curl; (ii) an inexact differential with an integrating factor (3.331b) for non-zero curl and zero helicity; and (iii) an inexact differential without an integrating factor (3.331c) for non-zero helicity:

$$X_1 dx_1 + X_2 dx_2 + X_3 dx_3 = \begin{cases} d\Phi & \text{if} \quad \vec{\Omega} = 0, & (3.331a) \\ \lambda^{-1} d\Phi & \text{if} \quad \vec{\Omega} \neq 0 = H, & (3.331b) \\ \lambda^{-1} d\Phi + d\Psi & \text{if} \quad H \neq 0. & (3.331c) \end{cases}$$

In cases (i, ii) there is (3.331a, b) a family of surfaces orthogonal to the vector field (Figure 3.7), but not in case (iii). In the latter case (3.331c) there are on any surface (Figure 3.9) curves orthogonal to the vector field.

NOTE 3.15: First-Order Differentials in Any Dimension

A differential of first-order in N variables (3.332a) may have vector of coefficients (3.283a) with non-zero (3.332b) curl (3.283b) and non-zero (3.332c) helicity (3.282):

$$0 = \sum_{n=1}^{N} X_n \, dx_n: \qquad \Omega_{mn} = \partial_m X_n - \partial_n X_m \neq 0,$$

$$H_{mns} = \sum_{m,n,s}^{cycl} X_s \left(\partial_m X_n - \partial_n X_m \right).$$

$$(3.332a\text{–}c)$$

Thus there are three cases: (i) an exact differential if the curl is zero (3.333a) implying (3.305b) integrability conditions; (ii) an inexact differential with an integrating factor (3.333b) if the curl is not zero but the helicity is zero, implying (3.306b) integrabity conditions; and (iii) an inexact differential without integrating factor (3.333c) if the helicity is not zero, in which case $N - 2$ constraints can be imposed:

$$\sum_{n=1}^{N} X_n dx_n = \begin{cases} d\Phi & \text{if} \quad \Omega_{mn} = 0, & (3.333a) \\ \lambda^{-1} d\Phi & \text{if} \quad \Omega_{mn} \neq 0 = H_{mns}, & (3.333b) \\ \lambda^{-1} d\Phi + \sum_{m=1}^{N-2} d\Phi_m & \text{if} \quad H_{mns} \neq 0. & (3.333c) \end{cases}$$

In cases (i, ii) there is (3.333a, b) a family of hypersurfaces orthogonal to the vector field of coefficients (3.283a) but not in case (iii). In the latter case $N - 2$ functions (3.292a, b) can be specified such that the matrix of partial derivatives has maximum possible rank (3.292c); this implies that the $N - 2$ hypersurfaces intersect on a surface, and that on that surface exists a curves orthogonal to the vector field. The first-order differentials (sections 3.8–3.9 and notes 3.1–3.15) [first-order differential equations (sections 3.1–3.7 and 5.1–5.5)] are integrable in special homogeneous cases [notes 3.16–3.20 (section 3.7)]. Methods of integration quicker than the preceding apply to special differentials such as the homogeneous type (note 3.16) in two (more than two) variables [note 3.17 (3.18)] as shown by several examples (notes 3.19–3.20).

NOTE 3.16: Homogeneous Functions and Differential

*A function $P(x,y)$ is said to be **homogeneous of degree q** if it satisfies (3.334a):*

$$P(x,y) = x^q P\left(1, \frac{y}{x}\right); \quad P \in D(|R^2|): \quad x\frac{\partial P}{\partial x} + y\frac{\partial P}{\partial y} = qP, \quad (3.334a\text{–}c)$$

and if it is differentiable (3.334b) the Euler's theorem (3.334c) applies; the latter can be proved from:

$$x\frac{\partial P}{\partial x} + y\frac{\partial P}{\partial y} = qx^q P\left(1, \frac{y}{x}\right) - \frac{y}{x^2}x^{q+1}\frac{\partial P}{\partial y} + \frac{y}{x}x^q\frac{\partial P}{\partial y} = qP. \quad (3.335)$$

If the functions X, Y are homogeneous of the same degree q, they satisfy the identity:

$$Y\left(x\frac{\partial X}{\partial x} + y\frac{\partial X}{\partial y}\right) = YqX = X\left(x\frac{\partial Y}{\partial x} + y\frac{\partial Y}{\partial y}\right), \quad (3.336a, b)$$

that is equivalent to:

$$\frac{\partial}{\partial y}\left\{\frac{X}{xX + yY}\right\} = \frac{\partial}{\partial x}\left\{\frac{Y}{xX + yY}\right\}. \quad (3.336c)$$

The proof of (3.336c) follows from:

$$(xX + yY)^2\left[\frac{\partial}{\partial y}\left(\frac{X}{xX + yY}\right) - \frac{\partial}{\partial x}\left(\frac{Y}{xX + yY}\right)\right]$$

$$= (xX + yY)\left(\frac{\partial X}{\partial y} - \frac{\partial Y}{\partial x}\right) - X\left(x\frac{\partial X}{\partial y} + y\frac{\partial Y}{\partial y} + Y\right) + Y\left(x\frac{\partial X}{\partial x} + y\frac{\partial Y}{\partial x} + X\right) \quad (3.337)$$

$$= Y\left(x\frac{\partial X}{\partial x} + y\frac{\partial X}{\partial y}\right) - X\left(x\frac{\partial Y}{\partial x} + y\frac{\partial Y}{\partial y}\right) = 0.$$

    The identities (3.336a, b) ≡ (3.336c) for homogeneous functions (x, y) of the same degree (3.335), are used next (note 3.17) to integrate homogeneous first-order differentials.

NOTE 3.17: Integration of Homogeneous Differentials

The identity (3.336c) ≡ (3.153c) implies that *if the functions (3.338a, b) are homogeneous (3.334a) with the same degree q:*

$$X(x,y) = x^q X\left(1, \frac{y}{x}\right), \quad Y(x,y) = x^q Y\left(1, \frac{y}{x}\right), \quad (3.338a, b)$$

then: *(i) the vector with components (3.339a, b) leads to an exact differential in two variables (3.153b) ≡ (3.339c) ≡ (3.339d):*

$$\{A,B\} = \frac{\{X,Y\}}{xX+yY}: \quad d\Phi = A\,dx + B\,dy = \frac{X\,dx+Y\,dy}{xX+yY}; \quad\quad\quad (3.339a\text{--}d)$$

(ii) the (standard LXXXVI) first-order differential (3.340a) has integrating factor (3.340b) given by (3.340c):

$$d\Psi = X\,dx + Y\,dy = \lambda^{-1}d\Phi: \quad \frac{1}{\lambda} = xX + yY; \quad\quad\quad (3.340a\text{--}c)$$

and (iii) if the first-order differential in two variables (3.340a) ≡ (3.341b) is (standard LXXXVII) exact and homogeneous of degree (3.341a) the integral is (3.341c):

$$q \neq -1: \quad X\,dx + Y\,dy = d\Phi \quad \Rightarrow \quad xX + yY = C. \quad\quad\quad (3.341a\text{--}c)$$

The proof of (3.340a–c) follows from the identity (3.340b, c) ≡ (3.339c, d). To prove (3.341a–c) the condition that (3.341a) is an exact differential (3.153c) ≡ (3.342a) is used in (3.342b):

$$\frac{\partial X}{\partial y} = \frac{\partial Y}{\partial x}: \quad d(xX+yY) - X\,dx - Y\,dy = x\,dX + y\,dY$$

$$= x\left(\frac{\partial X}{\partial x}dx + \frac{\partial X}{\partial y}dy\right) + y\left(\frac{\partial Y}{\partial x}dx + \frac{\partial Y}{\partial y}dy\right)$$

$$= x\left(\frac{\partial X}{\partial x}dx + \frac{\partial Y}{\partial x}dy\right) + y\left(\frac{\partial X}{\partial y}dx + \frac{\partial Y}{\partial y}dy\right) \quad\quad (3.342a\text{--}c)$$

$$= \left(x\frac{\partial X}{\partial x} + y\frac{\partial X}{\partial y}\right)dx + \left(x\frac{\partial Y}{\partial x} + y\frac{\partial Y}{\partial y}\right)dy$$

$$= q(X\,dx + Y\,dy),$$

where the property of homogeneity (3.334c) was used in (3.342c). Rewriting (3.342c) ≡ (3.343):

$$d(xX + yY) = (q+1)(X\,dx + Y\,dy), \quad\quad\quad (3.343)$$

it follows that if (3.341a) holds then (3.343) implies (3.341c).

NOTE 3.18: Extension to Homogeneous Differentials in N Variables

The preceding results extend to first-order differentials in N variables: *if the functions (3.344a) are homogeneous of degree* q:

$$X_n(x_1,...,x_N) = (x_1)^q X_n\left(1, \frac{x_2}{x_1},...,\frac{x_N}{x_1}\right) \quad \Leftrightarrow \quad \sum_{m=1}^{n} x_m \frac{\partial X_n}{\partial X_m} = qX_n, \qquad \text{(3.344a, b)}$$

from (3.344b) it follows that: (i) the (standard LXXXVIII) first-order differential (3.345a):

$$d\Psi = \sum_{m=1}^{N} X_n \, dx_n : \qquad \frac{1}{\lambda} = \sum_{m=1}^{N} X_n x_n, \qquad \text{(3.345a, b)}$$

has integrating factor (3.345b); and (ii) if the first-order differential (3.345a) is exact (3.346b) and (standard LXXXIX) the vector of coefficients is homogeneous (3.343a) of degree (3.346a):

$$q \neq -1: \qquad d\Phi = \sum_{n=1}^{N} X_n \, dx_n \quad \Rightarrow \quad \sum_{n=1}^{N} X_n x_n = C, \qquad \text{(3.346a–c)}$$

then the general integral is (3.346c).

NOTE 3.19: Integration of an Homogeneous First-Order Differential of Degree Two

As an example, consider the first-order differential in two variables (3.347a), that is homogeneous of degree two:

$$\left(x^2 + y^2\right)dx - xy \, dy = 0; \qquad \frac{1}{\lambda} = x\left(x^2 + y^2\right) - xy^2 = x^3; \qquad \text{(3.347a, b)}$$

its integrating factor is (3.347b). Dividing (3.347a) by (3.347b) leads to an exact differential (3.348a):

$$0 = \frac{dx}{x} + \frac{y^2}{x^3} dx - \frac{y}{x^2} dy = d\left(\log x - \frac{y^2}{2x^2}\right): \qquad \log x = C + \frac{y^2}{2z^2}, \qquad \text{(3.348a, b)}$$

that specifies the general integral (3.348b) of (3.347a).

NOTE 3.20: Exception for a Homogeneous First-Order Differential of Degree -1

The first-order differential:

$$0 = \left(3x^2 + y^2\right)dx + 2xy\, dy = d\left(x^3 + xy^2\right) \equiv d\Phi, \tag{3.349}$$

is homogeneous of degree two and exact; it can be checked that it satisfies the integrability condition (3.350a):

$$\frac{\partial}{\partial y}\left(3x^2 + y^2\right) = 2y = \frac{\partial}{\partial x}(2xy), \quad x\left(3x^2 + y^2\right) + 2xy^2 = 3\Phi = C, \tag{3.350a, b}$$

and that the general integral is (3.350b). The exact homogeneous first-order differential (3.351a):

$$0 = \frac{y}{x^2}dx - \frac{dy}{x} = -d\left(\frac{y}{x}\right), \quad y = Cx, \tag{3.351a, b}$$

has a general integral (3.351b); it does not coincide with (3.352a, b):

$$\{X,Y\} = \left(\frac{y}{x^2}, -\frac{1}{x}\right), \quad C = xX + yY = 0, \tag{3.352a, b}$$

because the degree of (3.351a) is $q = -1$, and the theorem (3.346b, c) fails.

NOTE 3.21 Higher-Orders, Sub-Spaces, and Tensors

The differentials may be generalized from the first order (sections 3.8–3.9 and notes 3.1–3.20) to higher orders (note 3.21) with associated invariance properties (note 3.22). A method of solution of higher-order differentials is the factorization into lower-orders (note 3.23), for example a second-order differentials may be factorized into two first-order differentials using a difference of squares (note 3.24). The simplest *second-order differential in two variables (standard XIX) is*:

$$0 = A(x,y)(dx)^2 + B(x,y)(dy)^2 + D(x,y)dx\,dy, \tag{3.353}$$

that may be solved (standard XC) as a quadratic equation for the slope:

$$0 = B(x,y)y'^2 + D(x,y)y' + A(x,y), \tag{3.354}$$

with roots:

$$2B(x,y)y' = -D(x,y) \pm \left[\left[D(x,y) \right]^2 - 4A(x,y)B(x,y) \right]^{1/2}. \qquad (3.355)$$

As in (3.9a, b) with $M = 2$, the roots (3.355) specify a set of points in the plane (Figure 3.3) where two families of integral curves (3.356a, b) intersect:

$$\Phi(x,y) = C_1, \qquad \Psi(x,y) = C_2. \qquad (3.356a, b)$$

The general case is a **differential of order M in N variables** (3.357b):

$$i_1,...,i_M = 1,...,N: \qquad 0 = X_{i_1...i_M} dx^{i_1} ... dx^{i_M}, \qquad (3.357a, b)$$

where: (i) each of the M indices takes N values (3.357a), where N is the dimension of the space and a repeated index is summed (3.275a, b) from 1 to N; (ii) the coefficients form a **symmetric tensor of order M**:

$$X_{i_1...i_{r-1}i_r i_{r+1}...i_{s-1}i_s i_{s+1}...i_M} = X_{i_1...i_{r-1}i_s i_{r+1}...i_{s-1}i_r i_{s+1}...i_M}. \qquad (3.357c)$$

and (iii) the general integral of (3.357b), if it exists, is a **sub-space** of dimension $N - M$, that is the intersection of M hypersurfaces (3.358a, b):

$$m = 1,...;M: \qquad \Phi_m(x_1,...,x_N) = C_m, \qquad Ra\left(\frac{\partial \Phi_m}{\partial x_n}\right) = M, \qquad (3.358a-c)$$

with matrix of partial derivatives (3.358c) of maximum possible rank M, meaning that the normal vectors are linearly independent. The differential of order M involves (3.357a, b) a symmetric tensor (3.357c) of order M, and is distinct from a **differential form** of order M (notes III.9.1-III.9.52) that involves a skew-symmetric tensor of order M or M-vector.

NOTE 3.22: Curvilinear Coordinates, Invariance, and Transformation Law

A **curvilinear coordinate transformation** is specified (note III.9.6) by a set of N continuously differentiable functions (3.359a) with finite, non-zero Jacobian (3.359b) with the differentials of coefficients related by (3.359c):

$$x^{i'} \in C^1\left(|R^N\right): \qquad Det\left(\frac{\partial x^{i'}}{\partial x^i}\right) \neq 0, \infty, \qquad dx^{i'} = \frac{\partial x^{i'}}{\partial x^i} dx^i. \qquad (3.359a-c)$$

*The differential (3.357b) is **invariant**; that is, it takes the same form (3.360) in each allowable coordinate system (3.359a–c):*

$$0 = X_{i_1...i_M} dx^{i_1} ....dx^{i_M}, \qquad (3.360)$$

*if and only if the coefficients have the **transformation law**:*

$$X_{i_1'...i_M'} = X_{i_1...i_M} \frac{\partial x^{i_1}}{\partial x^{i_1'}} ... \frac{\partial x^{i_M}}{\partial x^{i_M'}}, \qquad (3.361)$$

*corresponding to a **covariant tensor of order M** that is symmetric (3.357c) in all indices.* The invariance properties of differentials are addressed in the tensor calculus (notes III.9.1-III.9.52).

NOTE 3.23: Factorization of Differentials of Degree Two

The consideration of differentials (notes 3.1–3.22) will be concluded with some cases of second degree in two or three variables (notes 3.23–3.24). *The second-order differential in two variables (3.353) ≡ (3.362a) can be always factorized into the product of two first-order differentials (3.363b):*

$$0 = B(dy - \lambda_+ \, dx)(dy - \lambda_- \, dx), \qquad 2\lambda_\pm B = -D \pm \sqrt{D^2 - 4AB}. \qquad (3.362a, b)$$

This is not generally the case with (standard XCI) the second-order differential in three variables:

$$0 = A(dx)^2 + B(dy)^2 + C(dz)^2 + 2D\,dx\,dy + 2E\,dx\,dz + 2F\,dy\,dz. \qquad (3.363)$$

If the condition (3.364a) is met, the second-order differential (3.363) can be substituted in the perfect square (3.364b):

$$C \neq 0: \quad (C\,dz + E\,dx + F\,dy)^2$$
$$= E^2(dx)^2 + F^2(dy)^2 + 2EF\,dx\,dy + C\left[C(dz)^2 + 2(E\,dx + F\,dy)dz\right]$$
$$= E^2(dx)^2 + F^2(dy)^2 + 2EF\,dx\,dy - C\left[A(dx)^2 + B(dy)^2 + 2D\,dx\,dy\right]$$
$$= (E^2 - CA)(dx)^2 + (F^2 - BC)(dy)^2 + 2(EF - CD)dx\,dy; $$
$$\qquad (3.364a, b)$$

thus the differential (3.363) reduces to a difference of squares if the r.h.s. of (3.364b) is a perfect square (3.365a), implying (3.365b):

$$0 = (E^2 - CA)(F^2 - CB) - (EF - CD)^2: \quad ABC + 2DEF - AF^2 - BE^2 - CD^2 = 0. $$
$$\qquad (3.365a, b)$$

It has been shown that *if the differential (3.363) of the second order in three variables satisfies (standard XCI) the condition (3.365b), then it can be written as difference of squares (6.366):*

$$\left(C\,dz + E\,dx + F\,dy\right)^2 - \left(\sqrt{E^2 - CA}\,dx + \sqrt{F^2 - CB}\,dy\right)^2 = 0, \qquad (3.366)$$

so that it can be split into the product of two differential of first-degree in three variables:

$$C\,dz + \left\{E \pm \sqrt{E^2 - CA}\right\}dx + \left\{F \pm \sqrt{F^2 - CB}\right\}dy = 0, \qquad (3.367)$$

which can be solved by the methods presented before (section 3.9).

NOTE 3.24: Differential of Order Two as Difference of Squares

As an example, consider the differential of second order in three variables:

$$x^2\left(dx\right)^2 + y^2\left(dy\right)^2 - z^2\left(dz\right)^2 + 2xy\,dx\,dy = 0, \qquad (3.368)$$

whose coefficients:

$$\{A, B, C, D, E, F\} = \left\{x^2, y^2, -z^2, xy, 0, 0\right\}, \qquad (3.369)$$

satisfy (3.365a) $\equiv$ (3.365b) and lead (3.367) to the factorization (3.370a):

$$\left(xz\,dx + yz\,dy + z^2\,dz\right)\quad\left(xz\,dx + yz\,dy - z^2 dz\right) = 0; \qquad (3.370a)$$

this factorization follows readily from (3.368):

$$0 = \left(x\,dx + y\,dy\right)^2 - \left(z\,dz\right)^2 = \left(x\,dx + y\,dy + z\,dz\right)\left(x\,dx + y\,dy - z\,dz\right). \qquad (3.370b)$$

Thus the problem reduces to two differentials of first-degree in three variables (3.371a):

$$0 = x\,dx + y\,dy \pm z\,dz, \qquad x^2 + y^2 \pm z^2 = C_\pm, \qquad (3.371a, b)$$

whose solutions (3.372b) are: (i) a sphere with centre at the origin and radius $\sqrt{C_+}$, whose section by any plane through the origin $\alpha x + \beta y + \lambda z = 0$ is a circle; and (ii) a hyperboloid of revolution with axis OZ, cutting the OZ-axis at the point $z = \pm z_0$ with $z_0 = \sqrt{-C_-}$, whose sections by planes $x = $ const or $y = $ const ($z = $ const with $|z| > z_0$) are hyperbolas (circles). Either expression (3.371a, b) satisfies the differential of second-order (3.368), and the general integral:

$$\left(x^2 + y^2 + z^2 - C\right)\left(x^2 + y^2 - z^2 - C\right) = 0, \qquad (3.372)$$

is the product (3.372) involving only one constant C of integration, because (subsection 5.4.4) first-order differentials (3.371a, b) have a general integral involving only one constants of integration.

Conclusion 3

A first-order differential equation relates a function (or dependent variable) to its variable (or independent variable) and to its first-order derivative. The solution or general integral is obtainable by specific methods tailored to eight classes: (i) separable; (ii/iii) linear unforced (forced); (iv/v) non-linear Bernoulli (Riccati); (vi) homogeneous; and (vii/viii) exact (inexact) differentials. Of the eight there are: (a) six always reducible to integrations or quadratures; (b) one that requires finding an integrating factor (viii); and (c) one is not generally solvable (v) though it can be solved in many cases, for example if one particular integral is available. The latter remark is one of the cases in Table 3.1 indicating the number q of quadratures needed to find the general integral of a linear first order (non-linear Riccati) equation if p linearly independent (distinct but not necessarily linearly independent) particular integrals are known, with the sum being $p + q = 2(p + q = 3)$. The exception is $p = 0$ because the Riccati equation is not generally integrable; it is related to the linear second-order differential equation with variable coefficients that is also not generally integrable, though it is solvable in many important specific cases like special functions (chapter 9).

A first-order differential equation specifies one slope at each point of the plane (Figure 3.1) if it is explicit; that is, has only one root for the slope. The curve with this slope is an integral curve, or particular integral of the differential equation. Since there is an integral curve through every regular point, they form a family of curves (Figure 3.2) involving an arbitrary constant; that is, the general integral of the differential equation. Choosing a particular value of the arbitrary constant leads back to the particular integral passing through a given point taken as the boundary condition (Figure 3.4). If the differential equation leads to more than one slope at particular points these are multiple points through which the curve passes several times in different directions, for example twice through the double points (Figure 3.3). If the differential equation is implicit and there is more than one slope at all points, there are several families of integral curves, for example two families (Figure 3.5) for a differential equation quadratic in the slope.

The first-order differential equation is equivalent to a first-order differential in two variables whose coefficients specify vector field, to which the integral curves are orthogonal (Figure 3.6). The first-order differential may be generalized to any number of variables, and is no longer equivalent to a single first-order differential equation. For example, a first-order differential is specified by a vector field of coefficients: (i) that has a family of orthogonal surfaces (Figure 3.7) if it is integrable, that is, has zero helicity; and (ii) if the helicity is not zero, there is no family of orthogonal surfaces, but there is a family of orthogonal curves on any surface (Figure 3.9). The helicity is constant for a circular cylindrical helix with constant slope (Figure 3.8). The extension of differentials to order M in an N-dimensional space relates (Table 3.2) to the number of independent components of the bivector curl and trivector helicity of a vector field.

4

Unsteady, Non-Linear, and Chaotic Systems

The basic linear second-order system with coefficients independent of time is the damped or amplified and forced harmonic attractor or repeller (chapter 2) that can be generalized in three directions. The first generalization is to consider all singularities of a linear system of two first-order differential equations with constant coefficients as regards the trajectory in the physical plane (section 4.2) and the paths in the phase plane (section 4.1); the corresponding singularities that is the points that are not (i) ordinary, are the (ii) contact, (iii) nodal, (iv) focal, (v) saddle, and (vi) asymptotic or spiral points, plus the (vii) center. A second generalization is to allow the coefficients of the linear second-order system to depend on time; for example, in the case of a harmonic oscillator with vibrating support (section 4.3) the energy input into the oscillator through the support causes a parametric resonance (section 4.3) that is distinct from ordinary resonance (sections 2.7–2.8). Parametric resonance occurs only for excitation frequencies close to the first harmonic of the natural frequency, and also more weakly for its sub-multiples, and it leads to exponential rather than linear growth of amplitude with time, in the absence of damping. The effect of damping on parametric resonance is to introduce a threshold amplitude for the vibration of the boundary while the amplitude still grows exponentially with time; in ordinary resonance with damping there is no threshold amplitude and the oscillations have constant amplitude.

The third generalization concerns a non-linear second-order system with constant coefficients, corresponding to non-linear restoring and friction forces (sections 4.4–4.6). An example is a soft or hard spring corresponding to a non-linear restoring force (section 4.4) leading to elliptic integrals; the non-linear restoring force generates harmonics, that is, causes oscillations at multiples of the fundamental frequency (section 4.5). The forcing of a non-linear oscillator can lead to non-linear resonance with two distinct amplitudes, with amplitude jumps and hysteresis loops extracting or feeding energy (section 4.6). The non-linear damping or amplification lead respectively to decay or growth of fields; an example of the latter is the self-excited dynamo (section 4.7) that generates the magnetic field of rotating bodies like the earth and sun, and other planets and stars. The preceding examples include several cases of bifurcations; that is, the existence of multiple solutions of a differential equation beyond a bifurcation value of a parameter (section 4.8); a large number of successive bifurcations can lead to a "chaotic" state, in the sense that it is strongly affected by small changes of initial conditions.

The evolution of a system subject to perturbations determines its stability (section 4.9); that is, it is stable (unstable) if the perturbations always decay (grow at least in some cases).

4.1 Streamlines of a Flow Near a Stagnation Point

Consider a steady flow in the plane with a stagnation point where the velocity is zero; in a neighborhood, the components of velocity may be linear functions of position, leading to a system of two coupled first-order differential equations with constant coefficients. Eliminating time leads to a non-linear differential equation for the trajectory, with a singularity of degree unity at the stagnation point; eliminating one of the variables leads to a second-order linear differential equation, representing a generalized oscillator. The trajectories near the stagnation point may be interpreted as due to the coupling of two identical, orthogonal, generalized oscillators, and in the case (i) of critical damping the (case I) singularity is a contact point (subsection 4.1.4) where the trajectories (Figure 4.1a) have an inflexion tangent to a straight principal line. In the case (ii) of overcritical damping the trajectory may be: (II) a nodal point (Figure 4.1b), through which all paths pass with the same tangent; (III) a focus (Figure 4.1c), crossed by paths in every direction; (IV) a simple point (Figure 4.1d), through which passes only one of a family of parallel lines (subsection 4.1.6); or (V) a saddle point (Figure 4.1e), where converge two principal lines to which all paths asymptote (subsection 4.1.7). In the case (iii) of subcritical damping the paths have neither common tangents nor asymptotes and there are two sub-cases: (VI) the case of zero spiral decay degenerates into a set of ellipses (subsection 4.1.8) with a common center (Figure 4.1f); and (VII) an asymptotic point as the limiting point (subsection 4.1.9) of a spiral (Figure 4.1g). Other shapes of streamlines are possible if the vanishing of the velocity in the neighborhood of the stagnation point is quadratic (example 10.9) or of higher degree.

4.1.1 Distinction between Path and Trajectory

For a **steady motion** in the (x, y) plane the components of velocity (u, v) do not depend on time t and depend only on the position:

$$\frac{dx}{dt} = u(x,y), \qquad\qquad \frac{dy}{dt} = v(x,y). \qquad (4.1a, b)$$

Integration of the system (4.1a, b) of two coupled first-order differential equations specifies the **trajectory**:

$$\{x_0, y_0\} \equiv \{x(t_0), y(t_0)\}: \qquad x(t) = f(t; x_0, y_0), \quad y(t) = g(t; x_0, y_0), \quad (4.2a\text{–}d)$$

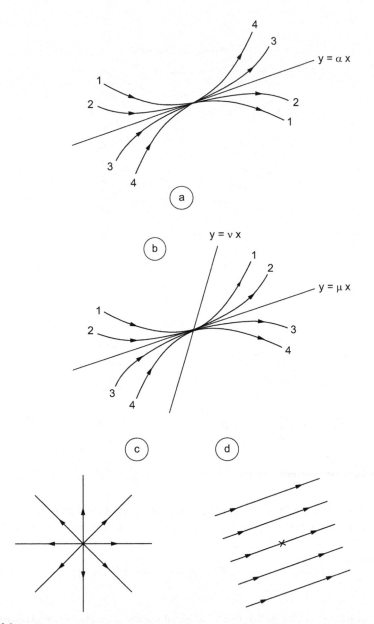

FIGURE 4.1

At a plane stagnation point of first-degree both components of the velocity have a simple zero, leading to seven possible trajectories in its neighborhood: (i) a simple point through which passes only one trajectory (d); (ii/iii) a tangent (b) [inflexion (a)] point through which pass an infinite number of trajectories with the same tangent direction without (with) inflexion; (iv) a center through which pass trajectories in all directions (c); (v) an asymptotic point (g) where all trajectories start (end); (vi) a center through which no trajectories pass with closed paths around it (f); (vii) a saddle point (e) through which no trajectories pass, corresponding to the intersection of two asymptotic lines, separating four regions with open paths.

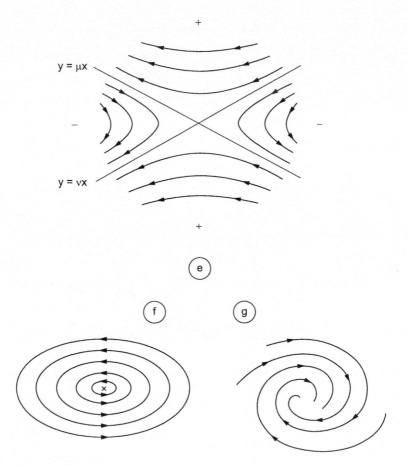

FIGURE 4.1 (CONTINUED)

that is, the position (4.2c, d) at all times *t*, given the initial position (4.2a, b). Eliminating time between (4.1a, b) leads to the equation of the **path** (4.3a):

$$\frac{dy}{dx} = \frac{u(x,y)}{v(x,y)} = \tan\theta: \qquad F(x,y) = \text{const} = F(x_0, y_0), \qquad (4.3\text{a, b})$$

whose solution (4.3b) is the curve described by a particle, or **streamline** of the flow. The trajectory (4.2c, d) indicates the location of the particle along the path (4.3c) at each instant. The path (4.3c) indicates the tangent to the direction of the motion at any point; that is, it makes an angle θ in (4.3a) with the *x*-axis; but it tells nothing about the modulus of the velocity, that is multiplying (4.1a, b) by λ does not change (4.3a); thus the path does not indicate the position of the particle at any time. The path (trajectory) near a stagnation point will be considered first (second) in the section 4.1 (4.2).

4.1.2 Stagnation Point and Degree of the Singularity

The differential equation (4.3a) is **regular** at a point (x_0, y_0) if the derivative is uniquely defined there; the point (x_0, y_0) is a **singularity** of the differential equation (4.3a) if the derivative is indeterminate or infinite, for example $y' = 0/0$. The singularity of the differential equation (4.3a) for a steady flow corresponds to a **stagnation point** where both components of velocity vanish; taking the origin as the stagnation point, if the velocity is a differentiable function of position (4.4a), its components are linear functions (4.4b) in the neighborhood of the stagnation point:

$$u, v \in D^1\left(|R^2\right): \qquad \begin{bmatrix} u(x,y) \\ v(x,y) \end{bmatrix} = \begin{bmatrix} \dfrac{\partial u}{\partial x} & \dfrac{\partial u}{\partial y} \\ \dfrac{\partial v}{\partial x} & \dfrac{\partial v}{\partial y} \end{bmatrix}_{x=0=y} \begin{bmatrix} x \\ y \end{bmatrix} = \begin{bmatrix} a & b \\ c & d \end{bmatrix}\begin{bmatrix} x \\ y \end{bmatrix},$$

$$(4.4a, b)$$

where terms of $O\left(x^2, xy, y^2\right)$ have been neglected, assuming that the origin is a simple zero. In this case (4.4b), the origin is a **singular point of degree one** of the differential equation (4.3a) $\equiv$ (4.5b):

$$\frac{c}{a} \neq \frac{d}{b}: \qquad \frac{dy}{dx} = \frac{cx + dy}{ax + by}, \qquad (4.5a, b)$$

that is non-trivial if the condition (4.5a) is met. In the more general case of the first order differential equation (4.6b):

$$N = \max(m, n, p, q): \qquad \frac{dy}{dx} = \frac{cx^m + dy^n}{ax^p + by^q}, \qquad (4.6a, b)$$

the origin would be a singularity of degree (4.6a) equal to the largest of the numbers n, m, p, q. First-order differential equations with singularities of degree higher, than the first are considered in the EXAMPLE 10.9; the first-order singularity is addressed next (subsections 4.1.3–4.1.9).

4.1.3 Singularity of First Degree and Homogeneous O.D.E.

The first-order differential equation (4.5a, b) is of the homogeneous type (3.127) since the r.h.s. is a function of the ratio of variables (3.131a) $\equiv$ (4.7a) alone. Using the latter variable (4.7a), the first-order singular differential equation of degree one (4.5b) becomes (4.7b):

$$z = \frac{y}{x}: \qquad x\frac{dz}{dx} = \frac{d}{dx}(xz) - z = \frac{c + dz}{a + bz} - z = -\frac{bz^2 + (a-d)z - c}{a + bz}. \qquad (4.7a, b)$$

It is clear that a particular solution of (4.7b) is (4.8a); that is, a straight line (4.8b):

$$z = k \equiv const: \qquad y = kx, \qquad 0 = g(k) \equiv bk^2 + (a - d)k - c, \qquad \text{(4.8a–c)}$$

whose slope k is a root of (4.8c); the binomial (4.8c) is related to the characteristic polynomial of the system of two coupled first-order linear differential equations with constant coefficients:

$$\frac{dx}{dt} = ax + by, \qquad \frac{dy}{dt} = cx + dy, \qquad \text{(4.9a, b)}$$

obtained from (4.1a, b) by linearization (4.4b) about the stagnation point of degree unity at the origin; the characteristic polynomial appears in connection with the trajectory (subsection 4.2.1). Since (4.8c) is a quadratic equation with real coefficients, three cases arise, namely; (i) a real double root (subsection 4.1.4); (ii) two real distinct roots (subsections 4.1.5–4.1.7); and (iii) a pair of complex conjugate roots (subsections 4.1.8–4.1.9).

4.1.4 One Principal Line and Contact Point

The differential equation (4.7b), involving the quadratic expression (4.8c), is separable (4.10a):

$$\frac{dx}{x} = -\frac{a + bz}{g(z)} dz, \qquad \log x = \log C - \int^{y/x} \frac{a + bz}{g(z)} dz, \qquad \text{(4.10a, b)}$$

and has solution (4.10b) where C is an arbitrary constant of integration. The integration in (4.10b) depends on the form of the quadratic function (4.8c); in the case (i) of a double real root the discriminant (4.11a) is zero:

$$\Delta \equiv (a - d)^2 + 4cb = 0: \qquad g(z) = b(z - \alpha)^2, \qquad \alpha = \frac{d - a}{2b}, \qquad y = \alpha x, \qquad \text{(4.11a–d)}$$

and the double root (4.11b) of (4.8c) is (4.11c) and specifies one **principal line** (4.11d) that is a straight line, satisfying the differential equation (4.5b). The complete set of integral curves are solutions of (4.10b; 4.11b):

$$
\begin{aligned}
x &= C \exp\left\{ -\int^{y/x} \frac{z + a/b}{(z - \alpha)^2} dz \right\} \\
&= C \exp\left\{ -\int^{y/x} \left[\frac{1}{z - \alpha} + \frac{\alpha + a/b}{(z - \alpha)^2} \right] dz \right\} \qquad \text{(4.12)} \\
&= \frac{C}{y/x - \alpha} \exp\left(\frac{\alpha + a/b}{y/x - \alpha} \right);
\end{aligned}
$$

thus the general integral of the differential equation (4.5b) in the case (4.11a) is (4.13c):

$$\gamma \equiv \alpha + \frac{a}{b} = \frac{a+d}{2b}: \qquad \log\left(\frac{y-\alpha x}{C}\right) = \frac{\gamma x}{y-\alpha x}, \qquad (4.13\text{a–c})$$

involving the constant (4.13a), where (4.11c) was used leading to (4.13b). The principal line (4.11d) is the particular integral obtained from (4.13c), setting the arbitrary constant of integration equal to zero: $C = 0$. In order to interpret the integral curves (4.13c) for general $C \neq 0$, note that: (i) as $x \to 0$ then $y \to 0$ with $y = \alpha x$; that is, all integral curves are tangent to the principal line at the origin; and (ii) if the sign of x changes, then (4.13c) implies that the sign of y also changes; that is, there is a point of inflexion. Thus *is the case (i), sub-case I of a double root (4.11a) of the binomial (4.8c; 4.11b), the singular point of the differential equation (4.5b) is (problem 70) a* **contact point***; that is, it is an inflexion point of all paths (4.13a–c) that are tangent to the principal line (4.11d) at the origin (Figure 4.1a).*

4.1.5 Two Real and Distinct Principal Lines

In the case (ii) of positive discriminant (4.14a), the binomial (4.8c) has two real distinct roots v, μ (4.14b):

$$\Delta \equiv (a-d)^2 + 4cb > 0: \qquad g(z) = b(z-v)(z-\mu), \qquad (4.14\text{a, b})$$

given by (4.15a) in terms of (4.11c; 4.15b):

$$v, \mu = \alpha \pm \beta: \qquad \beta \equiv \frac{\sqrt{\Delta}}{2b} = \alpha \frac{|\Delta|^{1/2}}{d-a} = \alpha\left|1 + \frac{4bc}{(d-a)^2}\right|^{1/2}; \qquad (4.15\text{a, b})$$

in (4.15b) where (4.11c) and (4.14a) were used. In this case (ii) in (4.14a) there are two principal lines (4.16a):

$$\frac{y}{x} = \alpha \pm \beta = \tan\theta_\pm, \qquad \theta_* \equiv \theta_+ - \theta_- = arc\tan\left\{\frac{\sqrt{\Delta}}{b-c}\right\}, \qquad (4.16\text{a, b})$$

with inclinations $\theta_\pm$, cutting each other at an angle (4.16b) determined by:

$$\tan\theta_* = \tan(\theta_+ - \theta_-) = \frac{\tan\theta_+ - \tan\theta_-}{1 + \tan\theta_+ \tan\theta_-} = \frac{2\beta}{1+\alpha^2 - \beta^2} = \frac{\sqrt{\Delta}}{b-c}; \qquad (4.17)$$

in (4.17) were used:

$$\beta^2 - \alpha^2 = \frac{\alpha^2 \, 4bc}{(d-a)^2} = \frac{c}{b}, \qquad \frac{2\beta}{1+\alpha^2-\beta^2} = \frac{\sqrt{\Delta}/b}{1-c/b} = \frac{\sqrt{\Delta}}{b-c}, \qquad \text{(4.18a, b)}$$

that follow from (4.15b; 4.11c). If the discriminant is zero $\Delta = 0$ in (4.11a), then $\beta = 0$ in (4.15b) and the angle between the principal lines is zero $\theta_* = 0$ in (4.16b), showing that they coincide $\nu = \alpha = \mu$, and there is only one principal line (subsection 4.1.3) in case (i) or sub-case I. In the case (ii) of two real and distinct principal lines there are four subcases (subsections 4.1.6–4.1.7).

4.1.6 Saddle Point or "Mountain Pass"

The complete set of integral curves (4.10b) in the case (ii) of (4.14a) two distinct real roots (4.14b) is given by (4.19):

$$\frac{x}{C} = \exp\left\{ -\int^{y/x} \frac{z + a/b}{(z-\nu)(z-\mu)} dz \right\}$$

$$= \exp\left\{ -\int^{y/x} \left(\frac{A}{z-\nu} + \frac{B}{z-\mu} \right) dz \right\} \qquad \text{(4.19)}$$

$$= \left(\frac{y}{x} - \nu \right)^{-A} \left(\frac{y}{x} - \mu \right)^{-B},$$

where A (B) are the residues of the function in curved brackets at the simple poles $z = \nu(z = \mu)$; they are obtained (I.15.24b) $\equiv$ (1.205) from the function, omitting $z - \nu(z - \mu)$ and setting $z = \nu(z = \mu)$ in the other terms:

$$\{A, B\} = \frac{\{\nu + a/b, -\mu - a/b\}}{\mu - \nu} = \frac{\{\alpha + \beta + a/b, -\alpha + \beta - a/b\}}{2\beta}$$

$$= \frac{(\beta + \gamma, \beta - \gamma)}{2\beta} = \frac{1}{2} \pm \frac{\gamma}{2\beta}, \qquad \text{(4.20)}$$

where (4.15a, 4.13a) were used. From (4.20) follows:

$$A + B = 1: \qquad\qquad \vartheta \equiv -\frac{B}{A} = \frac{\gamma - \beta}{\gamma + \beta} = \frac{a + d - \sqrt{\Delta}}{a + d + \sqrt{\Delta}}, \qquad \text{(4.21a, b)}$$

where (4.13a; 4.15b) were used in (4.21b); the latter is simplified further using (4.14a):

$$\left(a+d+\sqrt{\Delta}\right)^{2}\vartheta = \left(a+d\right)^{2}-\Delta = \left(a+d\right)^{2}-\left(a-d\right)^{2}-4bc = 4\left(ad-bc\right). \quad (4.22)$$

Using (4.21a, b), the general integral (4.19) is given by (4.23a, b) where (4.23c) is another arbitrary constant:

$$y - vx = D\left(y - \mu x\right)^{\vartheta}: \qquad \vartheta = 4\frac{ad-bc}{\left(a+d+\sqrt{\Delta}\right)^{2}}, \qquad D \equiv C^{1/A}. \quad (4.23a\text{--}c)$$

The principal lines $y = vx(y = \mu x)$ are the particular integrals obtained by setting the arbitrary constant of integration $D = 0(D = \infty)$. If $\vartheta < 0$ in (4.23b); that is, $ad < bc$ in (4.5b), then $x = 0 = y$ is not a solution of (4.23a) for $D \neq 0, \infty$, and thus no integral curves pass through the origin, except for the principal lines. As $y \to vx$ it follows from (4.23a) that $y - \mu x \to \infty$, and vice-versa $y \to \mu x$ implies $y - vx \to \infty$; all integral curves have as asymptotes the two principal lines at infinity. Thus in *the case (ii) of two real distinct roots (4.15a, b; 4.14a) of the binomial (4.8c)* ≡ *(4.14b) the singular point of the differential equation (4.5b) with bc > ad (sub-case V) is (problem 71) a **saddle point**, through which pass the two principal lines $y = \lambda x, \mu x$ that are the asymptotes of all integral curves (4.23a) at infinity, for $\vartheta < 0$ in (4.23b); the saddle point may be visualized (Figure 4.1e) as a col or mountain pass through which pass two level curves $y = vx, \mu x$, separating the mountains (valleys), identified as + regions (− regions) above (below) the col.*

4.1.7 Nodal and Simple Points and Focus

The case (ii) of the two real distinct roots (4.14a) includes sub-cases other than the saddle point for $\vartheta < 0$ in (4.23b) or $ad < bc$ in (4.5b). If $\vartheta > 0$, then $x \to 0$ implies $y \to 0$ in (4.23a), and all integral curves pass through the origin. Their slope:

$$\lim_{x \to 0}\left(\frac{dy}{dx} - v\right) = D\vartheta \lim_{x \to 0}\left(y - \mu x\right)^{\vartheta-1}\left(\frac{dy}{dx} - \mu\right), \quad (4.24)$$

is zero for $\vartheta > 1$, showing that they are tangent to the principal line $y = vx$ at the origin $x = 0 = y$; if $\vartheta < 1$ they would be tangent to the other principal line $y = \mu x$. In both cases of the differential equation (4.5b), with (i) positive discriminant (4.14a) for $\vartheta > 1(\vartheta < 1)$ in (4.23b) has a singular point that is (sub-case II) a **nodal point** (Figure 4.1b), where (problem 72) all integral curves are tangent to a principal line $y = vx(y = \mu x)$ without inflexion, since changing the sign of x need not change to sign of y in (4.23a) for $\vartheta > 0$.

The two instances of case (ii) still not considered are $\vartheta = 0$ and $\vartheta = 1$. In the case $\vartheta = 1$, it is still true that all integral curves (4.23a) pass through the origin, but in this case (4.25a) they are straight lines (3.25b):

$$\vartheta = 1: \qquad\qquad y = mx, \qquad\qquad m \equiv \frac{v - \mu D}{1 - D}, \qquad\qquad (4.25\text{a–c})$$

with arbitrary slope (4.25c); thus the singular point is (sub-case III) a **focal point** (Figure 4.1c), through (problem 73) which integral curves pass isotropically in all directions. The last remaining instance of case (ii) is $\vartheta = 0$, which separates the instances of no $(\vartheta < 0; sub-caseV)$ [all $(\vartheta > 0$; sub-cases II, and IV)] integral curves passing through the origin; the instance (sub-case IV) of $\vartheta = 0$ corresponds (4.23b) to $ad = bc$, so that (4.5b) reduces to a constant and the integral curves are parallel straight lines. It can be checked from (4.23a) that in the case (4.26a) [(4.26b)] the integral curves are straight lines (4.26c) [(4.26d)]:

$$\vartheta = 0, \infty: \qquad\qquad y = vx + D, \mu x, \qquad\qquad (4.26\text{a–d})$$

parallel to the principal line $y = vx (y = \mu x)$, and only one passes through the origin (Figure 4.1d), that (problem 74) is (case IV) a **simple point**.

It has been shown that *in the case (ii) of two real and distinct roots (4.15a, b; 4.11c) of the binomial (4.8c)* $\equiv$ *(4.14b) with (4.14a), the integral curves (4.23a), besides (problem 71) the instance $\vartheta < 0$ in (4.23b) of a (case V) saddle point (subsection 4.1.6), lead to three more instances: (sub-case II) if $1 < \vartheta < \infty (0 < \vartheta < 1)$ the integral curves (4.23a) are tangent to the principal line $y = vx (y = \mu x)$ at the singular point that is (problem 72) a nodal point (Figure 4.1b); (sub-case III) the separating instance $\vartheta = 1$, corresponds (problem 73) to a **focus** (4.25a–c) through which integral curves pass in all directions (Figure 4.1c); (sub-case IV) the extreme instances $\vartheta = 0 (\vartheta = \infty)$ correspond (4.26a–d) to straight integral lines parallel to the principal line $y = vx (y = \mu x)$ that is the only one passing through the origin (Figure 4.1d) that is (problem 74) a simple point.*

4.1.8 Imaginary Principal Lines and Center

In the last case (iii) of negative (4.27a) discriminant (4.11a):

$$\Delta \equiv (a - d)^2 + 4bc < 0: \qquad\qquad v, \mu = \alpha \pm i\beta, \qquad\qquad (4.27\text{a, b})$$

the roots of the binomial (4.8c) $\equiv$ (4.14b) are a complex conjugate pair (4.27b) with (α, β) given by (4.11c; 4.15b); the corresponding characteristic polynomial is (4.8c) $\equiv$ (4.14b) $\equiv$ (4.27c):

$$g(z) = C(z - \alpha - i\beta)(z - \alpha + i\beta) = C\left[(z - \alpha)^2 + \beta^2\right]. \qquad\qquad (4.27\text{c})$$

Thus the principal lines are imaginary; that is, no real straight lines through the origin coincide with integral curves.

The integral curves are given (4.10b, 4.27c) by (4.28):

$$\frac{x}{C} = \exp\left\{ -\int_{y/x}^{y/x} \frac{z+a/b}{(z-\alpha)^2+\beta^2} \right\} = \exp\left\{ -\int_{}^{y/x} \frac{z-\alpha+\gamma}{(z-\alpha)^2+\beta^2}\,dz \right\}$$

$$\tag{4.28}$$

$$= \left| 1 + \left(\frac{y/x-\alpha}{\beta} \right)^2 \right|^{-1/2} \exp\left[-\frac{\gamma}{\beta} \text{ arc tan}\left(\frac{y/x-\alpha}{\beta} \right) \right],$$

where (4.13a) was used. In the integration (4.28) ≡ (4.29b) the change of variable (4.29a) was used:

$$\zeta \equiv \frac{z-\alpha}{\beta}: \qquad \frac{x}{C} = \exp\left\{ -\int^{\frac{y/x-\alpha}{\beta}} \frac{\zeta+\gamma/\beta}{1+\zeta^2}\,d\zeta \right\}$$

$$= \exp\left[-\frac{1}{2}\log\left(1+\zeta^2\right) - \frac{\gamma}{\beta} \text{ arc tan } \zeta \right]_{\zeta=\frac{z-\alpha}{\beta}} \tag{4.29a, b}$$

$$= \left[\left|1+\zeta^2\right|^{-1/2} \exp\left(-\frac{\gamma}{\beta} \text{ arc tan } \zeta \right) \right]_{\zeta=\frac{z-\alpha}{\beta}},$$

and the derivative of the circular arc tangent (II.7.112b).

Only the first term on the r.h.s. of (4.28) remains for $\gamma = 0$ or (4.30a) by (4.13a) implying (4.30b) from (4.27a) leading to (4.30c):

$$a+d = 0: \qquad a^2 + bc < 0, \qquad C^2 = (y-\alpha x)^2 + \beta^2 x^2, \tag{4.30a–c}$$

that is the simplest *instance (sub-case VI) of the case (iii) of a complex conjugate pair of roots (4.27b; 4.11c; 4.15b) of the binomial (4.8c) ≡ (4.27c) with (4.27a), in which (problem 75) the integral curves of the differential equation (4.5b) with (4.30a) and (4.30b) are a family of ellipses (4.30c), with the singular point as their common center (Figure 4.1f).*

4.1.9 Asymptotic Point and Spiral Path

In the general case of (4.28), that is (4.31a) included in (4.30a–c), the complex variable (4.31b) is introduced:

$$\gamma \neq 0: \qquad re^{i\theta} = \beta x + i(y-\alpha x); \tag{4.31a, b}$$

its modulus r and argument θ:

$$r = \left| \left(y - \alpha x \right)^2 + \beta^2 x^2 \right|^{\frac{1}{2}}, \qquad \tan \theta = \frac{y - \alpha x}{\beta x}, \qquad (4.32a, b)$$

appear in (4.28), which becomes (4.33b) in the case (4.33a):

$$a + d \neq 0: \qquad r = C \exp(-\eta \theta), \qquad \eta \equiv \frac{\gamma}{\beta} = \frac{a + d}{\sqrt{\Delta}}, \qquad (4.33a\text{–}c)$$

with (4.33c) evaluated by (4.13a; 4.15b). *The integral curves (4.33c) are* **spirals** *(Figure 4.1g), with (problem 76) the singular point as an* **asymptotic point**, *in the sub-case VII in (4.33a) of the differential equation (4.5b), for the case (iii) when the binomial (4.8c)* $\equiv$ *(4.27c) with (4.27a), has a conjugate pair of roots (4.27b; 4.11c; 4.15b), specifying the distance r in (4.32a) and angle θ in (4.32b). The streamlines of a flow near a stagnation point of degree one are summarized in Table 4.1, where the discriminant (4.27a)* $\equiv$ *(4.34):*

$$\Delta \equiv \left(a - d \right)^2 + 4bc = \left(a + d \right)^2 + 4\left(bc - ad \right), \qquad (4.34)$$

is used in the distinction between the sub-cases I to VII. The paths near some cases of stagnation points of degree (4.6a, b) higher than the first (4.5a, b) are considered (problems 79–80) in example 10.3.

4.2 Trajectories of a Particle Near a Singularity

The same problem of the flow near a stagnation point of degree one may be solved [section 4.1 (4.2)] in terms of paths (trajectories). The trajectory (subsection 4.2.1) is specified by a system of two coupled linear first-order differential equations with constant coefficients that has the same characteristic polynomial as the equivalent second-order system (subsection 4.2.2). Two linearly independent particular integrals supply a pair of decoupled coordinates (subsection 4.2.3) that are used in the same three cases (seven sub-cases) of the characteristic polynomial: (i) the double root (subsection 4.2.4) corresponds to one principal line (subsection 4.1.4) and the critical damping leads to the sub-case I of an inflexion point (Figure 4.1a); (ii) the distinct real roots (subsection 4.2.5) correspond to two real and distinct principal lines (subsections 4.1.5–4.1.7) and the supercritical damping leads to the sub-cases II/III/IV/V of respectively nodal/focal/simple/saddle points (Figures 4.1b/c/d/e); and (iii) the complex conjugate roots (subsection 4.2.6) correspond to

TABLE 4.1

Streamlines Near a Stagnation Point of Degree One (Singularities of a Linear First-Order Autonomous Dynamical System)

Characteristic Polynomial Case	Singularity Sub-case	Figure 4.1	Path 4.1.1–4.1.3	Trajectory 4.2.1–4.2.3	Parameter $\dfrac{dy}{dx}=\dfrac{ax+by}{cx+dy}$	Eigenvalues $0 = D^2 - (a+d)D + ad - bc$ $= (D-\xi_+)(D-\xi_-)$
(i) *double root* 4.1.4	I: contact point	Figure 4.1a	4.1.4: (4.13a–c;4.11a,c)	4.2.4:(4.48a,b;4.49a,b;4.54b)	$\Delta \equiv (a-d)^2 + 4bc = 0$	$\xi_+ = \xi_- = -\lambda$
(ii) *distinct real roots* 4.1.5	II: nodal point	Figure 4.1b	4.1.7: (4.24)	4.2.5: (4.59;4.56a,b;4.55b; 4.47) $rs>0,\, r \neq s$	$(a+d)^2 > \Delta > 0$	$\xi_+ = r,\, \xi_- = s$
	III: focal point	Figure 4.1c	4.1.7: (4.25a–c)	4.2.5: (4.59;4.56a,b;4.55b; 4.47) $r = s \neq 0$	$a+d \neq 0 = \Delta$	$\xi_+ = \xi_- = r = s$
	IV: simple point	Figure 4.1d	4.1.7: (4.26a–d)	4.2.5: $r = 0 = s$	$bc - ad = 0 < \Delta$	$\xi_\pm = 0 = \xi_-$
	V: saddle point	Figure 4.1e	4.1.6:(4.23a,b;4.15a,b; 4.11c;4.14a)	4.2.5: (4.59;4.56a,b;4.55b; 4.47) $rs<0,\, r \neq s$	$\Delta > (a+d)^2 > 0$	$\xi_+ = r,\, \xi_- = s$
(ii) *complex conjugate pair* 4.1.8	VI: center	Figure 4.1f	4.1.8: (4.30a–c; 4.15b;4.11c)	4.2.6: (4.72;4.71a,b;4.68a,b; 4.47) $\lambda = 0$	$a+d = 0 > \Delta$	$\xi_\pm = \pm i\delta$
	VII: asymptotic point	Figure 4.1g	4.1.9: (4.33a–c; 4.32a,b;4.15b;4.11c)	4.2.6: (4.72;4.71a,b;4.68a,b; 4.47) $\lambda \neq 0$	$a+d \neq 0 > \Delta$	$\xi_\pm = -\lambda \pm i\delta$

Note: The trajectories near a stagnation point of first-degree involve three cases and seven subcases illustrated in the Figures 4.1a–g.

imaginary principal lines (subsections 4.1.8–4.1.9) and the subcritical damping (no damping) leads to the sub-case VII (VI) of [Figure 4.1f(g)], an asymptotic point (center). The three cases (i, ii, iii) have respectively $1+4+2=7$ sub-cases that correspond to the superposition of damped/undamped/ amplified monotonic/oscillatory motions in orthogonal/oblique directions. The sub-case VI can be extended to two orthogonal oscillators with the same frequency, distinct amplitudes, and an arbitrary phase shift (subsection 4.2.7), leading to a trajectory that is an ellipse oblique to the coordinate axis (subsection 4.2.8).

4.2.1 Coupled Linear First-Order System

The system of two coupled first-order ordinary differential equations with constant coefficients $(4.4b) \equiv (4.35b, c)$:

$$\dot{z} \equiv \frac{dz}{dt}: \qquad\qquad \dot{x} = ax + by, \qquad \dot{y} = cx + dy, \qquad\qquad (4.35a\text{–}c)$$

where dot denotes derivative with regard to time (4.35a) may be considered by two methods: (i) elimination of time leading to the non-linear homogeneous first-order differential equation (4.5b) specifying the paths (section 4.1); and (ii) direct solution for the trajectory (section 4.2) implying that it is equivalent to a single second-order linear differential equation with constant coefficients obtained by elimination between (4.35b, c):

$$\ddot{x} = a\dot{x} + b\dot{y} = a\dot{x} + b(cx + dy) = a\dot{x} + bcx + d(\dot{x} - ax). \qquad (4.36)$$

Thus the system of two first-order equations (4.35b, c) is of order two, since it is equivalent to a single differential equation of order two:

$$\left\{ P_2\left(\frac{d}{dt}\right)\right\} x, y(t) = \left\{ \frac{d^2}{dt^2} - (a+d)\frac{d}{dt} + (ad - bc)\right\} x, y(t) = 0, \quad (4.37a, b)$$

which is the same for the two variables; it is sufficient to solve (4.37a) for x [or (4.37b) for y], since the other variable $y(x)$ is determined by substitution in (4.35b) [(4.35c)]. An alternative way to obtain the result (4.37a, b) is to seek solutions for (x, y) as identical exponentials of time (4.38a) with distinct amplitudes:

$$\{x(t), y(t)\} = e^{\vartheta t}\{A, B\}: \qquad \begin{bmatrix} \vartheta - a & -b \\ -c & \vartheta - d \end{bmatrix}\begin{bmatrix} A \\ B \end{bmatrix} = 0; \qquad (4.38a, b)$$

substituting (4.38a) in (4.35b, c) leads to (4.38b). A non-trivial solution (4.39a) requires that the determinant of the matrix of coefficients be zero (4.39b):

$$\{A,B\} \neq \{0,0\}: \qquad 0 = \begin{vmatrix} \vartheta - a & -b \\ -c & \vartheta - d \end{vmatrix}$$

$$= (\vartheta - a)(\vartheta - d) - bc$$

$$= \vartheta^2 - (a+d)\vartheta + ad - bc = P_2(\vartheta),$$

(4.39a–c)

leading to the characteristic polynomial (4.39c) $\equiv$ (4.37a, b) = (4.40a, b):

$$D \equiv \frac{d}{dt}: \quad P_2(D) = D^2 - (a+d)D + (ab - bc) = D^2 + 2\lambda D + \omega_0^2 = (D - \xi_+)(D - \xi_-).$$

(4.40a, b)

Thus *the characteristic polynomial (4.40a, b) of the first-order system of linear differential equations with constant coefficients (4.35a–c) can be obtained: (i) either directly (4.39c) from the vanishing of the determinant (4.39b) matrix of coefficients (4.38b) for the exponential solutions (4.38a); or (ii) indirectly, eliminating (4.36) for a linear second-order differential equation with constant coefficients satisfied by each of the variables (4.37a, b).* The characteristic polynomial of the second degree (4.40a, b) leads to an analogy with the equivalent second-order linear system with constant coefficients (chapter 2) corresponding to a damped harmonic oscillator (subsection 4.2.2).

4.2.2 Equivalent Damped Harmonic Oscillator

The characteristic polynomial corresponds to that (2.81b) $\equiv$ (4.40b) for a "damped oscillator" with:

$$\lambda = -\frac{a+d}{2} = -b\gamma, \qquad \omega_0^2 = ad - bc \neq 0, \qquad (4.41a, b)$$

where: (i) (4.13a) was used in (4.41a); (ii) the condition $\omega_0 \neq 0$ in (4.41b) implies (4.5a) that the differential equation (4.5b) is non-trivial $dy/dx \neq const$. *The characteristic polynomial (4.40a, b) corresponds to **a generalized oscillator**, since (problem 79) the two pairs of terms are interpreted as follows: (i) the (4.42a) first two terms (4.42c) specify (4.41b) a velocity (4.42b) that is amplified (decays) with time for $\lambda < 0 (\lambda > 0)$ in (4.42d) and is constant for $\lambda = 0$:*

$$ad = bc; \qquad v \equiv \dot{x}: \qquad 0 = \ddot{x} + 2\lambda \dot{x} = \dot{v} + 2\lambda v, \qquad v(t) = v(0)e^{-2\lambda t};$$

(4.42a–d)

(ii) the first and third (4.43a; 4.44a; 4.45a) terms lead (4.41b) to oscillatory (4.43c, d)
[monotonic (4.44c, d)] motion if the condition (4.43b) [(4.44b)] is met:

$$a + d = 0, \quad ad > bc: \qquad \ddot{x} + |\omega_0|^2 x = 0, \qquad x(t) = x(0)\exp(\pm i|\omega_0|t), \quad (4.43a\text{–}d)$$

$$a + d = 0, \quad ad < bc: \qquad \ddot{x} - |\omega_0|^2 x = 0, \qquad x(t) = x(0)\exp(\pm|\omega_0|) \qquad (4.44a\text{–}d)$$

$$a + d = 0, \quad ad = bc: \qquad\qquad \ddot{x} = 0, \qquad x(t) = x(0) + t\dot{x}(0), \qquad (4.45a\text{–}d)$$

with the intermediate case (4.45b) corresponding to uniform motion (4.45c, d). Thus
the trajectories near a stagnation point of first degree (4.35a–c), correspond to
the composition of two identical generalized oscillators, which are orthogonal (x, y) and coupled and can be decoupled (subsection 4.2.3).

4.2.3 Decomposition into a Pair of Decoupled Coordinates

The linear second-order differential equation (4.37a, b) for the generalized
oscillator has two particular integrals $\zeta(t), \eta(t)$, that may be called **decoupled coordinates** because they satisfy differential equations that are decoupled, unlike (4.35a–c). The orthogonal coupled coordinates of the generalized
oscillators are linear combinations of the decoupled ones:

$$\begin{bmatrix} x(t) \\ y(t) \end{bmatrix} = \begin{bmatrix} C_1 & C_2 \\ D_1 & D_2 \end{bmatrix} \begin{bmatrix} \zeta(t) \\ \eta(t) \end{bmatrix}, \qquad (4.46)$$

where (C_1, C_2) are arbitrary constants of integration and (D_1, D_2) are functions of C_1, C_2 determined by compatibility with (4.35b, c); thus the decoupled
coordinates ζ, η specify rectilinear axis that are generally oblique. *The integral curves of a singular first-order differential equation of first-degree (4.5b) correspond to two identical, orthogonal coupled (4.35a–c) generalized oscillators (4.37a, b)
that may be decoupled into motions along oblique (4.46) axis (ζ, η) that are always
distinct since they are linearly independent particular integrals of the generalized
free oscillator.* The discriminant of the characteristic polynomial (4.40b):

$$\Delta = 4(\lambda^2 - \omega_0^2) = (a + d)^2 - 4(ad - bc) = (a - d)^2 + 4bc, \qquad (4.47)$$

coincides with that (4.47) ≡ (4.34) ≡ (4.11a) of the quadratic form or binomial
(4.8c) specifying the principal lines; that is, the decoupled coordinates lie
along the principal lines. Thus the same three cases arise: (i) double real root
of (4.8c) corresponding to a contact point (subsection 4.1.4) and generalized
oscillators with critical damping (subsection 4.2.4); (ii) two real distinct roots
of (4.8c) corresponding (subsection 4.1.5) to a saddle point (subsection 4.1.6)

or a focus, or a nodal or simple point (subsection 4.1.7) and generalized oscillators with supercritical damping (subsection 4.2.5); and (iii) a complex conjugate pair of roots of (4.8c), corresponding [subsection 4.1.8 (4.1.9)] to a center (asymptotic point) for undamped (subcritically damped) generalized oscillators (subsection 4.2.6).

4.2.4 Critical Damping and Double Root

In the case (i) of (problem 80) a real double root (4.48a) of the characteristic polynomial (4.48b):

$$0 = \Delta = 4(\lambda^2 - \omega_0^2): \qquad P_2(D) = D^2 + 2\lambda D + \lambda^2 = (D+\lambda)^2, \qquad (4.48a, b)$$

the linearly independent particular integrals (2.85a, b) $\equiv$ (4.49a, b):

$$\zeta(t) = e^{-\lambda t}, \qquad\qquad \eta(t) = te^{-\lambda t}, \qquad\qquad (4.49a, b)$$

satisfy the differential equations of first and second order respectively:

$$0 = (D+\lambda)\zeta = \dot{\zeta} + \lambda\zeta, \qquad 0 = (D+\lambda)^2 \eta = \ddot{\eta} + 2\lambda\dot{\eta} + \lambda^2\eta; \qquad (4.50a, b)$$

although the orders of the decoupled equations (4.50a, b) add $1 + 2 = 3$ to three, they are dependent, and the actual order of the system (4.35a–c) remains (4.36) two.

Substituting (4.49a, b) in the first line of (4.46) leads to (4.51a):

$$x(t) = (C_1 + C_2 t)e^{-\lambda t}; \qquad\qquad (4.51a)$$

$$by(t) = \dot{x} - ax = \left\{ C_2 - (\lambda + a)C_1 \right\}e^{-\lambda t} - C_2(\lambda + a)te^{-\lambda t}; \qquad (4.51b)$$

substitution of (4.51a) in (4.35b) leads to (4.51b). Comparison of (4.51b) with the second line of (4.46) specifies the constants:

$$\begin{bmatrix} D_1 \\ D_2 \end{bmatrix} = \frac{1}{b}\begin{bmatrix} -\lambda - a & 1 \\ 0 & -\lambda - a \end{bmatrix}\begin{bmatrix} C_1 \\ C_2 \end{bmatrix}. \qquad (4.52)$$

Substitution of the decoupled coordinates (4.49a, b) in (4.51a, b) specifies the directions of the oblique decoupled axis $\zeta, \eta,$ relative to the orthogonal coupled axis x, y:

$$\begin{bmatrix} x(t) \\ y(t) \end{bmatrix} = \frac{1}{b}\begin{bmatrix} C_1 b & C_2 b \\ C_2 - C_1(\lambda + a) & -C_2(\lambda + a) \end{bmatrix}\begin{bmatrix} \zeta(t) \\ \eta(t) \end{bmatrix}. \qquad (4.53)$$

Eliminating the parameter time t between the decoupled coordinates (4.49a, b) leads to:

$$-\lambda t = \log \zeta : \qquad -\lambda \eta(t) = \zeta(t) \log\{\zeta(t)\}, \qquad (4.54a, b)$$

*the integral curves (4.54b), correspond (problem 80) to (4.13c) that are tangent to the principal line ζ at a (sub-case I) the **contact point** of inflexion (Figure 4.1a); this applies to the decoupled oblique coordinates (4.49a, b) and to coupled orthogonal generalized oscillators (4.53), with critical damping in (4.48a) or $\omega_0 = \lambda$ in (4.41a, b).*

4.2.5 Supercritical Damping and Distinct Real Roots

In the case (ii) of real distinct roots of the characteristic polynomial (4.40b):

$$\Delta > 0: \qquad P_2(D) = (D-r)(D-s), \qquad r, s = \frac{a+d \pm \sqrt{\Delta}}{2}, \qquad (4.55a, b)$$

the particular integrals (4.56a, b):

$$\zeta(t) = e^{rt}, \eta(t) = e^{st} : \qquad \dot{\zeta} = r\zeta, \qquad \dot{\eta} = s\eta, \qquad (4.56a-d)$$

satisfy decoupled, real, first-order differential equations (4.56c, d). Substituting (4.56a, b) in the first line of (4.46) leads to (4.57a):

$$x(t) = C_1 e^{rt} + C_2 e^{st}; \quad by(t) = \dot{x} - ax = C_1(r-a)e^{rt} + C_2(s-a)e^{st}; \quad (4.57a, b)$$

substitution of (4.57a) in (4.35b) leads to (4.57b). Comparison of (4.57b) with the second-line of (4.46) specifies the constants (D_1, D_2) in (4.58a):

$$\begin{bmatrix} D_1 \\ D_2 \end{bmatrix} = \frac{1}{b} \begin{bmatrix} r-a & 0 \\ 0 & s-a \end{bmatrix} \begin{bmatrix} C_1 \\ C_2 \end{bmatrix}; \quad \begin{bmatrix} x(t) \\ y(t) \end{bmatrix} = \frac{1}{b} \begin{bmatrix} C_1 b & C_2 b \\ C_1(r-a) & C_2(s-a) \end{bmatrix} \begin{bmatrix} \zeta(t) \\ \eta(t) \end{bmatrix};$$

$$(4.58a, b)$$

also, (4.57a, b) $\equiv$ (4.58b) indicate the direction of the oblique, decoupled axis (ζ, η) relative to the orthogonal, coupled axis (x, y).

Eliminating the time t between (4.56a, b):

$$\zeta(t) = \left(e^t\right)^r = \left(\eta^{1/s}\right)^r = \{\eta(t)\}^{r/s}, \qquad (4.59)$$

specifies *the integral curves (4.59) in oblique decoupled axis ζ, η, corresponding to (4.23a, b) in orthogonal coupled axis x, y, leading to the following four instances of*

coupled generalized oscillators with super-critical damping: (sub-case II) if (4.60a)
[(4.61a)] the integral curves are tangent to the η-axis (ζ-axis) at the origin (4.60a)
*[(4.61b)]; that is, (problem 81) a **nodal point** (Figure 4.1b):*

$$r > s: \qquad \lim_{\eta \to 0} \frac{d\zeta}{d\eta} = \lim_{\eta \to 0} \frac{r}{s} \left[\eta(t) \right]^{r/s-1} = 0, \qquad (4.60a, b)$$

$$s > r: \qquad \lim_{\eta \to 0} \frac{d\eta}{d\zeta} = \lim_{\eta \to 0} \frac{s}{r} \left[\zeta(t) \right]^{s/r-1} = 0; \qquad (4.61a, b)$$

(sub-case III) if (4.62a) the (problem 82) integral curves are straight lines (4.62b)
*passing in every direction through the origin; that is, a **focal point** (Figure 4.1c):*

$$r = 0 \neq s: \quad \zeta(t) = 1 \neq \eta(t); \quad s = 0 \neq r: \quad \eta(t) = 0 \neq \zeta(t); \quad (4.62a, b; 4.63a, b)$$

(sub-case IV) if (4.63a) the integral curves are (4.63b) straight lines parallel to the
*η-axis, and only one passes through the origin; that is (problem 83), a **simple point***
(Figure 4.1d); (sub-case V) in all preceding instances (II, III, IV) it was assumed
that r, s had the same sign rs > 0, whereas (problem 84) if they have opposite signs
(4.64a):

$$rs < 0: \qquad \lim_{\eta \to 0} \zeta = \infty, \qquad \lim_{\zeta \to 0} \eta = \infty, \qquad \{\zeta, \eta\} \neq \{0, 0\}, \qquad (4.64a\text{--}d)$$

all integral curves have (4.64b, c) as asymptotes the ζ, η-axis, and none passes (4.65d)
*through the origin; that is, a **saddle point** (Figure 4.1e).*

4.2.6 Subcritical Damping and Complex Roots

In the case (iii) of complex conjugate roots of the characteristic polynomial:

$$\Delta < 0: \qquad P_2(D) = (D + \lambda)^2 + \delta^2, \qquad \delta = \frac{\sqrt{-\Delta}}{2}, \qquad (4.65a\text{--}c)$$

the particular integrals (4.66a, b):

$$\zeta, \eta(t) = \exp(-\lambda t \pm i\delta t): \qquad \dot{\zeta} = (i\delta - \lambda)\zeta, \qquad \dot{\eta} = -(i\delta + \lambda)\eta, \qquad (4.66a\text{--}d)$$

satisfy decoupled equations of the first-order, with complex conjugate coefficients (4.66c, d). Real decoupled coordinates may be used instead of (4.66a, b):

$$\phi(t) = e^{-\lambda t} \cos(t\delta), \qquad \psi(t) = e^{-\lambda t} \sin(t\delta). \qquad (4.67a, b)$$

Substituting (4.67a, b) in the first line of (4.46) leads to (4.68a):

$$x(t) = e^{-\lambda t}\{C_1 \cos(t\delta) + C_2 \sin(t\delta)\};$$

(4.68a)

$$by(t) = \dot{x} - ax = e^{-\lambda t}\{[-(\lambda + a)C_1 + C_2\delta]\cos(t\delta) - [C_1\delta + (\lambda + a)C_2]\sin(t\delta)\};$$

(4.68b)

substitution of (4.68a) in (4.35b) leads to (4.68b), implying that the directions of the decoupled oblique axis (ϕ, ψ) relative to the orthogonal coupled axis (x, y) are specified by:

$$\begin{bmatrix} x(t) \\ y(t) \end{bmatrix} = \frac{1}{b} \begin{bmatrix} C_1 b & C_2 b \\ C_2\delta - (\lambda + a)C_1 & -C_1\delta - C_2(\lambda + a) \end{bmatrix} \begin{bmatrix} \phi(t) \\ \psi(t) \end{bmatrix}.$$

(4.69)

The comparison of (4.46) ≡ (4.69) leads to:

$$\begin{bmatrix} x(t) \\ y(t) \end{bmatrix} = \begin{bmatrix} C_1 & C_2 \\ D_1 & D_2 \end{bmatrix} \begin{bmatrix} \phi(t) \\ \psi(t) \end{bmatrix}; \qquad \begin{bmatrix} D_1 \\ D_2 \end{bmatrix} = \frac{1}{b} \begin{bmatrix} -\lambda - a & \delta \\ \delta & -\lambda - a \end{bmatrix} \begin{bmatrix} C_1 \\ C_2 \end{bmatrix}.$$

(4.70a, b)

Partial elimination of t between (4.67a, b) leads to:

$$\{\phi(t)\}^2 + \{\psi(t)\}^2 = \exp(-2\lambda t).$$

(4.71)

*The integral curves (4.67a, b) in decoupled oblique axis (ϕ, ψ), correspond to (4.33a–c) in coupled orthogonal axis (x, y) for generalized oscillators with sub-critical damping, representing: (sub-case VII) in (problem 85) the case $\lambda > 0 (\lambda < 0)$ of damping (amplification), a **spiral** fall $\phi, \psi \to 0$ (**exit** $\phi, \psi \to \infty$) as $t \to \infty$ towards the (away from) the origin which is as $t \to \infty (t \to -\infty)$ an **asymptotic point** (Figure 4.1g); (sub-case VI) in the case $\lambda = 0$ when (problem 86) no damping is present the coupled identical undamped oscillators have for integral curves a circle (ellipse) in oblique ϕ, ψ (orthogonal x, y) axis, with the origin as common **center** (Figure 4.1f).*

4.2.7 Superposition of Two Orthogonal Undamped Oscillators

As an extension of the sub-case VI, consider two orthogonal undamped oscillators:

$$x(t) = a\cos(\omega t), \quad y(t) = b\cos(\omega t - \varphi) = b[\cos(\omega t)\cos\varphi + \sin(\omega t)\sin\varphi],$$

(4.72a–c)

with distinct amplitudes and the same frequency ω; initial time $t = 0$ is chosen so as to have zero phase in the first oscillator (4.72a) so that φ in the second oscillator (4.72b) is the phase shift between the two. From the second oscillator (4.72c) in the form (4.73):

$$\{y(t) - b\cos(\omega t)\cos\varphi\}^2 = b^2\sin^2(\omega t)\sin^2\varphi = b^2\sin^2\varphi\left[1 - \cos^2(\omega t)\right], \quad (4.73)$$

time can be eliminated using the first (4.72a):

$$\left[y(t) - \frac{b}{a}x(t)\cos\varphi\right]^2 = b^2\left\{1 - \frac{x(t)^2}{a^2}\right\}\sin^2\varphi; \quad (4.74)$$

this leads to the equation of the path:

$$a^2 y^2 + b^2 x^2 - 2abxy\cos\varphi = a^2 b^2\sin^2\varphi. \quad (4.75)$$

If (problem 87) there is no phase shift (4.76a):

$$\varphi = 0: \qquad 0 = (ay - bx)^2, \qquad y = \frac{b}{a}x = x\tan\theta; \quad (4.76a\text{--}c)$$

that is, the oscillators pass through the origin at the same time (Figure 4.2a) and the path (4.76b) is a straight line (4.76c) with slope determined by the ratio of amplitudes. If the two oscillators are (problem 88) exactly out-of-phase (4.77a):

$$\varphi = \frac{\pi}{2}: \qquad 0 = a^2 y^2 + b^2 x^2 - a^2 b^2, \qquad \frac{x^2}{a^2} + \frac{y^2}{b^2} = 1; \quad (4.77a\text{--}c)$$

that is, one oscillator is at the maximum displacement when the other is at zero displacement (4.77b), the trajectory is (Figure 4.2b) an ellipse (4.77c), with axis along the coordinate axis, and half-axis equal to the amplitudes.

4.2.8 In-Phase, Out-Of-Phase, and Intermediate Cases

In the case of intermediate phase shifts (4.78a), the quadratic (4.75) represents an ellipse because: (i) the trajectory is finite $|x(t)| < a$ and $|y(t)| < b$ as follows from (4.72a, b); (ii) the only finite centered quadric is an ellipse and degenerate cases, like a straight line (4.76a–c). The ellipse need not be aligned (Figure 4.2c) with the coordinate axis (4.77b) ≡ (4.77c) unless the two motions (4.72a, b) are out-of-phase (4.77a). This can be confirmed by considering

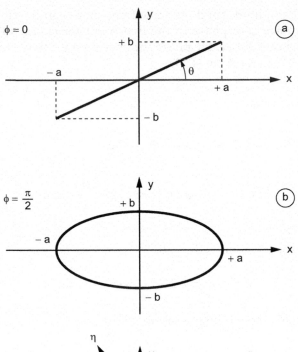

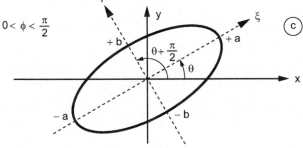

FIGURE 4.2
A two-dimensional harmonic oscillator consists of the superposition of motions in orthogonal directions with the same natural frequency distinct amplitudes (*a, b*) and a phase shift ϕ, leading to: (i) for zero phase shift a straight line (a) with slope *b/a* passing through the origin; (ii) in the out-of-phase case $\phi = \pi/2$ to an ellipse with half-axis (*a, b*) along the coordinate axis (b) and center at the origin; or (iii) in all other cases $0 < \phi < \pi/2$ the center remains at the origin (c) and the axis of the ellipse is rotated (Figure 4.3).

(Figure 4.3) a rotation by an angle θ of the Cartesian reference frame (4.78a) to the Cartesian reference frame (4.78b) related by (4.78c):

$$z \equiv x + iy, \qquad \zeta \equiv \xi + i\eta: \qquad\qquad \zeta = e^{-i\theta}z, \qquad z = e^{i\theta}\zeta, \qquad (4.78\text{a–d})$$

in complex coordinates:

$$x + iy = (\cos\theta + i\sin\theta)(\xi + i\eta) \quad \Leftrightarrow \quad \xi + i\eta = (\cos\theta - i\sin\theta)(x + iy); \quad (4.78\text{e, f})$$

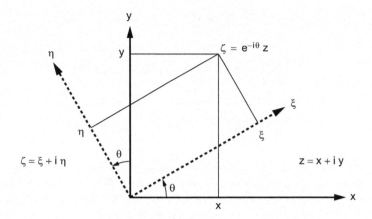

FIGURE 4.3
A point in the plane is represented by a complex number and a rotation corresponds to a change of phase.

the corresponding transformation in real coordinates:

$$x = \xi \cos\theta - \eta \sin\theta, \qquad y = \xi \sin\theta + \eta \cos\theta, \qquad (4.79a, b)$$

$$\xi = x \cos\theta + y \sin\theta, \qquad \eta = -x \sin\theta + y \cos\theta, \qquad (4.79c, d)$$

is a **rotation** *by an angle* θ *in the positive or clockwise direction (4.79a, b)* $\equiv$ *(I.16.56a, b) whose inverse, namely the rotation in the negative on clockwise direction (4.79c, d) is obtained by exchanging* θ *with* $-\theta$. *Substitution of (4.79a, b) in (4.75) leads to:*

$$a^2 b^2 \sin^2\varphi = \xi^2 \left[a^2 \sin^2\theta + b^2 \cos^2\theta - ab\cos\varphi\sin(2\theta) \right]$$
$$+ \eta^2 \left[a^2 \cos^2\theta + b^2 \sin^2\theta + ab\cos\varphi\sin(2\theta) \right] \qquad (4.80)$$
$$+ \xi\eta \left[(a^2 - b^2)\sin(2\theta) - 2ab\cos\varphi\cos(2\theta) \right].$$

The axis of the ellipse are aligned (Figure 4.2b) with the (ξ, η) coordinate axis (Figure 4.3) when the coefficient of $\xi\eta$ in (4.80) vanishes; that is, for the rotation angle (4.81a):

$$2\cos\varphi\cot(2\theta) = \frac{a^2 - b^2}{ab} = \frac{a}{b} - \frac{b}{a}: \qquad \left(\frac{\xi}{A}\right)^2 + \left(\frac{\eta}{B}\right)^2 = 1, \qquad (4.81a, b)$$

leading to the equation (4.81b) of an ellipse with axis (4.82a, b):

$$A^2 = \frac{a^2 b^2 \sin^2 \varphi}{a^2 \sin^2 \theta + b^2 \cos^2 \theta - ab \cos \varphi \cos(2\theta)}, \tag{4.82a}$$

$$B^2 = \frac{a^2 b^2 \sin^2 \varphi}{a^2 \cos^2 \theta + b^2 \sin^2 \theta + ab \cos \varphi \cos(2\theta)}. \tag{4.82b}$$

It has been shown that *two orthogonal undamped oscillators (4.72a, b) with the same frequency* ω *and amplitudes (a, b) lead as a path to: (i) for (problem 89) arbitrary phase shift (4.83a) to (Figure 4.2c) an ellipse (4.81b; 4.79c, d)* $\equiv$ *(4.83b) with half-axis (4.82a, b) along the lines with (4.81a) angles* θ *and* $\theta + \pi/2$:

$$0 < \varphi < \frac{\pi}{2}: \qquad \left(\frac{x \cos\theta + y \sin\theta}{A} \right)^2 + \left(\frac{x \sin\theta - y \cos\theta}{B} \right)^2 = 1; \tag{4.83a, b}$$

(ii) if (problem 88) they are out of phase (4.77a) and (4.81a) implies $\theta = 0$, *the axis of the ellipse (4.82a, b) coincides (Figure 4.2b) with the coordinate axis (4.77b) and the half-axis equal to the amplitudes (4.77c); (iii) if (problem 87) they are in phase (4.76a) the ellipse degenerates (Figure 4.2a) into a straight line (4.76b) with slope equal to the ratio of amplitudes (4.76c).*

4.3 Variable Coefficients and Parametric Resonance

The linear second-order differential equation with constant coefficients (4.37a, b) has two generalizations: (i) coefficients depending on time for variable systems, for example periodic coefficients (section 9.3), leading to parametric resonance (section 4.3); and (ii) non-linear systems and equations (sections 9.1–9.2), leading to a variety of phenomena including hysteresis and bifurcations (sections 4.4–4.9). The parametric resonance occurs in the absence of forcing for second-order mechanical system (electric circuit) whose the mass (induction), damping (resistance) or spring (capacity) may depend on time. An example is an undamped oscillator with mass and spring resilience varying with time (subsection 4.3.1) that corresponds to an oscillator with "natural frequency" a function of time; another example is a mass suspended from a linear spring with vibrating support (subsection 4.3.2), leading to the Mathieu equation. For this or any other linear second-order differential equations with periodic coefficients (subsection 4.3.3) the solution is a periodic function multiplied by an exponential of time (subsection 4.3.4); the latter (subsection 4.3.5) grows (decays) for an unstable (stable) system and has constant amplitude for an indifferent system. In the case of an undamped variable system only the indifferent and unstable cases are possible (subsection 4.3.6), with the latter corresponding to parametric resonance.

The particular case of a constant mass suspended from a linear spring with oscillating support (subsection 4.3.2) leads to the Mathieu equation that: (i) has four even or odd solutions (subsection 4.3.8) with period π or 2π specified by trigonometric series (subsection 4.3.7), (ii) the recurrence formula for the coefficients involves the eigenvalues (subsection 4.3.10) that are the roots of a continued fraction (subsection 4.3.9); and (iii) from (ii) follow the excitation frequencies that lead to instability (subsection 4.3.11) and the motion due to parametric resonance (subsection 4.3.10). The parametric resonance is considered using approximate solutions, first for the case of excitation frequency (subsection 4.3.12) close to the first harmonic that is twice the natural frequency (subsection 4.3.13); the second-order approximation (subsection 4.3.14) is first applied to excitation at a frequency close to the second harmonic. The excitation at a frequency close to the fundamental (subsection 4.3.15) is also a second-order effect. There is also parametric resonance at submultiples of the first harmonic (subsection 4.3.18), and adding damping introduces a threshold amplitude of excitation (subsection 4.3.16) that increases at higher-orders. The threshold amplitude of excitation of parametric resonance (subsection 4.3.17) in the presence of damping is such that the energy supply through vibration of the support overcomes dissipation. The case of a pendulum with vertically oscillating support is an example of parametric resonance (subsection 4.3.19). There are at least 11 differences (subsection 4.3.20) between parametric (ordinary) resonance [section(s) 4.3 (2.6–2.7)].

4.3.1 Oscillator with Variable Mass and Resilience

Considering an undamped linear oscillator (Figure 4.4) the equation of motion balances: (i) the restoring force (4.84a) that is a linear function of the displacement with resilience that may be a function of time; and (ii) the

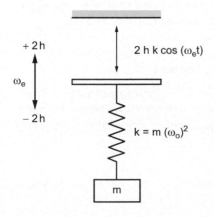

$+2h$

ω_e

$-2h$

$2\,h\,k\,\cos(\omega_e t)$

$k = m\,(\omega_0)^2$

m

FIGURE 4.4
The parametric resonance occurs for a linear system whose coefficients are periodic functions of time, for example a mass suspended from a spring whose support oscillates at an excitation frequency ω_e that may differ or not from the natural frequency ω_0. In the case of the harmonic oscillator, the parametric resonance occurs only for frequencies that are sub-multiples of the first harmonic of the natural frequency and is modified by damping (Figure 4.5).

inertia force (4.84a) equal to the time derivative of the linear momentum (4.84b) that equals the product of the velocity by the mass (4.84c) that may also be a function of time:

$$-k(t)x(t) = F(t), \qquad F(t) = \frac{dp}{dt}, \qquad p(t) = m(t)\frac{dx}{dt}; \qquad \text{(4.84a–c)}$$

thus the equation of motion is:

$$m(t)\frac{d^2x}{dt^2} + \frac{dm}{dt}\frac{dx}{dt} = \frac{d}{dt}\left(m(t)\frac{dx}{dt}\right) = -k(t)x(t). \qquad \text{(4.84d)}$$

Introducing the modified time (4.85a) ≡ (4.85b) ≡ (4.85c):

$$\frac{m(t)}{dt} = \frac{1}{dT}: \qquad \frac{dt}{dT} = m(t), \quad T = \int \frac{dt}{m(t)}, \quad X(T) \equiv x(t), \qquad \text{(4.85a–d)}$$

the equation of motion (4.84d; 4.85d) takes the form (4.85e):

$$-k(t)X(T) = \frac{dT}{dt}\frac{d}{dT}\left(\frac{dt}{dT}\frac{dX}{dt}\right) = \frac{1}{m(t)}\frac{d^2X}{dT^2}. \qquad \text{(4.85e)}$$

Thus *an undamped linear oscillation with variable mass and spring resilience (4.84d) satisfies (problem 90) a pseudo-harmonic oscillator equation (4.85e) ≡ (4.86b):*

$$\omega(T) = \sqrt{k(t)m(t)}: \qquad \left\{\frac{d^2}{dT^2} + \left[\omega(T)\right]^2\right\}X(T) = 0, \qquad \text{(4.86a, b)}$$

involving the modified time (4.85a) ≡ (4.85b) ≡ (4.85c) and pseudo-natural frequency (4.86a); the pseudo-natural frequency (4.86a) does not reduce to the natural frequency (2.54a) because the mass also appears in the modified time (4.85b).

4.3.2 Harmonic Oscillator with Vibrating Support

In the case of constant mass and spring resilience the natural frequency is (4.87a) ≡ (2.54a) and if the support of the spring (Figure 4.4) performs oscillations with amplitude 2h and excitation frequency ω_e, this is equivalent to a resilience of the spring varying in time (4.87b):

$$\omega_0 = \sqrt{\frac{k_0}{m}}: \qquad k(t) = k_0\left[1 + 2h\cos(\omega_e t)\right], \qquad \text{(4.87a, b)}$$

and leads (problem 91) to the equation of motion (4.84d) in the original time:

$$\ddot{x} + \omega_0^2 \left[1 + 2h \cos(\omega_e t) \right] x = 0. \tag{4.88}$$

*The latter corresponds to a **Mathieu (1873) equation** (4.89):*

$$\frac{d^2 w}{d\theta^2} + \left[a - 2q \cos(2\theta) \right] w = 0, \tag{4.89}$$

with: (i) independent variable (4.90a); (ii) dependent variable (4.90d); and (iii) parameters (4.90b, c):

$$\theta \equiv \frac{\omega_e t}{2}, \quad a \equiv \left(\frac{2\omega_0}{\omega_e} \right)^2, \quad q \equiv -ha = -h \left(\frac{2\omega_0}{\omega_e} \right)^2, \quad w(\theta) \equiv x(t). \tag{4.90a–d}$$

The general properties of linear differential equations of second order with periodic coefficients (subsections 4.3.3–4.3.6) are used to solve (subsections 4.3.7–4.3.12) the Mathieu equation (4.89) exactly; these are illustrated by approximate solutions of the harmonic oscillator with vibrating support (4.88) without (with) damping [subsections 4.3.13–4.3.15 (4.3.16–4.3.18)]. An example is a pendulum with support oscillating vertically (subsections 4.3.19–4.3.20).

4.3.3 Linear Differential Equation with Periodic Coefficients (Floquet 1883)

The Mathieu equation (4.89) is a particular case of the unforced linear ordinary differential equation (1.34b; 1.36c) ≡ (4.91b) with periodic coefficients (4.91a):

$$A_n(t+\tau) = A_n(t): \qquad \sum_{n=0}^{N} A_n(t) \frac{d^n x}{dt^n} = 0, \tag{4.91a, b}$$

considered next (in the section 9.3) for order two (order N). In the second-order case:

$$N = 2: \qquad A_2(t)\ddot{x}(t) + A_1(t)\dot{x}(t) + A_0(t)x(t) = 0, \tag{4.92a, b}$$

the general integral is a linear combination (1.36b) ≡ (4.92d) of two linearly independent particular integrals with the non-zero Wronskian (1.40c) ≡ (4.92c):

$$0 \neq W(x_1 x_2) = x_1 \dot{x}_2 - \dot{x}_1 x_2: \qquad x(t) = C_1 x_1(t) + C_2 x_2(t), \tag{4.92c, d}$$

where (C_1, C_2) are arbitrary constants. The differential equation (4.92b) is unchanged (4.91a) by the substitution $t \rightarrow t + \tau$ and thus $x_{1,2}(t + \tau)$ are also solutions; since a linear second-order differential equation can have only two linearly independent solutions, both $x_{1,2}(t + \tau)$ must be linear combinations (4.93b) of $x_{1,2}(t)$:

$$C_{11}C_{22} \neq C_{12}C_{21}: \qquad \begin{bmatrix} x_1(t+\tau) \\ x_2(t+\tau) \end{bmatrix} = \begin{bmatrix} C_{11} & C_{12} \\ C_{21} & C_{22} \end{bmatrix} \begin{bmatrix} x_1(t) \\ x_2(t) \end{bmatrix}; \qquad \text{(4.93a, b)}$$

the matrix of coefficients has non-zero determinant (4.93a) because the system (4.93b) must be invertible, since $x_{1,2}(t)$ are also linear combinations of $x_{1,2}(t + \tau)$. The matrix of coefficients has eigenvalues that are the roots of 2×2 determinant:

$$0 = \begin{vmatrix} C_{11} - s & C_{12} \\ C_{21} & C_{22} - s \end{vmatrix} = (C_{11} - s)(C_{22} - s) - C_{12}C_{21}, \qquad \text{(4.94a)}$$

corresponding to the quadratic polynomial:

$$0 = s^2 - \left(C_{11} + C_{22}\right)s + C_{11}C_{22} - C_{12}C_{21} = (s - s_+)(s - s_-). \qquad \text{(4.94b)}$$

If the eigenvalues are distinct (4.95a), the corresponding eigenfunctions have the proportional property (4.95b, c):

$$s_+ \neq s_-: \qquad x_\pm(t+\tau) = s_\pm x_\pm(t), \qquad \text{(4.95a–c)}$$

$$s_+ = s_- \equiv s_0: \qquad x_+(t+\tau) = s_0 x(t), \qquad x_-(t+\tau) = s_0\left[x_-(t) + x_+(t)\right], \qquad \text{(4.96a–c)}$$

whereas if the eigenvalues coincide (4.96a), the first eigenfunction is the same (4.96b) $\equiv$ (4.97b) and the second is modified (4.96c) $\neq$ (4.95c).

4.3.4 Characteristic Exponents and Periodic Solutions

In the first case of distinct eigenvalues (4.95a) from (4.95b, c) it follows that:

$$e^{-b_\pm(t+\tau)}x_\pm(t+\tau) = e^{-b_\pm\tau}s_\pm e^{-b_\pm t}x_\pm(t) = e^{-b_\pm t}x_\pm(t) = \varphi_\pm(t) = \varphi_\pm(t+\tau), \qquad \text{(4.97)}$$

is a periodic function of with period τ provided that (4.98a, b) is met:

$$s_+ \neq s_-: \qquad s_\pm = \exp(b_\pm\tau); \qquad x_\pm(t) = \exp(b_\pm t)\varphi_\pm(t); \qquad \text{(4.98a–c)}$$

thus the solution (4.98c) of (4.92b) is the product of a periodic function (4.97) by an exponential of time (4.98a). In the second case of coincident eigenvalues (4.96a) ≡ (4.99a) it follows that (4.99b):

$$s_+ = s_- \equiv s_0 = \exp(b_0\tau): \qquad\qquad x_\pm(t) = \exp(b_0 t)\varphi_\pm(t), \qquad\qquad (4.99\text{a, b})$$

satisfies (4.96b, c) ≡ (4.100a, b):

$$\varphi_+(t+\tau) = e^{-b_0\tau}e^{-b_0 t}x_+(t+\tau) = s_0 e^{-b_0\tau}e^{-b_0 t}x_+(t) = e^{-b_0 t}x_+(t) = \varphi_+(t), \qquad (4.100\text{a})$$

$$e^{-b_0 t}\left[x_-(t) + x_+(t)\right] = \varphi_-(t) + \varphi_+(t); \qquad\qquad (4.100\text{b})$$

and the functions $\varphi_\pm$ are again periodic. It has been shown that *a linear second-order (4.92a) ordinary differential equation (4.91b) with periodic coefficients (4.91a) has (problem 92) solutions that are the product of a periodic function by an exponential of time (4.98c) with distinct* **characteristic exponents** *(4.98b) that are the* **eigenvalues** *(4.98a) of (4.94a). If the eigenvalues coincide (4.99a), the single characteristic exponent (4.99a) ≡ (4.101a) appears (4.99b) in the solutions (4.96b, c) ≡ (4.101b, c):*

$$s_+ = s_- = s_0 = \exp(b_0\tau): \qquad x_+(t) = e^{b_0 t}\varphi_+(t), \quad x_-(t) = e^{b_0 t}\left[\varphi_+(t) + \varphi_-(t)\right],$$
$$(4.101\text{a–c})$$

where again $\varphi_\pm(t)$ are periodic functions with period τ.

4.3.5 Stability, Instability, and Indifference Boundary

In the preceding solutions the periodic functions are bounded (4.102a), implying that the growth for large time is determined by the exponential factor:

$$\left|\varphi_\pm(t)\right| = \left|\varphi_+(t+\tau)\right| \le M: \qquad\qquad \left|x_\pm(t)\right| \le \left|e^{bt}\right|\left|\varphi_\pm(t)\right| \le M\left|e^{bt}\right|. \qquad (4.102\text{a, b})$$

The characteristic exponent may be complex (4.94a) ≡ (4.94b) and only the real part affects the modulus of the exponential (4.103b):

$$b = \alpha + i\beta: \qquad\qquad \left|e^{bt}\right| = \left|e^{\alpha t}\right|\left|e^{i\beta t}\right| = e^{\alpha t} = \exp[t\,\text{Re}\,b]. \qquad (4.103\text{a, b})$$

Thus *a linear second-order (4.92a) differential equation (4.91b) with periodic coefficients (4.91a) has general integral (4.104b) [(4.105b)] if the characteristic exponents are distinct (4.104a) [equal (4.105a)]:*

$$b_+ \ne b_-: \qquad x(t) = C_+ x_+(t) + C_- x_-(t) = C_+ e^{b_+ t}\varphi_+(t) + C_- e^{b_- t}\varphi_-(t), \quad (4.104\text{a, b})$$

$$b_+ = b_- \equiv b_0: \qquad x(t) = C_+ x_+(t) + C_- x_-(t) = e^{b_0 t}\left[(C_+ + C_-)\varphi_+(t) + C_-\varphi_-(t)\right],$$
$$(4.105\text{a, b})$$

where $\varphi_{\pm}(t)$ *are periodic functions (4.102a) with period* τ. *The limit for large time:*

$$\mathrm{Re}(b_+) \geq \mathrm{Re}(b_-): \quad \lim_{t\to\infty}|x(t)| = \begin{cases} = \infty & \text{if} \quad \mathrm{Re}(b_+) > 0: & \text{instability} \\ \neq 0, \infty & \text{if} \quad \mathrm{Re}(b_+) = 0: & \text{indiferent}, \\ = 0 & \text{if} \quad \mathrm{Re}(b_+) < 0: & \text{stability}, \end{cases}$$

$$(4.106a\text{--}d)$$

shows that the exponent with larger real part (4.106a) determines: (i) **stability** *if it is negative (4.106d): (ii)* **instability** *if it is positive (4.106b); and (iii)* **indifference** *if the value is zero (4.106c).* The results are reconsidered next for the variable undamped linear oscillator (subsection 4.3.2) showing that (subsection 4.3.6) only the cases (i) and (iii) occur.

4.3.6 Stability of the Linear Variable Oscillator

The preceding results (4.104a, b; 4.105a, b) apply to the undamped (4.86b) linear variable oscillator, for which the two terms in the case of distinct eigenvalues (4.107a) may be written (4.107c) using the eigenvalues (4.107b):

$$b_+ \neq b_-: \qquad \exp(b_{\pm}t)\varphi_{\pm}(t) = s_{\pm}^{t/\tau}\varphi_{\pm}(t), \qquad s_{\pm} = \exp(b_{\pm}\tau). \qquad (4.107a\text{--}c)$$

The general integral for the displacement is (4.108b) [(4.109b)] for distinct (4.108a) [equal (4.109a)] eigenvalues:

$$s_+ \neq s_-: \qquad x(t) = C_+ s_+^{t/\tau}\varphi_+(t) + C_- s_-^{t/\tau}\varphi_-(t) = C_+ x_+(t) + C_- x_-(t), \qquad (4.108a, b)$$

$$s_+ = s_- \equiv s_0: \quad x(t) = s_0^{t/\tau}\left[(C_+ + C_-)\varphi_+(t) + C_-\varphi_-(t)\right] = C_+ x_+(t) + C_- x_-(t). $$

$$(4.109a, b)$$

Since $x_{\pm}$ satisfy the differential equation (4.86b) $\equiv$ (4.110a, b) follows (4.110c):

$$\ddot{x}_{\pm} + \omega^2 x_{\pm} = 0: \qquad 0 = x_+\ddot{x}_- - x_-\dot{x}_+ = \frac{d}{dt}(x_+\dot{x}_- - x_-\dot{x}_+), \qquad (4.110a\text{--}c)$$

implying that the Wronskian (1.40c) is constant (4.111a):

$$const \equiv E = x_+(t)\dot{x}_-(t) - x_-(t)\dot{x}_+(t)$$

$$= x_+(t+\tau)\dot{x}_-(t+\tau) - x_-(t+\tau)\dot{x}_+(t+\tau) = s_+ s_- E;$$

$$(4.111a\text{--}c)$$

the Wronskian is unchanged (4.111b) replacing t by $t + \tau$, and (4.95b, c) lead to (4.111c). It follows (4.111c) $\equiv$ (4.112a) that the product of eigenvalues is unity:

$$s_+ s_- = 1; \qquad \{s_+, s_-\} = \{s_+^*, s_-^*\} \quad \text{or} \quad \{s_+, s_-\} = \{s_-^*, s_+^*\}; \qquad \text{(4.112a–c)}$$

also, since the differential equation (4.86b) has real coefficients, both $x_\pm$ and x_+^* are solutions and thus the two pairs of eigenvalues $s_\pm$ must coincide with $s_\pm^*$ leading to either (4.112b) or (4.112c). In the first case $s_\pm$ are real (4.113a) and if they are distinct one is larger than unity (4.113b, c) so that (4.108b) diverges for large t leading (4.113d) to instability:

$$s_\pm = s_\mp^*: \qquad s_- < 1, \qquad s_+ = \frac{1}{s_-} > 1, \qquad \lim_{t \to \infty} s_+^{t/\tau} = \infty, \quad b_+ > 0 > b_-, \qquad \text{(4.113a–e)}$$

corresponding (4.107c) to (4.113e) $\equiv$ (4.106b). In the second case, $s_\pm$ are complex conjugate (4.114a) and hence (4.112c) have unit modulus (4.114b, c) so that (4.108b) is finite (4.114d) for large t:

$$s_\pm = s_\pm^*: \quad 1 = s_- s_+ = s_+^* s_+ = |s_+|^2 = |s_-|^2, \quad \lim_{t \to \infty} |s_\pm|^{t/\tau} = 1, \quad \mathrm{Re}(b_\pm) = 0, \qquad \text{(4.114a–e)}$$

corresponding (4.107c) to the indifference (4.114e) $\equiv$ (4.106c). In the second case the eigenvalues (4.114a) with unit modulus (4.114b) can be written in the form (4.144f) leading to the solution (4.108b) $\equiv$ (4.114g) that is oscillatory and bounded:

$$s_\pm = \exp(\pm i\psi): \qquad x(t) = C_+ \varphi_+(t) \exp(i\psi t/\tau) + C_- \varphi_-(t) \exp(-i\psi t/\tau). \quad \text{(4.114f, g)}$$

If the eigenvalues coincide it follows from (4.113a–c) that they are both unity $s_\pm = 1$ and (4.107c) implies $b_0 = 0$ hence indifference (4.106c). If has been shown that *for the linear undamped variable oscillator (4.86b) the displacements (4.108b) [(4.109b)] with (problem 93) distinct (equal) eigenvalues imply that there are two possible cases: (i) instability (4.113a–e) if the eigenvalues or characteristic exponents are real and distinct; (ii) indifference otherwise, that is for complex conjugate (4.114a–e) or coincident eigenvalues or characteristic exponents.* The determination of the characteristic exponents is illustrated (subsection 4.3.8) in the process of obtaining the exact solutions (subsection 4.3.9) of the Mathieu equation (4.89) involving trigonometric series (subsection 4.3.7).

4.3.7 Expansion of Mathieu (1873) Functions in Trigonometric Series

The solution of the Mathieu equation (4.89) may be sought in the form of a trigonometric series:

$$w(\theta) = \sum_{m=0}^{\infty} A_m \cos(m\theta) + \sum_{m=1}^{\infty} B_m \sin(m\theta), \tag{4.115}$$

that is similar to a Fourier series (II.5.188b) ≡ (4.115) but is more general than the Fourier series (II.5.188b) since the coefficients may not satisfy the formulas for the coefficients (II.5.189a, b; II.5.190a, b). Substituting the trigonometric series (4.115) in the Mathieu equation (4.89) leads to:

$$0 = \sum_{m=0}^{\infty} \left[(a - m^2) \cos(m\theta) - 2q \cos(m\theta) \cos(2\theta) \right] A_m$$

$$+ \sum_{m=1}^{\infty} \left[(a - m^2) \sin(m\theta) - 2q \sin(m\theta) \cos(2\theta) \right] B_m. \tag{4.116}$$

Using the identities (II.5.88a) ≡ (4.117a) and (II.5.88c) ≡ (4.117b):

$$2\cos(m\theta)\cos(2\theta) = \cos\left[(m+2)\theta \right] + \cos\left[(m-2)\theta \right], \tag{4.117a}$$

$$2\sin(m\theta)\cos(2\theta) = \sin\left[(m+2)\theta \right] + \sin\left[(m-2)\theta \right], \tag{4.117b}$$

leads to:

$$0 = \sum_{m=-2}^{\infty} \left[(a - m^2) A_m - q(A_{m+2} + A_{m-2}) \right] \cos(m\theta)$$

$$+ \sum_{m=-1}^{\infty} \left[(a - m^2) B_m - q(B_{m+2} + B_{m-2}) \right] \sin(m\theta). \tag{4.118}$$

In order for the trigonometric series to vanish, all terms in square brackets must vanish, leading to the recurrence formulas for the coefficients.

4.3.8 Even/Odd Mathieu Functions with Period $\pi/2\pi$

The vanishing of the coefficients in (4.118) leads to four cases: (i) the coefficients A_m (B_m) of the cosines (sines) lead to solutions that are even (odd) functions of the variable θ; and (ii) the coefficients are related by jumps of two

$m \rightarrow m \pm 2$, so that the solution with even $m = 2r$ (odd $m = 2r + 1$) coefficients have period $\pi(2\pi)$:

$$w_1(\theta) = \sum_{r=0}^{\infty} A_{2r} \cos(2r\theta); \quad r > 1: \quad (a - 4r^2) A_{2r} = q(A_{2r+2} + A_{2r-2}),$$

$$(4.119\text{a–c})$$

$$w_2(\theta) = \sum_{r=0}^{\infty} A_{2r+1} \cos[(2r+1)\theta]; \quad r \geq 1: \quad [a - (2r+1)^2] A_{2r+1} = q(A_{2r+3} + A_{2r-1}),$$

$$(4.120\text{a–c})$$

$$w_3(\theta) = \sum_{r=1}^{\infty} B_{2r} \sin(2r\theta); \quad r > 1: \quad (a - 4r^2) B_{2r} = q(B_{2r+2} + B_{2r-2}),$$

$$(4.121\text{a–c})$$

$$w_4(\theta) = \sum_{r=0}^{\infty} B_{2r+1} \sin[(2r+1)\theta]; \quad r \geq 1: \quad [a - (2r+1)^2] B_{2r+1} = q(B_{2r+3} + B_{2r-1}).$$

$$(4.122\text{a–c})$$

Thus *the Mathieu equation (4.89) has solutions (problem 91) that are even (4.119a; 4.120a) [odd (4.121a; 4.122a)] in (4.123a) [(4.123b)] the variable* θ:

$$w_{1,2}(\theta) = w_{1,2}(-\theta), \qquad w_{3,4}(\theta) = -w_{3,4}(-\theta), \qquad (4.123\text{a, b})$$

$$w_{1,3}(\theta) = w_{1,3}(\theta + \pi), \qquad w_{2,4}(\theta) = w_{2,4}(\theta + 2\pi), \qquad (4.124\text{a, b})$$

and with (4.119a; 4.121a) [(4.120a; 4.122a)] period $\pi(2\pi)$ *in (4.124a) [(4.124b)]. The recurrence formulas for the coefficients (4.119c; 4.121c) [(4.120c; 4.122c)] hold for* $m = 2r \geq 4(m = 2r + 1 \geq 3)$; *the first three relations for the* $A_m(B_m)$ *in (4.116) not included in (4.119a; 4.120a) [(4.121a; 4.122a)] are (4.125a–c) [(4.126a, b)]:*

$$aA_0 = qA_2, \qquad (a-1)A_1 = q(A_1 + A_3), \qquad (a-4)A_2 = q(2A_0 + A_4),$$

$$(4.125\text{a–c})$$

$$(a-1)B_1 = q(B_3 - B_1), \qquad (a-4)B_2 = qB_4. \qquad (4.126\text{a, b})$$

In all equations, e.g. (4.118; 4.119c; 4.120c; 4.121c; 4.122c) it is assumed that the coefficients with negative indices (4.127a) are zero (4.127b, c):

$$m = 1, 2, 3, \ldots: \qquad A_{-m} = 0 = B_{-m}. \qquad (4.127\text{a–c})$$

The proof of (4.125a–c) [4.126a, b)] follows from the vanishing of the coefficients of cosines (4.128a) [sines (4.128b)] of lowest order in (4.116):

$$0 = \left[a - 2q\cos(2\theta)\right]A_0 + \left[(a-1)\cos\theta - 2q\cos\theta\cos(2\theta)\right]A_1$$

$$+ \left[(a-4)\cos(2\theta) - 2q\cos^2(2\theta)\right]A_2$$

$$+ \left[(a-9)\cos(3\theta) - 2q\cos(3\theta)\cos(2\theta)\right]A_3$$

$$+ \left[(a-16)\cos(4\theta) - 2q\cos(4\theta)\cos(2\theta)\right]A_4 + ...$$

$$= a\,A_0 - 2q\,A_0\cos(2\theta) + (a-1)\,A_1\cos\theta - qA_1\left[\cos\theta + \cos(3\theta)\right]$$

$$+ (a-4)\,A_2\cos(2\theta) - qA_2\left[1 + \cos(4\theta)\right] + (a-9)\,A_3\cos(3\theta)$$

$$- q\,A_3\left[\cos\theta + \cos(5\theta)\right] + (a-16)\,A_4\cos(4\theta) - q\,A_4\left[\cos(2\theta) + \cos(6\theta)\right] + ...$$

$$= a\,A_0 - q\,A_2 + \left[(a-1)\,A_1 - q(A_1 + A_3)\right]\cos\theta$$

$$+ \left[(a-4)\,A_2 - q(2\,A_0 + A_4)\right]\cos(2\theta) + ... \tag{4.128a}$$

$$0 = \left[(a-1)\sin\theta - 2q\sin\theta\cos(2\theta)\right]B_1$$

$$+ \left[(a-4)\sin(2\theta) - 2q\sin(2\theta)\cos(2\theta)\right]B_2$$

$$+ \left[(a-9)\sin(3\theta) - 2q\sin(3\theta)\cos(2\theta)\right]B_3$$

$$+ \left[(a-16)\sin(4\theta) - 2q\sin(4\theta)\cos(2\theta)\right]B_4$$

$$\tag{4.128b}$$

$$= (a-1)\,B_1\sin\theta - qB_1\left[\sin(3\theta) - \sin\theta\right] + (a-4)\,B_2\sin(2\theta)$$

$$- q\,B_2\sin(4\theta) + (a-9)\,B_3\sin(3\theta) - q\,B_3\left[\sin(5\theta) + \sin\theta\right]$$

$$+ (a-16)\,B_4\sin(4\theta) - qB_4\left[\sin(6\theta) + \sin(2\theta)\right] + ...$$

$$= \left[(a-1)\,B_1 + q\,(B_1 - B_3)\right]\sin\theta + \left[(a-4)\,B_2 - qB_4\right]\sin(2\theta) + ...$$

The series expansions (4.119a)/(4.120a)/(4.121a)/(4.122a) satisfy the Mathieu differential equation (4.89) by virtue of the recurrence formulas for the coefficients respectively (4.119b, c; 4.125a, c)/(4.120b, c; 4.125b)/(4.121b, c; 4.126b)/(4.122b, c; 4.126a). For the series to converge the Cauchy (subsection I.29.2.1) necessary condition (I.29.16a) must be met, requiring that the coefficients tend to zero for large order:

$$0 = \lim_{r\to\infty} A_{2r} = \lim_{r\to\infty} A_{2r+1} = \lim_{r\to\infty} B_{2r} = \lim_{r\to\infty} B_{2r+1}. \tag{4.128c–f}$$

These conditions are not met for all values of the parameter a, but only for an denumerably infinite set of eigenvalues, that are determined next.

4.3.9 Denumerable Set of Eigenvalues as Roots of Continued Fractions

The triple recurrence formulas for the coefficients (4.119c)/(4.120c)/(4.121c) /(4.122c) can be written respectively in the form of two-term fractions (4.129b)/(4.130b)/(4.131b)/(4.132b) starting with (4.129a) ≡ (4.125a)/(4.130a) ≡ (4.125b)/(4.131a) ≡ (4.126b)/(4.132a) ≡ (4.126a):

$$aA_0 = qA_2: \qquad \frac{A_{2r}}{A_{2r-2}} = \cfrac{1}{\dfrac{a-4r^2}{q} - \dfrac{A_{2r+2}}{A_{2r}}}, \qquad \text{(4.129a, b)}$$

$$(a-1)A_1 = q(A_1 + A_3): \qquad \frac{A_{2r+1}}{A_{2r-1}} = \cfrac{1}{\dfrac{a-(2r+1)^2}{q} - \dfrac{A_{2r+3}}{A_{2r+1}}}, \qquad \text{(4.130a, b)}$$

$$(a-4)B_2 = qB_4: \qquad \frac{B_{2r}}{B_{2r-2}} = \cfrac{1}{\dfrac{a-4r^2}{q} - \dfrac{B_{2r+2}}{B_{2r}}}, \qquad \text{(4.131a, b)}$$

$$(a-1)B_1 = q(B_3 - B_1): \qquad \frac{B_{2r+1}}{B_{2r-1}} = \cfrac{1}{\dfrac{a-(2r+1)^2}{q} - \dfrac{B_{2r+3}}{B_{2r+1}}}. \qquad \text{(4.132a, b)}$$

Thus the triple recurrence formulas (4.129b)/(4.130b)/(4.131b)/ (4.132b) lead (II.1.91a, b; II.1.93a, b) respectively to the continued fractions (4.133)/(4.134)/(4.135)/(4.136) starting with (4.129a)/(4.130a)/(4.131a)/(4.132a):

$$\frac{a}{q} - \frac{1}{(a-4)/q-} \frac{1}{(a-16)/q-} \cdots \frac{1}{(a-4r^2)/q-}$$

$$= \frac{a}{q} - \frac{q}{a-4-} \frac{q^2}{a-16-} \cdots \frac{q^r}{a-4r^2-} \cdots = 0 = \prod_{n=1}^{\infty} [a - c_{1,n}(q)], \qquad \text{(4.133)}$$

$$\frac{a-1}{q} - 1 - \frac{1}{(a-9)/q-} \frac{1}{(a-25)/q-} \cdots \frac{1}{[a-(2r+1)^2]/q-}$$

$$= \frac{a-1}{q} - 1 - \frac{q}{a-9-} \frac{q^2}{a-25-} \cdots \frac{q^r}{a-(2r+1)^2-} \cdots = 0 = \prod_{n=1}^{\infty} [a - c_{2,n}(q)], \qquad \text{(4.134)}$$

$$\frac{a-4}{q} - \frac{1}{(a-16)/q-} \frac{1}{(a-36)/q-} \cdots \frac{1}{\left(a-4r^2\right)/q-}$$

$$= \frac{a-4}{q} - \frac{q}{a-16-} \frac{q^2}{a-36-} \cdots \frac{q^r}{a-4r^2-} \cdots = 0 = \prod_{n=1}^{\infty} \left[a - c_{3,n}(q) \right], \tag{4.135}$$

$$0 = \frac{a-1}{q} + 1 - \frac{1}{(a-9)/q-} \frac{1}{(a-25)/q-} \cdots \frac{1}{\left[a-(2r+1)^2\right]/q-} \cdots$$

$$= \frac{a-1}{q} + 1 - \frac{q}{a-9-} \frac{q^2}{a-25-} \cdots \frac{q^r}{a-(2r+1)^2-} \cdots = 0 = \prod_{n=1}^{\infty} \left[a - c_{4,n}(q) \right]. \tag{4.136}$$

The roots of the continued fractions (4.133)/(4.134)/(4.135)/(4.136) specify the parameter a as a function of q respectively in the recurrence relations (4.119b, c)/(4.120b, c)/(4.121b, c)/(4.122b, c) with (4.125a–c; 4.126a, b). It has been shown that *the Mathieu equation (4.89) has (problem 91): (i) four solutions as trigonometric series (4.119a; 4.120a; 4.121a; 4.122a) that are even (4.123a) [odd (4.123b)] functions with (4.124a) [(4.124b)] period $\pi(2\pi)$; (ii) in the corresponding recurrence formulas for the coefficients (4.119b, c; 4.125a, c)/(4.120b, c; 4.125b)/(4.121b, c; 4.126b)/(4.122b, c; 4.126a) appear the parameters (a, q); (iii) the convergence of the series requires that the parameters be related by the roots (4.137b) with (4.137a) of respectively the continued fraction (4.133)/(4.134)/(4.135)/(4.136);*

$$p = 1,2,3,4 : a = c_{p,n}(q); w_{1-4}(\theta)$$

$$= \left\{ A_0 ce_{2n}(\theta;q), A_1 ce_{2n+1}(\theta;q), B_2 se_{2n} 2n(\theta;q), B_1 se_{2n+1}(\theta;q) \right\}, \tag{4.137a–f}$$

and (iv) thus there is a denumerably infinite set of solutions depending only on the variable θ and parameter q, namely the **Mathieu cosine (sine) functions** *of period π in (4.137c) [(4.137e)] and period 2π in (4.137d) [(4.137f)].*

4.3.10 Pairs of Mathieu Cosine and Sine Functions

Besides the two pairs of Mathieu cosine and sine functions (4.137c, d) with period π, and (4.137e, f) with period 2π, it is possible to obtain pairs with period $p\pi$ for all positive integer p. All these solutions are incompatible: (i) for different periods; and (ii) between Mathieu cosine and sine functions. The reason is that in each case the set of eigenvalues (4.137a, b) is distinct. Thus only one of the preceding solutions can be used as a particular integral of the Mathieu equation. The general integral requires a second linearly independent particular integral that is obtained subsequently (subsections 9.7.12–9.7.17) in the context of the general theory (section 9.3) of linear differential equations (4.91b) with periodic coefficients (4.91a).

The exact solutions of the Mathieu differential equation (subsections 4.3.7–4.3.9) suggest (subsection 4.3.10) an approximate method (subsections 4.3.11–4.3.20) suitable to study parametric resonance (subsections 4.3.1–4.3.6). For example, consider the Mathieu cosine (4.120a–c; 4.137d) ≡ (4.138a–d) [sine (4.122a–c; 4.137f) ≡ (4.139a–d)] function with period 2π:

$$w_{2,n}(\theta;q) = A_1 \cos\theta + A_3 \cos(3\theta) + ... = A_1 \left[\cos\theta + \left(\frac{a-1}{q} - 1\right)\cos(3\theta) + ... \right]$$

$$= A_1 \left[\cos\theta + \left[\frac{c_{2,n}(q)-1}{q} - 1\right]\cos(3\theta) + ... \right] = A_1 \, ce_{2n+1}(\theta;q),$$

$$(4.138a\text{–}d)$$

$$w_{4,n}(\theta;q) = B_1 \sin\theta + B_3 \sin(3\theta) + ... = B_1 \left[\sin\theta + \left(1 + \frac{a-1}{q}\right)\sin(3\theta) + ... \right]$$

$$= B_1 \left\{ \sin\theta + \left[1 + \frac{c_{4,n}(q)-1}{q}\right]\sin(3\theta) + ... \right\} = B_1 se_{2n+1}(\theta;q),$$

$$(4.139a\text{–}d)$$

where: (i) the coefficient $A_2(A_3)$ is related to $A_1(B_1)$ by (4.125b) [(4.126a)], and all $A_{2n+1}(B_{2n+1})$ are proportional to $A_1(B_1)$ leading to (4.138a) ≡ (4.138b) [(4.139a) ≡ (4.139b)]; (ii) the eigenvalues $c_{2n}(c_{4,n})$ are functions of q specified by the roots of the continued fraction (4.134) [(4.136)] that appear in (4.138c) [(4.139c)]; and (iii) each root corresponds to a Mathieu cosine (sine) function of period 2π in (4.138d) [(4.139d)] to within a multiplying constant $A_1(B_1)$ that can be set to unity.

The lowest-order approximation to the first root of the continued fraction (4.134) [(4.136)] is obtained truncating after the first two terms (4.140a) with upper (lower) sign leading to (4.140b) ≡ (4.140a):

$$\frac{a-1}{q} = \pm 1 \quad \Leftrightarrow \quad a - 1 = \pm q = \mp ha, \qquad (4.140a\text{–}c)$$

where (4.90c) was substituted leading to (4.140c). Substituting (4.90b) in (4.140c) ≡ (4.140d) leads to (4.140e):

$$\frac{1}{1\pm h} = a = \left(\frac{2\omega_0}{\omega_e}\right)^2 \quad \Leftrightarrow \quad \omega_e = 2\omega_0\sqrt{1\pm h} = 2\omega_0 \pm \omega_0 h + O(\omega_0 h^2), \quad (4.140d\text{–}f)$$

showing that the excitation frequency is close to twice the natural frequency (4.140f). Also, the condition (4.140a) cancels the second terms in (4.138b; 4.139b).

It has been shown that *a linear undamped harmonic oscillator (4.88) with natural frequency* ω_0, *excited (problem 94) with a vibrating support with small (4.141a) amplitude 2h, has the strongest parametric resonance when the excitation frequency is twice the natural frequency (4.140b) ≡ (4.141b) leading (4.90a) to an oscillation at the natural frequency (4.141c) for the even (4.141d) [odd (4.141e)] mode:*

$$h^2 \ll 1: \qquad \omega_e = 2\omega_0 \pm \omega_0 h, \qquad \theta = \frac{\omega_e t}{2} = \omega_0 t \left(1 \pm \frac{h}{2}\right); \qquad \text{(4.141a–c)}$$

$$x_2(t) = w_2(\theta) = A_1 \cos\left(\omega_0 t + \frac{1}{2}\omega_0 h t\right), \qquad \text{(4.141d)}$$

$$x_4(t) = w_4(\theta) = A_2 \sin\left(\omega_0 t - \frac{1}{2}\omega_0 h t\right). \qquad \text{(4.141e)}$$

The Mathieu equation (4.88) has solutions with period $k\pi$ for all positive integer k and this leads to parametric resonance not only for excitation frequency equal to the first harmonic of the natural frequency $\omega_e = 2\omega_0$ but also to submultiples and multiples (subsection 4.3.11).

4.3.11 Excitation Frequencies that Lead to Parametric Resonance

The leading two terms of the even (4.119a–c) [odd (4.121a–c)] solution with period π of the Mathieu equation (4.89) are (4.142a) [(4.142b)]:

$$w_{1,n}(\theta;q) = A_0 + A_2 \cos(2\theta) + A_4 \cos(4\theta) + \dots$$

$$= A_0 \left[1 + \frac{a}{q}\cos(2\theta) + \left(\frac{a^2 - 4a}{q^2} - 2\right)\cos(4\theta) + \dots\right]$$

$$= A_0 \left\{1 + \frac{1}{q}c_{1,n}(q)\cos(2\theta) + \left[c_{1,n}(q)\frac{c_{1,n}(q) - 4}{q^2} - 2\right]\cos(4\theta) + \dots\right\}$$

$$\equiv A_0 \, ce_{2n}(\theta;q), \qquad \text{(4.142a–d)}$$

$$w_{3,n}(\theta;q) = B_2 \sin(2\theta) + B_4 \sin(4\theta) + \dots = B_2 \left[\sin(2\theta) + \frac{a-4}{q}\sin(4\theta) + \dots\right]$$

$$= B_2 \left[\sin(2\theta) + \frac{c_{3,n}(q) - 4}{q^2}\sin(4\theta) + \dots\right] \equiv B_2 \, se_{2n}(\theta;q), \qquad \text{(4.143a–d)}$$

where: (i) the coefficients (A_2, A_4) are [coefficient B_3 is] related to $A_0(B_2)$ by (4.125a, c) [(4.126b)] and all $A_{2n}(B_{2n})$ are proportional to $A_0(B_2)$ leading to (4.142a) ≡ (4.142b) [(4.143a) ≡ (4.143b)]; (ii) the eigenvalues associated with the coefficients $A_{2n}(B_{2n})$ are functions of q specified by the roots of the continued faction (4.133) [(4.135)] that appear in (4.142c) [(4.143c)]; and (iii) each root corresponds to a Mathieu cosine (sine) function with period 2π in (4.142d) [(4.143d)] to within an arbitrary multiplying constant $A_0(B_2)$ that can be set to unity.

The Mathieu equation (4.89) has solutions with period $\pi(2\pi)$, namely (4.119a–c; 4.121a–c; 4.124a; 4.142a–d; 4.143a–d) [(4.120a–c; 4.122a–c; 4.124b; 4.138a–d; 4.139a–d)] and also with period $k\pi$ where k is any positive integer. This implies that for an harmonic oscillator (4.88) with natural (excitation) frequency $\omega_0(\omega_e)$ their harmonics of all orders $n(m)$ interact (II.5.88a) ≡ (4.144):

$$2\cos(m\omega_e t)\cos(n\omega_0 t) = \cos\left[(m\omega_e + n\omega_0)t\right] + \cos\left[(m\omega_e - n\omega_0 t)\right], \quad (4.144)$$

leading (4.145a, b) to the sum and difference frequencies (4.145c):

$$n, m = 1, 2, \ldots : \qquad\qquad \omega_{n,m}^\pm = m\omega_e \pm n\omega_0. \qquad\qquad (4.145\text{a–c})$$

If one of the frequencies (4.145c) coincides with the natural frequency (4.146a):

$$\omega_{n,m}^\pm = \omega_0 : \qquad\qquad \omega_e^\pm = \frac{1 \mp n}{m}\omega_0, \qquad\qquad (4.146\text{a, b})$$

there is parametric resonance, corresponding to the excitation frequencies (4.146b).

The parametric resonance can occur (problem 95) when the excitation frequency (4.87b) at the support is related to the natural frequency (4.87a) of the harmonic oscillator by (4.146b) where (4.145a, b) are positive integers. The smallest values (4.147a, b) of (4.145a, b) lead (4.147c) to the strongest parametric resonance that occurs (4.147c) for excitation frequency equal to twice the natural frequency (4.140b):

$$m = 1 = n : \qquad \omega_e^+ = 2\omega_0 ; \qquad m = 2, n = 1 : \qquad \omega_e^+ = \omega_0, \qquad (4.147\text{a–f})$$

larger values of (4.145a, b) such as (4.147d, e) lead to weaker parametric resonance (4.146b), in this case at the natural frequency (4.147f), and generally at multiples or sub-multiples (4.146b). The case (4.147f) corresponds (4.90b) to the lowest root $a = 4$ in (4.135) that cancels the second term in (4.143b). The preceding results obtained from the exact solution of the Mathieu equation (subsections 4.3.1–4.3.11) will be confirmed in the sequel (subsections 4.3.12–4.3.20) by a simpler approximate method that specifies: (i) the ranges of excitation frequency leading to parametric resonance; (ii) the characteristic exponents in the unstable cases; and (iii) the effects of damping or dissipation.

4.3.12 Wave Envelope with Amplitude Modulation

Two differences between the undamped normal (parametric) resonance [section 2.6 (subsections 4.3.1–4.3.11)] are: (i) the amplitude grows linearly (exponentially) in time; and (ii) the oscillation occurs at one **applied frequency** that must be equal to the natural frequency (a discrete set of **excitation frequencies** that are multiples or submultiples of the natural frequency). The differences between the ordinary (parametric) resonance relate to the oscillating support, and are illustrated next by approximate solutions of (4.88). In the case of fixed support the oscillations would be sinusoidal with constant amplitude, and in the case vibrating support the **method of variation of parameters** is used (subsection 1.3.6; notes 1.2–1.4; section 2.9; section 3.3) with the amplitudes taken as functions of time:

$$x(t) = A(t)\cos(\omega t) + B(t)\sin(\omega t), \tag{4.148}$$

and the frequency of oscillation ω is also to be determined. The approximate solution (4.148) allows for the two main features of parametric resonance: (i) periodic oscillation with sinusoidal form as the lowest-order approximation (subsections 4.3.7–4.3.11); and (ii) constant amplitude or **wave envelope** with **amplitude modulation,** in this case amplitude increasing or decreasing exponentially with time (subsections 4.3.1–4.3.6).

The equation of the parametric oscillator is $(4.76) \equiv (4.134)$:

$$\ddot{x} + \omega_0^2 x = -2h\omega_0^2 \cos(\omega_e t)x(t), \tag{4.149}$$

where the term on the r.h.s. is: (i) is specified *a priori* for ordinary resonance (2.165a) but not for parametric resonance (4.144) because the displacement also appears as a factor in the r.h.s. of (4.149); (ii) a forced oscillation (2.165b) occurs even if the initial displacement and velocity of the free motion are zero; and (iii) the parametric response does not occur if the initial displacement and velocity are both zero and thus requires a disturbance from the state of rest. The l.h.s. of (4.149) involves the acceleration or second-order time derivative of (4.148) specified by (4.150a):

$$\ddot{x} = \left(\ddot{A} - \omega^2 A + 2\omega\dot{B}\right)\cos(\omega t) + \left(\ddot{B} - \omega^2 B - 2\omega\dot{A}\right)\sin(\omega t), \tag{4.150a}$$

$$2x(t)\cos(\omega_e t) = 2\cos(\omega_e t)\left[A\cos(\omega t) + B\sin(\omega t)\right]$$

$$= A\left\{\cos\left[(\omega_e + \omega)t\right] + \cos\left[(\omega_e - \omega)t\right]\right\} + B\left\{\sin\left[(\omega_e + \omega)t\right] - \sin\left[(\omega_e - \omega)t\right]\right\}, \tag{4.150b}$$

and the r.h.s. of (4.149) involves (4.150b) where (II.5.88a, c) were used.

Substituting (4.150a, b) in (4.88) ≡ (4.149) gives:

$$\left[\ddot{A}+\left(\omega_0^2-\omega^2\right)A+2\omega\dot{B}\right]\cos(\omega t)+\left[\ddot{B}+\left(\omega_0^2-\omega^2\right)B-2\omega\dot{A}\right]\sin(\omega t)$$

$$=-h\,\omega_0^2 A\left\{\cos\left[(\omega_e+\omega)t\right]+\cos\left[(\omega_e-\omega)t\right]\right\} \tag{4.150c}$$

$$-h\,\omega_0^2\,B\left\{\sin\left[(\omega_e+\omega)t\right]-\sin\left[(\omega_e-\omega)t\right]\right\}.$$

There are only two amplitudes $\{A,B\}$ available to satisfy (4.150c); the coefficients of all cosines and sines with different frequencies must vanish, leading to six conditions, unless ω and $\omega_e\pm\omega$ coincide (4.151a):

$$\omega=\omega_e\pm\omega: \qquad\qquad \omega_e=0 \quad\text{or}\quad \omega_e=2\omega\sim2\omega_0, \tag{4.151a–c}$$

implying: (i) for the upper sign in (4.151a) that the excitation frequency is zero (4.151b) ≡ (4.152a) and there is no parametric resonance, only an harmonic oscillation (4.152b) with the natural frequency (4.152c):

$$\omega_e=0: \qquad\qquad m\ddot{x}+k(1+2h)x=0, \qquad \omega=\sqrt{\frac{k}{m}(1+2h)}; \tag{4.152a–c}$$

(ii) for the lower sign in (4.151a), an excitation frequency (4.151c) equal to twice the frequency ω, so that parametric resonance is strongest at the first harmonic; (iii) the frequency ω must be close to the natural frequency (4.87a) for the amplitudes (A, B) to vary, slowly leading to the cancellation of some terms of frequency ω in (4.150c); and (iv) in this case remain in (4.150c) terms with frequency $\omega_e+\omega=3\omega$ that are of higher order and hence negligible. The statements (iii) [(iv)] will be proved in detail next [subsections 4.3.13 (4.3.14)].

4.3.13 Excitation at the First Harmonic and Parametric Resonance at the Fundamental

In the case of excitation frequency close to the first harmonic (4.151c) ≡ (4.153a), the parametric resonance leads to an oscillation close to the natural frequency (4.153b):

$$\omega_e=2\omega_0+2\varepsilon: \qquad x_0(t)=A(t)\cos\left[(\omega_0+\varepsilon)t\right]+B(t)\sin\left[(\omega_0+\varepsilon)t\right], \tag{4.153a, b}$$

corresponding to (4.154a) and leading to (4.154b):

$$\omega=\omega_0+\varepsilon: \qquad \omega_0^2-\omega^2=(\omega_0-\omega)(\omega_0+\omega)=-\varepsilon(\omega+\omega_0). \tag{4.154a, b}$$

Substituting (4.153a, b; 4.154a, b) in (4.150c) gives:

$$\left\{\ddot{A}+\left[h\omega_0^2-\varepsilon(\omega+\omega_0)\right]A+2\omega\dot{B}\right\}\cos(\omega t)$$

$$+\left\{\ddot{B}-\left[h\omega_0^2+\varepsilon(\omega+\omega_0)\right]B-2\omega\dot{A}\right\}\sin(\omega t) \tag{4.155}$$

$$=-h\omega_0^2\left[A\cos(3\omega t)+B\sin(3\omega t)\right].$$

For a frequency close (4.156a) to the natural frequency (4.154a) it may be expected that: (i/ii) the first (second)-order derivatives of the amplitudes are small (4.156b) [negligible (4.156c)]:

$$\varepsilon^2 \ll \omega_0^2: \qquad \left\{\dot{A},\dot{B}\right\}\sim\varepsilon\left\{A,B\right\}, \qquad \left\{\ddot{A},\ddot{B}\right\}\sim\varepsilon^2\left\{A,B\right\}; \tag{4.156a–c}$$

and (iii) the terms on the r.h.s. of (4.155) with triple frequency 3ω are also of higher order (subsection 4.3.14) and hence (4.155) simplifies to:

$$\left[2\dot{B}+(\omega_0h-2\varepsilon)A\right]\cos(\omega t)-\left[2\dot{A}+(\omega_0h+2\varepsilon)B\right]\sin(\omega t)\sim O(h^2). \tag{4.157}$$

The vanishing of the coefficients of the terms of frequency ω on the l.h.s. of (4.157) leads to the coupled system of first-order differential equations with constant coefficients for the amplitudes:

$$2\dot{A}+(\omega_0h+2\varepsilon)B=0=2\dot{B}+(\omega_0h-2\varepsilon)A. \tag{4.158a, b}$$

Elimination between (4.158a, b) leads to a linear second-order differential equation with constant coefficients:

$$4\ddot{A}=-(\omega_0h+2\varepsilon)2\dot{B}=(\omega_0h+2\varepsilon)(\omega_0h-2\varepsilon)A=\left(\omega_0^2h^2-4\varepsilon^2\right)A, \tag{4.159}$$

that applies (4.160b) $\equiv$ (4.159) both to A and B in the form (4.160a, c):

$$b^2\equiv\left(\frac{\omega_0h}{2}\right)^2-\varepsilon^2: \qquad 0=\left\{\frac{d^2}{dt^2}+\varepsilon^2-\left(\frac{\omega_0h}{2}\right)^2\right\}A,B(t)=\left\{\frac{d^2}{dt^2}-b^2\right\}A,B(t).$$

$$\tag{4.160a–c}$$

The characteristic polynomial (4.160c) has real roots for (4.161a):

$$b^2>0: \qquad A(t),B(t)\sim\exp(bt)=\exp(\pm|b|t), \qquad \delta_0\equiv2\varepsilon_0\equiv|\omega_e-2\omega_0|<\omega_0h,$$

$$\tag{4.161a–d}$$

implying that the amplitude of oscillation (4.153b) near the natural frequency (4.154a) grows exponentially in time (4.161b, c) for excitation frequency in a narrow band (4.161d) around the first harmonic. Outside (4.162a) the frequency band (4.161d) the amplitudes are oscillatory (4.162b, c), hence have constant modulus (4.162d, e) and there is no parametric resonance:

$$|\omega_e - 2\omega_0| > \omega_0 h: \qquad A(t), B(t) \sim \exp(\pm i|b|t), \quad |A(t)| \sim const \sim |B(t)|.$$

$$(4.162a\text{–}e)$$

It has been shown that *a harmonic oscillator (4.88) with natural frequency (4.87a) whose support (problem 96) is excited with an amplitude 2h at a frequency close to the first harmonic (4.153a):*

$$\ddot{x} + \omega_0^2 x\left\{1 + 2h\cos\left[2(\omega_0 + \varepsilon)t\right]\right\} = 0, \qquad (4.163)$$

responds with an oscillation close to the natural frequency:

$$x(t) = \exp\left(t\sqrt{\left(\frac{\omega_0 h}{2}\right)^2 - \varepsilon^2}\right)\left\{C_1\cos\left[(\omega_0 + \varepsilon)t\right] + C_2\sin\left[(\omega_0 + \varepsilon)t\right]\right\}, \quad (4.164)$$

with an amplitude that: (i) oscillates in time (4.162b, c) with constant modulus (4.162d, e) outside (4.162a) the range (4.161d); (ii) within the range of frequencies of oscillation close to the first harmonic (4.161d) the amplitude increases exponentially with time (Figure 4.5a). The maximum growth (4.165c) is for excitation at twice the natural frequency (4.165a) ≡ *(4.165b):*

$$\varepsilon = 0, \quad \omega_e = 2\omega_0: \qquad x(t) = \exp\left(\frac{\omega_0 h t}{2}\right)\left[C_1\cos(\omega_0 t) + C_2\sin(\omega_0 t)\right].$$

$$(4.165a\text{–}c)$$

The result (4.164) was obtained by neglecting the r.h.s. of (4.155), on the assumption it is of higher order $O(\varepsilon^2)$; this assumption is proved next (subsection 4.3.14).

4.3.14 First- and Second-Order Parametric Resonance

The second-order approximation adds terms to (4.153b) at the triple of the fundamental frequency corresponding to the term on the r.h.s. of (4.155):

$$x_1(t) = x_0(t) + C(t)\cos\left[3(\omega_0 + \varepsilon)t\right] + D(t)\sin\left[3(\omega_0 + \varepsilon)t\right]. \qquad (4.166)$$

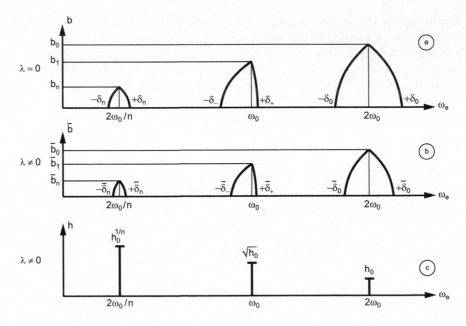

FIGURE 4.5

The parametric resonance (Figure 4.4) occurs (Figure 4.5) for excitation frequency close (a) to the first harmonic of the natural frequency $\omega_e = 2\omega_0$, and also for its submultiples $\omega_e = 2\omega_0/n = 2\omega_0, \omega_0, 2\omega_0/3,...$ with: (i) reduced amplitude; (ii) narrower range of frequencies of excitation; (iii) the range (ii) may be symmetric (or not), for example for $\omega_e = 2\omega_0 (\omega_e = \omega_0)$; and (iv) higher-order terms narrowing the range. Within each range of excitation frequency the characteristic exponent b of exponential amplification with time e^{bt} is maximum at the center of the range, decays to zero at the edges of the range (Figure 4.5a), and outside the range only oscillates with constant amplitude. In the presence of damping: (i) the range of frequencies of excitation is narrower and the peak value of the characteristic exponent smaller (Figure 4.5b); (ii) there is a threshold amplitude of excitation (Figure 4.5c) that increases for higher-order submultiples of the first harmonic, that is lower excitation frequency so that for a given excitation amplitude the parametric resonance is limited to the lowest orders.

In order to calculate the boundaries of parametric resonance the amplitudes are taken as constant when substituting (4.166; 4.153b) with (4.153a; 4.154a) in $(4.88) \equiv (4.149)$ leading to:

$$\left[\omega_0^2 - (\omega_0 + \varepsilon)^2\right]\{A \cos(\omega t) + B \sin(\omega t)\}$$

$$+\left[\omega_0^2 - 9(\omega_0 + \varepsilon)^2\right]\{C \cos(3\omega t) + D \sin(3\omega t)\}$$

$$= -2h\,\omega_0^2 \cos(2\omega t)\, x_1(t) \tag{4.167}$$

$$= -h\,\omega_0^2 \left\{A\left[\cos(3\omega t) + \cos(\omega t)\right] + B\left[\sin(3\omega t) - \sin(\omega t)\right]\right\}$$

$$-h\,\omega_0^2 \left\{C\left[\cos(5\omega t) + \cos(\omega t)\right] + D\left[\sin(5\omega t) + \sin(\omega t)\right]\right\},$$

where (II.5.88a, c) ≡ (4.117a, b) were used and: (i) the terms of frequency 5ω may be neglected because they are of higher order $O(h^3)$; (ii) the terms of frequency ω and 3ω are retained:

$$\left[-A\left(2\omega_0\varepsilon+\varepsilon^2\right)+h\omega_0^2\left(A+C\right)\right]\cos\left(\omega t\right)$$

$$+\left[-B\left(2\omega_0\varepsilon+\varepsilon^2\right)-h\omega_0^2\left(B-D\right)\right]\sin\left(\omega t\right)$$

$$+\omega_0^2\left[\left(hA-8C\right)\cos\left(3\omega t\right)+\left(hB-8D\right)\sin\left(3\omega t\right)\right] \qquad (4.168)$$

$$=-h\omega_0^2\left[C\cos\left(5\omega t\right)+D\sin\left(5\omega t\right)\right]=O\left(h^3\right).$$

Since all coefficients must vanish: (i) the amplitudes of the higher frequency 3ω terms are determined (4.169a, b) by those of lower frequency ω:

$$C=\frac{hA}{8}, \qquad D=\frac{hB}{8}; \qquad 0=\varepsilon^2+2\omega_0\varepsilon\mp h\omega_0^2-\frac{h^2\omega_0^2}{8}, \qquad (4.169\text{a--d})$$

(ii) substitution of (4.169a, b) in the coefficients in square brackets of terms of frequency ω in (4.168) leads to (4.169c, d). The latter (4.169c, d) is a binomial in ε with roots:

$$\varepsilon_\pm=-\omega_0\left[1\pm\sqrt{1\pm h+\frac{h^2}{8}}\right]$$

$$=-\omega_0\left\{1\pm\left[1+\frac{1}{2}\left(\pm h+\frac{h^2}{8}\right)-\frac{1}{8}\left(\pm h+\frac{h^2}{8}\right)^2+O\left(h^3\right)\right]\right\} \qquad (4.170\text{a, b})$$

$$=-\omega_0\left\{1\pm\left[1+\frac{h}{2}-\frac{h^2}{16}+O\left(h^3\right)\right]\right\}=-2\omega_0, \quad \frac{\omega_0 h}{2}-\frac{\omega_0 h^2}{16},$$

evaluated to $O(h^2)$ for small h. Only the positive root is relevant and specifies the range of frequencies around the first harmonic for parametric resonance:

$$\delta_1\equiv 2\varepsilon_1=2\varepsilon_-=\left|\omega_e-2\omega_0\right|<\omega_0 h-\frac{\omega_0 h^2}{8}=\omega_0 h\left(1-\frac{h}{8}\right)<\omega_0 h=2\varepsilon_0, \qquad (4.171)$$

that to second-order $O(h^2)$ is reduced relative to (4.161d) the first order $O(h)$. Thus *the excitation frequency range of parametric resonance (4.164) around the first harmonic of the natural frequency (4.165b), is reduced (problem 97) in the second (4.171) relative to the first (4.161d) order in the excitation amplitude (4.166); this is* *shown by (4.171) that agrees with (4.161d) to $O(h)$ and adds at the next order a correction $O(h^2)$ that narrows the range of excitation frequencies.*

4.3.15 Second-Order Excitation at the Natural Frequency

In the case of excitation at the natural frequency (4.172a):

$$\omega_e = \omega_0 + \varepsilon = \omega: \qquad x_2(t) = x_0(t) + E\cos\left[2(\omega_0 + \varepsilon)t\right] + F\sin\left[2(\omega_0 + \varepsilon)t\right] + G,$$

$$(4.172a, b)$$

the terms in (4.172b) are added to (4.153b). Substituting (4.172a, b; 4.153b) in $(4.88) \equiv (4.149)$ leads to:

$$\left[\omega_0^2 - (\omega_0 + \varepsilon)^2\right]\left[A\cos(\omega t) + B\sin(\omega t)\right]$$

$$+ \left[\omega_0^2 - 4(\omega_0 + \varepsilon)^2\right]\left\{E\cos(2\omega t) + F\sin(2\omega t)\right\} + \omega_0^2 G$$

$$= -2h\omega_0^2\cos(\omega t)\,x_2(t) = -h\omega_0^2\left\{A\left[1 + \cos(2\omega t)\right] + B\sin(2\omega t)\right\}$$

$$- h\omega_0^2\left\{E\left[\cos(3\omega t) + \cos(\omega t)\right] + F\left[\sin(3\omega t) + \sin(\omega t)\right]\right\} - 2hG\omega_0^2\cos(\omega t),$$

$$(4.173)$$

where the amplitudes were taken as constants to determine the frequency range of parametric resonance and (II.5.82a; II.5.63a; II.5.88a, c) were used. The terms in (4.173) may be separated into constant and with frequencies ω and 2ω, with those of frequency 3ω neglected:

$$\left(-2\varepsilon A + h\omega_0 E + 2h\omega_0 G\right)\cos(\omega t) + \left(-2\varepsilon B + h\omega_0 F\right)\sin(\omega t)$$

$$+ \omega_0\left(hA - 3E\right)\cos(2\omega t) + \omega_0\left(hB - 3F\right)\sin(2\omega t) + \omega_0\left(G + hA\right) \qquad (4.174)$$

$$= -h\omega_0^2\left[E\cos(3\omega t) + F\sin(3\omega t)\right] = O(h^3).$$

The vanishing of: (i) the constant term leads to (4.175a); (ii) the coefficients of frequency 2ω lead to (4.175b, c):

$$G = -hA, \quad E = \frac{hA}{3}, \quad F = \frac{hB}{3}: \quad \varepsilon^+ = \frac{h\omega_0 F}{2B} = \frac{h^2\omega_0}{6},$$

$$(4.175a\text{–}e)$$

$$\varepsilon^- = h\omega_0\frac{E + 2G}{2A} = -\frac{5h^2\omega_0}{6};$$

(iii) the coefficients of frequency ω lead to (4.175d, e) into which were substituted (4.175a–c), leading to:

$$\delta_+ \equiv \frac{\omega_0 h^2}{6} = \varepsilon^+ > \omega_e - \omega_0 > \varepsilon^- = -\frac{5h^2\omega_0}{6} = -\delta_- = -5\delta_+. \qquad (4.176)$$

It has been shown that *a harmonic oscillator with natural frequency (4.87a) excited with amplitude 2h at a frequency close to the natural frequency:*

$$\ddot{x} + \omega_0^2 x \left\{ 1 + 2h\cos\left[(\omega_0 + \varepsilon)t\right] \right\} = 0, \qquad (4.177)$$

responds (problem 98) with an oscillation at the same frequency with exponentially growing amplitude in the range of excitation frequencies (4.176) that: (i) is (Figure 4.5a) narrower $O\left(h^2\right)$ than for parametric resonance at the first harmonic when it is (4.161d) symmetric and broader O(h); and (ii) is unsymmetrical in the sense that its extent below the natural frequency is five times that above the natural frequency.

4.3.16 Effect of Damping on Parametric Resonance

In the case of weak damping (4.178a), the exponential growth with time e^{bt} with characteristic exponent b is modified by the damping factor $e^{-\lambda t}$ corresponding to the amplitudes (4.178b, c):

$$\lambda^2 \ll b^2: \qquad A(t) \sim \exp\left[(b - \lambda)t\right] \sim B(t). \qquad (4.178a–c)$$

Comparing (4.178b, c) with (4.161b, c), it follows that the introduction of damping is equivalent to the substitution (4.179a) in (4.160a) leading to (4.179b):

$$b \to b - \lambda: \qquad (b - \lambda)^2 = \left(\frac{\omega_0 h}{2}\right)^2 - \varepsilon_0^2. \qquad (4.179a, b)$$

The boundary of the frequency range of parametric resonance corresponds to (4.180a) leading to (4.180b):

$$b = 0: \qquad |\omega_e - 2\omega_0| < 2\varepsilon_0 = \sqrt{\omega_0^2 h^2 - 4\lambda^2} \equiv \bar{\delta}_0; \qquad h > \frac{2\lambda}{\omega_0} \equiv h_0, \qquad (4.180a–c)$$

the condition that (4.180b) is real implies (4.180c), showing that there is a minimum **threshold amplitude** of oscillation of the support for parametric resonance to occur.

It has been shown that *an harmonic oscillator (4.181) with natural frequency (4.87a) and damping λ excited with amplitude 2h at a frequency close to the first harmonic:*

$$\ddot{x} + 2\lambda\dot{x} + \omega_0^2 x \left\{ 1 + 2h\cos\left[2(\omega_0 + \varepsilon)t \right] \right\} = 0, \tag{4.181}$$

in the case of weak damping (4.182a) responds (problem 99) with an oscillation close to the natural frequency (4.182b):

$$\lambda^2 \ll \omega_0^2: \qquad x(t) = \exp\left[\left(-\lambda + \sqrt{\frac{\omega_0^2 h^2}{4} - \varepsilon^2} \right) t \right]$$

$$\left\{ C_1 \cos\left[(\omega_0 + \varepsilon)t \right] + C_2 \sin\left[(\omega_0 + \varepsilon)t \right] \right\}, \tag{4.182a, b}$$

with an amplitude that: (i) has a decay due to damping; and (ii) oscillates outside (grows exponentially inside) the range of frequencies (4.180b). Both the decay and growth are exponential functions of time, with fastest growth (4.183c) for (Figure 4.5b) excitation at the frequency of the first harmonic (4.183a, b):

$$\varepsilon = 0, \quad \omega_e = 2\omega_0: \qquad x(t) = \exp\left[\left(\frac{\omega_0 h}{2} - \lambda \right) t \right] \left\{ C_1 \cos(\omega_0 t) + C_2 \sin(\omega_0 t) \right\}. \tag{4.183a–c}$$

The oscillations occur only above (Figure 4.5c) a threshold amplitude (4.180c) that ensures that parametric resonance dominates damping in (4.182c) leading to an amplitude growing exponentially with time. The existence of a threshold amplitude is due to: (i) the excitation through the support that supplies energy to the oscillator; (ii) the damping that dissipates some of the energy; thus (iii) a minimum threshold amplitude of the vibration of the support is needed to overcome damping and excite parametric resonance. This is demonstrated next (subsection 4.3.17) by considering the energy balance. In the absence of damping there is no threshold amplitude because any energy input through the support leads to oscillations with growing amplitude and instability.

4.3.17 Energy Input through the Vibrating Support

In the presence of damping there is an extra term in (4.88) ≡ (4.149) that corresponds to (2.15) the equation of the linear damped oscillator:

$$m\ddot{x} + \mu\dot{x} + kx = -2khx\cos(\omega_e t), \tag{4.184}$$

with natural ω_0 and excitation ω_e frequencies. The **response** of the oscillator to weak damping (4.185a, b) is (4.185c):

$$b^2, \lambda^2 \ll \omega_0^2, \omega_e^2: \qquad\qquad x(t) = e^{(b-\lambda)t} A(t)\cos(\omega t - \phi); \qquad\qquad (4.185\text{a–c})$$

where the phase is constant and the amplitude a function of time. It is assumed that the amplitude is a slowly varying function of time; that is, the time scale is long compared to the period of oscillation (4.186a) so that the time derivative of (4.185c) applies (4.186c) only to the cosine term:

$$\frac{\dot{A}}{A}\langle\langle\omega\rangle\rangle b - \lambda: \qquad\qquad \dot{x}(t) = -e^{(b-\lambda)t}\omega A(t)\sin(\omega t - \phi), \qquad\qquad (4.186\text{a–c})$$

and the exponential term is also not differentiated on account of (4.186b). The exponential factor in (4.185c; 4.186c) is common to all terms of (4.184; 4.185c) and will be omitted in the sequel, for example when substituting (4.186c) in (4.184), leading to (4.187a):

$$m\ddot{x} + \mu\dot{x} + kx = -2khA\cos(\omega_e t)\cos(\omega t - \phi)$$
$$= -khA\left\{\cos\left[(\omega_e - \omega)t + \phi\right] + \cos\left[(\omega_e + \omega)t - \phi\right]\right\}, \qquad (4.187\text{a, b})$$

with (II.5.88a) $\equiv$ (4.117a) used in (4.187b). For the strongest parametric resonance to occur the excitation frequency must equal twice the natural frequency (4.174b) $\equiv$ (4.188a), leading from (4.187b) to (4.188b):

$$\omega_e - \omega = \omega: \qquad m\ddot{x} + \mu\dot{x} + kx = -khA\left[\cos(\omega t + \phi) + \cos(3\omega t - \phi)\right], \qquad (4.188\text{a, b})$$

where the oscillation at triple frequency in the second term on the r.h.s. of (4.188b) is a higher order effect discarded in the sequel.

The total energy (2.14) $\equiv$ (4.189a); that is, kinetic plus potential, has a rate of change with time (4.189b) specified by (4.189c) where (4.188b) was used with only the oscillation at frequency ω in the first term on the r.h.s.:

$$2E = m\dot{x}^2 + kx^2: \qquad \frac{dE}{dt} = \dot{x}(m\ddot{x} + kx) = -\mu\dot{x}^2 - khA\dot{x}\cos(\omega t + \phi). \qquad (4.189\text{a–c})$$

The first term on the r.h.s. of (4.189c) causes a decay of the total energy with time due to the dissipation function, whose average over a period is given (4.186c) by (4.190a–c), using (2.119b):

$$\theta = \omega t - \phi: \qquad \Psi \equiv \mu\langle\dot{x}^2\rangle = \mu\omega^2 A^2 \langle\sin^2\theta\rangle = \frac{1}{2}\mu\omega^2 A^2. \qquad (4.190\text{a–c})$$

The second term on the r.h.s. of (4.189c) is the work per unit time performed on the oscillator through the support, whose average over a period is given (4.186c) by (4.191a):

$$\dot{W} = kh\omega A^2 \left\langle \sin(\omega t - \phi)\sin(\omega t + \phi + \pi/2)\right\rangle, \qquad (4.191a)$$

and depends on the phase shift ϕ between the oscillation (4.185c) and the support (4.184).

Three extreme cases arise: (i) for zero phase shift (4.191c) no work is done on average over a period; (ii/iii) the average work over a period is maximum (minimum) for half out-of-phase lag $\phi = -\pi/4$ (lead $\phi = \pi/4$), and is positive (negative) implying (4.191b) [(4.191d)] that energy is supplied to (extracted from) the oscillator through the support:

$$2\dot{W} = \begin{cases} kh\omega A^2 \equiv \dot{W}_{max} > 0 & \text{if} \quad \phi = -\pi/4, & (4.191b) \\ 0 & \text{if} \quad \phi = 0, & (4.191c) \\ -kh\omega A^2 \equiv \dot{W}_{min} < 0 & \text{if} \quad \phi = \pi/4. & (4.191d) \end{cases}$$

The strongest parametric resonance occurs for the maximum work supplied through the support (4.191b), in which case the average over a period of the rate of change of the total energy with time (4.189c) is given by (4.192a) leading (4.190c) to (4.192b):

$$\left\langle \frac{dE}{dt}\right\rangle = \dot{W}_{max} - \Psi = \frac{\omega}{2}A^2(kh - \mu\omega). \qquad (4.192a, b)$$

Introducing the natural frequency (2.23c) $\equiv$ (4.87a) $\equiv$ (4.192a) and damping (2.23b) $\equiv$ (4.192d) leads to (4.192e) using (4.180c):

$$\frac{k}{m} = \omega_0^2 \sim \omega^2, \quad \mu = 2m\lambda: \quad 2\left\langle\frac{dE}{dt}\right\rangle = mA^2\omega_0^2(h\omega_0 - 2\lambda) = mA^2\omega_0^3(h - h_0) > 0.$$

$$(4.192c-e)$$

Thus: *(i) in the absence of damping, if the vibration of the support does maximum work it increases the energy of the oscillator, and causes parametric resonance for any excitation amplitude; (ii) in the presence of damping, the latter must be overcome by the energy input from the vibrating support, and the condition that (4.192e) be positive leads to the threshold amplitude (4.180c). These statements (i) to (ii) follow from the average over a period of the (4.192a) rate of growth with time (4.189c) of the total energy (4.189a) that equals the maximum difference between the work of excitation (4.191b) and the dissipation (4.190c).*

4.3.18 Parametric Resonance at Sub-Multiples of the First Harmonic

The parametric resonance is stronger (weaker) for excitation at twice (at) the natural frequency and the range of frequencies (4.161d) [(4.176)] is wider (narrower). Using similar methods [subsection 4.3.13 (4.3.14)] if follows that *an undamped harmonic oscillator (problem 100) with natural frequency excited at any submultiple (4.193a) of the first harmonic (4.193b) of the natural frequency (4.87a): (a) has parametric resonance (Figure 4.5a) in the frequency range (4.193c) that becomes narrower with increasing order; (ii) one-quarter of the width of the frequency range for parametric resonance specifies the characteristic exponent (4.193d) for the maximum exponential growth of the amplitude with time (4.193e):*

$$n = 0,1,2,...: \quad \omega_e = \frac{2\omega_0}{n+1}, \quad \left| \omega_e - \frac{2\omega_0}{n+1} \right| < \delta_n \sim \omega_0 h^{n+1} = 2b_n; \quad \left| x_n(t) \right| \sim \exp(b_n t);$$

$$(4.193a\text{–}e)$$

$$\left| \omega_e - \frac{2\omega_0}{n+1} \right| < \bar{\delta}_n \sim \sqrt{\omega_0^2 h^{2n+2} - 4\lambda^2} \sim 2\bar{b}_n, \quad \left| x_n(t) \right| \sim \exp\left[\left(\bar{b}_n - \lambda \right) t \right], \quad (4.194a\text{–}c)$$

in the presence (problem 101) of damping: (i) the frequency range (4.194a) of parametric resonance is narrower (Figure 4.5b); (ii) the characteristic exponent is reduced (4.194b) leading to an exponential growth of the amplitude with time (4.194c) that is slower and in addition reduced by damping; (iii) there is a minimum threshold amplitude of excitation (4.195a) that increases (Figure 4.5c) with the order (4.195b); thus (iv) modes beyond a certain order k are suppressed (4.195c) by damping:

$$h_n \sim O\left(h_0^{1/n} \right) \sim \sqrt[n]{\frac{2\lambda}{\omega_0}}: \quad 1 > \frac{2\lambda}{\omega_0} = h_0 < h_1 \sim h_0^{1/2} < h_2 \sim h_0^{1/3} < .... < h_n \sim h_0^{1/(n+1)},$$

$$(4.195a, b)$$

$$h_0 < h_1 < h_2 < .... < h_k < h < h_{k+1} < .... < h_n < ..... \qquad (4.195c)$$

The two stronger cases of parametric resonance with excitation at twice the (at the) natural frequency, and the general case of a sub-multiple, are summarized in the Table 4.2 and illustrated in the Figure 4.5 concerning: (i) the frequency range and peak characteristic exponent without (with) damping [Figure 4.5a(b)]; and (iii) the threshold excitation amplitude with damping (Figure 4.5c). The preceding results on parametric resonance (subsections 4.3.12–4.3.18) are illustrated next for a stable suspended pendulum (Figure 4.6) that may be rendered unstable by parametric resonance induced by oscillating vertically its support so as to supply energy to the oscillation (subsection 4.3.19).

TABLE 4.2

Parametric Resonance

Excitation Frequency	First Harmonic $\omega_e = 2\omega_0$	Natural Frequency $\omega_e = \omega_0$	Submultiples of First Harmonic $\omega_e = \dfrac{2\omega_0}{n+1}$
Subsection (s)	4.3.13–4.3.14	4.3.15–4.3.17	4.3.19
without damping frequency range	$\left|\omega_e - 2\omega_0\right| < \delta_0 = \omega_0 h\left(1 - \dfrac{h}{8}\right)$	$\omega_0 - \delta_- < \omega_e < \omega_0 + \delta_+$ $\delta_- = \dfrac{5\omega_0 h}{6},\ \dfrac{\omega_0 h}{6} = \delta_+$	$\left|\omega_e - \dfrac{2\omega_0}{n+1}\right| < \delta_n$ $\delta_n \sim O\!\left(\omega_0 h^{n+1}\right)$
maximum of characteristic exponent	$b_0 = \dfrac{\omega_0 h}{2}$	$2b_1 \sim O(\omega_0 h)$	$2b_n \sim O\!\left(\omega_0 h^{n+1}\right)$
with damping frequency range	$\bar\delta_0 = \sqrt{\omega_0^2 h^2 - 4\lambda^2}$	$\bar\delta_1 = \sqrt{\omega_0^2 h^4 - 4\lambda^2}$	$\bar\delta_n \sim \sqrt{\omega_0^2 h^{2n+2} - 4\lambda^2}$
maximum of characteristic exponent	$\bar b_0 = \dfrac{\omega_0 h}{2} - \lambda$	$\bar b_1 \sim \omega_0 h^2 - \lambda$	$\bar b_n \sim \omega_0 h^{n+1} - \lambda$
threshold amplitude	$h_0 > \dfrac{2\lambda}{\omega_0}$	$h_1 > \sqrt{\dfrac{2\lambda}{\omega_0}}$	$h_n > \left(\dfrac{2\lambda}{\omega_0}\right)^{1/(n+1)}$

Note: The parametric resonance occurs for a linear oscillator (damped or undamped) with oscillating support excited at the frequency of: (i) the first harmonic, that is, twice the natural frequency; (ii–iii) any submultiples, either (ii) the natural frequency or (iii) lower, with reduced magnitude.

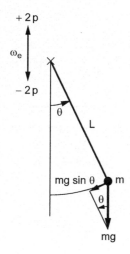

FIGURE 4.6
A simple pendulum whose support oscillates vertically (Figure 4.5) is an example (Figure 4.4) of parametric resonance. The energy supplied by the oscillation of the support may (or may not) destabilize the pendulum depending on the excitation frequency being close to twice the natural frequency or a submultiple (or not).

4.3.19 Pendulum with Support Oscillating Vertically

The oscillations of a pendulum of length L suspended in the gravity field are specified by (2.71b) ≡ (4.196a):

$$L\ddot{\theta} = -(g - \ddot{x})\sin\theta, \qquad x(t) = 2p\cos\left[2(\omega_0 + \varepsilon)t\right], \qquad \text{(4.196a, b)}$$

where (Figure 4.6) the acceleration $\ddot{x}$ of the support is subtracted from the acceleration of gravity g; the support oscillates (4.196b) with amplitude $2p$ at an excitation frequency (4.153a) close to the first harmonic of the natural frequency (2.72a) ≡ (4.197a):

$$\omega_0 = \sqrt{g/L}; \qquad \ddot{x}(t) = -8p\omega_0^2\cos\left[2(\omega_0 + \varepsilon)t\right]; \qquad \text{(4.197a, b)}$$

from the displacement (4.196b) follows the acceleration (4.197b) of the support, that substituted in (4.196a) leads to (4.198b) by linearization for small amplitude oscillations (4.198a):

$$\theta^2, p^2, p\theta \ll 1: \qquad \ddot{\theta} + \omega_0^2\theta = -\frac{8p}{L}\omega_0^2\theta\cos\left[2(\omega_0 + \varepsilon)t\right]. \qquad \text{(4.198a, b)}$$

The latter (4.198b) ≡ (4.149) is the equation of the harmonic parametric oscillator with excitation amplitude (4.199a) and frequency (4.199b):

$$h = \frac{4p}{L}: \qquad \omega_e = 2(\omega_0 + \varepsilon) = 2\sqrt{\frac{g}{L}} + 2\varepsilon, \qquad \text{(4.199a, b)}$$

leading to parametric resonance (4.171):

$$\left|\omega_e - 2\omega_0\right| = \left|\omega_e - 2\sqrt{\frac{g}{L}}\right| < \varepsilon < \omega_0 \frac{4p}{L}\left(1 - \frac{p}{2L}\right) = 4p\sqrt{\frac{g}{L^3}} - 2p^2\sqrt{\frac{g}{L^5}}, \qquad (4.200)$$

in the range (4.187).

Thus *a pendulum (Figure 4.6) with length L suspended in a uniform gravity field g with (problem 102) support vibrating vertically with amplitude 2p at an excitation frequency (4.199b) close to the first harmonic of the natural frequency (4.197a) for small oscillations (4.198a, b)* ≡ *(4.201a, b)*:

$$\theta^2 \ll 1: \qquad \qquad \ddot{\theta} + \frac{g}{L}\theta = -\frac{8pg}{L^2}\theta\cos\left[2\left(\sqrt{\frac{g}{\ell}} + \varepsilon\right)t\right], \qquad (4.201a, b)$$

responds (4.164; 4.199a; 4.197a) with an oscillation (4.202) at a frequency close to the natural frequency:

$$\theta(t) = \exp\left(t\sqrt{\frac{4p^2 g}{L^3} - \varepsilon^2}\right)\left\{C_1\cos\left[\left(\sqrt{\frac{g}{L}} + \varepsilon\right)t\right] + C_2\sin\left[\left(\sqrt{\frac{g}{L}} + \varepsilon\right)t\right]\right\}, \quad (4.202)$$

with an amplitude that oscillates outside (grows exponentially with time inside) the frequency range (4.200). The fastest amplitude growth (4.203b) occurs for excitation precisely at frequency of the first harmonic (4.203a):

$$\varepsilon = 0: \qquad \theta(t) = \exp\left(2pt\sqrt{\frac{g}{L^3}}\right)\left\{C_1\cos\left(t\sqrt{\frac{g}{L}}\right) + C_2\sin\left(t\sqrt{\frac{g}{L}}\right)\right\}. \quad (4.203a, b)$$

In the presence (problem 103) of linear damping with small amplitude (4.204a), the equation of motion (4.201b) adds a friction force leading to (4.204b):

$$\theta^2 \ll 1: \qquad \ddot{\theta} + 2\lambda\dot{\theta} + \frac{g}{L}\theta\left\{1 + 8p\cos\left[\left(2\sqrt{\frac{g}{L}} + \varepsilon\right)t\right]\right\} = 0; \qquad (4.204a, b)$$

for weak damping (4.205a) there is a decay (4.205c) that is dominated by the parametric resonance for the threshold amplitude (4.180c) = (4.205b) leading to exponential amplitude growth with time:

$$\lambda^2 \ll \omega_0^2 = \frac{g}{L}: \qquad \qquad p > \frac{h_0 L}{4} = \frac{\lambda L}{2\omega_0} = \frac{\lambda}{2}\sqrt{\frac{L^3}{g}}, \qquad (4.205a, b)$$

$$\theta(t) = \exp\left[t\left(2p\sqrt{\frac{g}{L^3}} - \lambda\right)\right]\left[C_1 \cos\left(t\sqrt{\frac{g}{L}}\right) + C_2 \sin\left(t\sqrt{\frac{g}{L}}\right)\right]. \quad (4.205c)$$

The parametric resonance also occurs for (problem 104) excitation frequencies close to submultiples of the first harmonic of the natural frequency (4.206a) in the frequency range (4.206b) that also specifies the characteristic exponent (4.206c) of the amplitude growth with time (4.206d):

$$\omega_e = \frac{2\omega_0}{n} = \frac{2}{n}\sqrt{\frac{g}{L}}: \quad \left|\omega_e - \frac{2}{n}\sqrt{\frac{g}{L}}\right| < O\left(\sqrt{\frac{g}{L}\left(\frac{4p}{L}\right)^{2n} - 4\lambda^2}\right) \sim 2\bar{b}_n, \quad (4.206a\text{--}c)$$

$$|x_n(t)| \sim \exp\left[(\bar{b}_n - \lambda)t\right]: \quad p_n = \frac{h_n L}{4} > O\left(\frac{L}{4}h_0^{1/(n+1)}\right) \sim \frac{L}{4}\left(\frac{2\lambda}{\omega_0}\right)^{1/(n+1)}$$

$$\sim 2^{-(2n+1)/(n+1)}\left(\frac{\lambda}{\sqrt{g}}\right)^{1/(n+1)} L^{(3+2n)/(2+2n)}. \quad (4.206\text{d, e})$$

The damping also introduces a threshold amplitude (4.195a) for the oscillation (4.206e). The case of the pendulum illustrates the differences between ordinary and parametric resonance (subsection 4.3.20).

4.3.20 Comparison between Ordinary and Parametric Resonance

The comparison of the linear damped harmonic oscillator in the cases of parametric (4.149) ≡ (4.207b) [ordinary (2.192a) ≡ (4.207a)] resonance:

$$\ddot{x} + 2\lambda\dot{x} + \omega_0^2 x = f_a e^{i\omega_a t}, \quad -2h\omega_0^2 x \cos\left[2(\omega_0 + \varepsilon)\frac{t}{n}\right], \quad (4.207a\text{, b})$$

shows [section(s) 4.3 (2.6–2.7)] a set (Table 4.3) of 12 interrelated differences: (i) it is due to oscillation of the support (an applied force) as an excitation (forcing) mechanism; (ii) for small amplitude, one-dimensional oscillations it leads to a linear second-order differential equation with periodic (constant) coefficients and without (with) forcing; (iii) the oscillations are dependent (independent) of the initial conditions in the sense that they occur in all cases except initial displacement and velocity both zero (occur without any exceptions); (iv) the parametric resonance occurs for excitation frequency close to any submultiple of the first

TABLE 4.3

Twelve Differences between Ordinary and Parametric Resonance

		Resonance	
Number Details in	Criterion Section (s)	Ordinary 2.6–2.7	Parametric 4.3
1	cause	external applied force	vibration of the support
2	differential equation for small oscillations	linear, forced with constant coefficients	linear, unforced with periodic coefficients
3	initial conditions	independent: not needed	dependent: initial perturbation needed
4	excitation frequency	natural frequency only	all submultiples of the first harmonic
5	frequency near resonance	beats not resonance	resonance (near) or oscillation (far)
6	damping	reduces the maximum amplitude	narrows frequency range of parametric resonance
7	damping	no threshold amplitude	introduces a threshold amplitude
8	no damping: amplitude grows with time	linearly	exponentially
9	damping: amplitude	constant	grows exponentially more slowly
10	dissipation	balances work of applied force	is less than the work input through vibration of the support
11	no damping: phase difference	out-of-phase lead $\phi = \dfrac{\pi}{2}$	half out-of-phase lag $\phi = -\pi/4$
12	with damping phase difference	intermediate phase lead $0 < \phi < \dfrac{\pi}{2}$	half out-of-phase lag: $\phi = -\pi/4$

Note: Comparison according to 12 criteria of the ordinary (parametric) resonance for a linear damped or undamped oscillator, due to forcing (oscillation of the support) at an applied (excitation) frequency equal to the natural frequency (any submultiple of the frequency of the first harmonic).

harmonic of the natural frequency (only at the natural frequency); (v) the frequency range for parametric resonance is narrower for higher orders (for applied frequency close to the natural frequency forcing leads to beats not resonance); (vi) damping further limits the range of frequencies for parametric resonance (does not change the natural frequency as the resonant frequency but reduces the maximum amplitude); (vii) damping causes a threshold amplitude for parametric resonance that increases at higher orders (there is no threshold amplitude); (viii) in the absence of damping the amplitude grows exponentially (linearly) with time; (ix) in the presence of damping the amplitude still grows exponentially with time beyond the threshold of excitation (is always constant); (x) in the presence of damping the dissipation is dominated by energy input beyond the threshold input amplitude (is exactly balanced by the work of the applied force); (xi-xii) the excitation (forcing) and oscillation are half out-of-phase lag $\phi = -\pi/4$ with or without damping (are lead out-of-phase $\phi = \pi/2$ in the absence of damping and have an intermediate phase $0 < \phi < \pi/2$ in the presence of damping). A third type of resonance, besides ordinary (sections 2.7–2.8) and parametric (section 4.3) is non-linear resonance (section 4.6) for large amplitude oscillations (sections 4.4–4.5).

4.4 Anharmonic Oscillator with Soft or Hard Springs

After linear systems with constant (sections 4.1–4.2) and time dependent (section 4.3) coefficients, are considered non-linear systems (sections 4.4–4.9), all of second-order (subsection 4.4.1). Starting with a non-linear restoring force (subsections 4.4.2) in the absence of damping (subsection 4.4.3), two types of motion are possible, namely a limit cycle oscillation (an open trajectory) around a center (from a saddle point) which is a point of stable (unstable) equilibrium. In the case of a non-linear (subsection 4.4.4) hard (soft) spring [subsections 4.4.5 (4.4.7)] spring, that is with a positive (negative) cubic term in the restoring force, only the center (also two saddle points) exist (subsection 4.4.6). The trajectory both in the cases of soft or hard springs can be specified: (a) exactly in terms a Jacobian elliptic function (subsection 4.4.9); (b) also exactly as a series of circular sines (subsection 4.4.10) or alternatively Chebychev polynomials of the second kind (subsection 4.4.12); (c) approximately as an oscillation at the natural frequency plus harmonics (subsection 4.4.15). The three solutions (a–c) demonstrate the effects of non-linearity on: (i) the ratio of the maximum velocity to the displacement (subsection 4.4.8); (ii) the ratio of the frequency of oscillation of the anharmonic oscillator to the natural frequency of the harmonic oscillator (subsection 4.4.11); (iii) the lowest-order non-linear corrections to the frequency (subsection 4.4.13) and the displacement (subsection 4.4.14) of the fundamental mode; (iv) the generation of the lowest-order harmonic (subsection 4.4.15).

4.4.1 Linear/Non-Linear Time Dependent/Independent Systems

The simplest second-order system (chapter 2; sections 4.1–4.2) is linear with constant coefficients:

$$\ddot{x} + 2\lambda\dot{x} + \omega_0^2 x = f(t), \tag{4.208a}$$

where the forcing is an arbitrary function of time. For a linear, time-dependent system the coefficients are the damping and frequency that may depend on time:

$$\ddot{x} + 2\lambda(t)\dot{x} + \left[\omega(t)\right]^2 x = f(t); \tag{4.208b}$$

an example is parametric resonance (section 4.3), for which the damping was constant but not the frequency. A non-linear system has non-linear damping (subsections 4.4.7–4.4.8) and/or restoring (subsections 4.4.4–4.4.6) forces:

$$\ddot{x} + h(\dot{x}) - j(x) = f(t). \tag{4.208c}$$

The combination of (4.208b) and (4.208c) is the unsteady and non-linear damping:

$$\ddot{x} + h(\dot{x},t) - j(x,t) = f(t). \tag{4.208d}$$

This is a particular case of the general non-linear second-order differential equation solvable for the acceleration:

$$\ddot{x} = F(\dot{x}, xt). \tag{4.209}$$

The non-linear oscillator (4.208c) will be considered in the sequence: (i) no forcing or damping (sections 4.4–4.5); (ii) forcing (section 4.6); (iii) damping (section 4.7); and (iv) more general cases (sections 4.8–4.9).

4.4.2 Non-Linear Restoring Force without Damping or Forcing

The study of non-linear second-order systems starts with a non-linear restoring force (sections 4.4–4.5), subsequently considering forcing (section 4.6) and non-linear damping or amplification (sections 4.7–4.8). An example is the one-dimensional motion of a particle with constant mass (4.210a) under a force that depends only on position. The non-linear restoring force (4.210b) derives from a potential:

$$m = \text{const}: \quad m\ddot{x} = h(x) = -\frac{d\Phi}{dx}, \qquad E = \frac{m}{2}\dot{x} + \Phi(x) = const, \tag{4.210a–c}$$

and appears in the total energy (4.210c), which is conserved, because (4.210a, b) implies (4.211):

$$0 = \dot{x}\left[m\ddot{x} - h(x)\right] = m\dot{x}\ddot{x} + m\frac{dx}{dt}\frac{d\Phi}{dx} = \frac{d}{dt}\left[\frac{m}{2}\dot{x}^2 + \Phi(x)\right]. \qquad (4.211)$$

Thus *for (problem 105) a non-linear second-order system with constant mass (4.210a) without damping or forcing (4.210b), the path in the phase plane of position and velocity is specified by the potential in the total energy (4.210c) that is conserved (4.211); the total energy is a **first integral or conserved quantity** that is constant during the motion and involves only the velocity, and thus supplies a first-order differential equation for the trajectory (4.210c) instead of a second-order differential equation for the Newton law (4.210b) stating the balance of forces. The constant of integration passing from (4.210b) to (4.210c) is the total energy. The trajectory in physical space is specified in inverse form, with time as a function of position, by (4.212b):*

$$x_0 \equiv x(t_0): \qquad t - t_0 = \int_{x_0}^{x}\frac{dx}{\dot{x}} = \int_{x_0}^{x}\left|\frac{2}{m}\left[E - \Phi(\xi)\right]\right|^{-1/2} d\xi, \qquad (4.212a, b)$$

taking (4.212a) the position $x = x_0$ at time $t = t_0$. In (4.212b) there are two constants of integration corresponding to the second-order differential equation (4.210b), namely: (i) the total energy (4.210c) that is conserved (4.211); and (ii) the position x_0 at (4.212a) a time t_0. The result (4.212a, b) follows solving (4.210c) for the velocity $\dot{x}$.

4.4.3 Separatrices of Limit Cycles and Open Paths

Since the kinetic energy is always positive $\dot{x}^2 > 0$, it follows from (4.210c) that *for an undamped, unforced non-linear oscillator (problem 106): (i) the motion is possible where the potential energy does not exceed the total energy $\Phi(x) \le E$, and their difference determines the velocity $\dot{x}$; (ii) the velocity is zero $\dot{x} = 0$ at the **stopping points** $a = 0$ where the potential equals the total energy $\Phi(a) = E$, that is, these points are the boundaries of the motion; (iii) a boundary of the motion is an equilibrium point if the acceleration is zero (4.210b), that is, the restoring force is zero and the potential is stationary; (iv) the equilibrium will be stable if the potential has a minimum; (v) the closed (open) paths or **limit cycles (unstable regimes)** are separated by **saddle points**; (iv) through a saddle point pass two **separatrices** having limit cycles or unstable regimes in alternating sectors.* The application of these remarks is illustrated by the potential in Figure 4.7a and the corresponding paths in the phase plane of Figure 4.7b: (i) the energy E_1 equals the potential energy only at the point a, and thus the path is open and extends to infinity from $x \ge a$; (ii) reducing the energy to E_2 leads to an unstable equilibrium point f

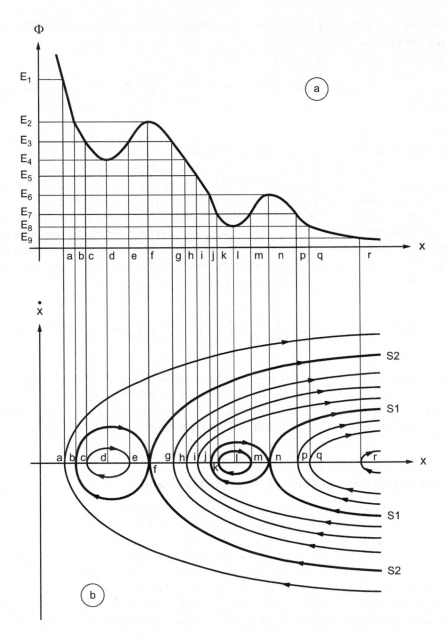

FIGURE 4.7

The total energy of a conservative system is (a) the sum of the kinetic and potential energies, and thus the velocity is zero (the motion is possible only) where the potential energy equals (is less than) the total energy (b), leading to three cases: (i) in (a) a **potential well** like E_3 and E_7 the trajectory (b) is closed that is a **periodic cycle** respectively (c, e) and (k, m); (ii) outside a potential well (a) like E_1 and E_9 the trajectories (b) are open, respectively a and s; and (iii) the **separatrices** between (i) and (ii) such (b) as S_1 and S_2 correspond to (a) maxima of the potential energy respectively E_2 and E_6, and are like (ii) open, that is, semi-infinite trajectories.

where the potential energy has a local maximum; (iii) the equilibrium point *f* is unstable because further reduction of energy to E_3 leads either to an open path for $x \geq f$ or closed path for $c \leq x \leq e$, neither of which returns to $x = f$; (iv) $x = f$ is a saddle point, separating unstable motion for $x > f$ from a limit cycle oscillation for $x < f$; (v) the limit cycle oscillation occurs around a center *d*, that is a point of stable equilibrium, where the energy is E_4 and the potential energy has a local minimum; (vi) for lower energy E_5 there is an open path starting at $x \geq i$; (vii) the lower energy E_6 corresponds to another saddle point $x = j$; (viii) the lower energy E_7 corresponds to a limit cycle $k \leq x \leq m$ and an open path $x \geq p$; (ix) the center $x = \ell$ of the limit cycle is a point of stable equilibrium corresponding to the energy E_8 and another local minimum of the potential energy; (x) the energy E_8 also corresponds to an open path $x \geq q$; (xi) the lower energy E_9 corresponds to an open path $x \geq r$. *The curves S1 and S2 in the Figure 4.7b denote paths through the saddle points that are the* **separatrices** *having in alternating sectors closed and open paths.* This qualitative picture of paths in the phase plane (Figure 4.7) for a general potential (subsection 4.4.3) relates to the **qualitative or topological theory of differential equations**; it is followed by quantitative analysis (subsections 4.4.4–4.4.15; section 4.5) for a bi-quadratic potential.

4.4.4 Non-Linear Soft and Hard Spring

Expanding the restoring force in (4.210a) in powers of the distance leads to (4.213a) and to the corresponding potential (4.213b):

$$m\dot{x} = j(x) = -k_0 - kx - k_2 x^2 - k\beta x^3 + O(x^4) \equiv -\frac{d\Phi}{dx}, \quad (4.213a)$$

$$\Phi(x) = \Phi(0) + k_0 x + \frac{k}{2}x^2 + \frac{k_2}{3}x^3 + \frac{k\beta}{4}x^4 + O(x^5), \quad (4.213b)$$

where: (i) since the potential is defined within an added constant that does not affect the force the first term on the r.h.s. of (4.213b) may be omitted (4.214a), (ii) the origin is an equilibrium point if the restoring force is zero, that is the potential is stationary (4.214b); (iii) the equilibrium position is stable if the restoring force points toward the equilibrium, that is, the potential has a minimum (4.214c), and thus the resilience *k* is positive:

$$\Phi(0) = 0, \quad -j(0) = \Phi'(0) = k_0 = 0, \quad k = -j'(0) = \Phi''(0) > 0, \quad 0 = k_2 = k_4 = \ldots = k_{2n};$$

$$(4.214a\text{–}d)$$

(iv) a restoring force must always point toward the equilibrium point at the origin, and thus have the opposite sign to the displacement, that is, *h* changes to $-h$ when *x* changes to $-x$, thus excluding even powers (4.214d) and allowing only odd powers in (4.215a); (v) since the potential is the integral of the restoring force, odd powers are excluded and only even powers appear in (4.215b):

$$m\ddot{x} = j(x) = -kx\left[1 + \beta x^2 + O\left(x^4\right)\right] = -j(-x),$$

$$\Phi(x) = \frac{k}{2}x^2 + \frac{k\beta}{4}x^4 + O\left(x^6\right) = \Phi(-x); \qquad \text{(4.215a, b)}$$

and (vi) the terms omitted are one (two) orders higher than the last retained in (4.213a, b) [(4.215a, b)]. Thus, *the undamped unforced non-linear oscillator is specified to lowest order by (4.216b):*

$$k > 0: \qquad m\ddot{x} = j(x) = -kx\left(1 + \beta x^2\right) = -\frac{d\Phi}{dx}, \qquad \text{(4.216a, b)}$$

where: (i) the resilience of the spring is positive (4.216a) ≡ (4.214a) to ensure stability of the equilibrium position at the origin; (ii) the quadratic coefficient β may be positive β > 0 (or negative β < 0) corresponding (Figure 4.8a) to a **hard (soft)spring** *for which the restoring force is larger (smaller) than for a linear spring as the displacement increases. For example (chapter I.8) the simple pendulum (Figure 4.5) corresponds [(2.71b) with upper + sign, or (4.198b) with p = 0] to (4.217a):*

$$\ddot{\theta} = -\omega_0^2 \sin\theta = -\omega_0^2\left[\theta - \frac{\theta^3}{6} + O\left(\theta^5\right)\right], \qquad \beta = -\frac{1}{6}, \qquad \text{(4.217a–c)}$$

to a soft spring (4.215a) ≡ (4.217b) with **non-linearity parameter** *(4.217e).*

4.4.5 Period of Limit Cycle Oscillation of Hard Spring

The hard spring (Figure 4.8b) can have only limit cycle oscillations (Figure 4.8c) around the center that is the point of stable equilibrium at the origin. From (4.210c, 4.215b) follows *the total energy (4.218b) of the oscillator with (problem 107) a hard spring (4.218a):*

$$\beta = |\beta| > 0: \quad 2E = m\dot{x}^2 + kx^2\left[1 + \frac{\beta}{2}x^2\right]; \quad \dot{x}(a) = 0: \quad k\beta a^4 + 2ka^2 - 4E = 0,$$

$$\text{(4.218a–d)}$$

the amplitude (4.218d) ≡ (4.219b) corresponds the points of zero velocity (4.218c) ≡ (4.219a):

$$\dot{x}(a) = 0: \qquad a_\pm^2 = \frac{1}{|\beta|}\left(-1 \pm \sqrt{1 + \frac{4E|\beta|}{k}}\right), \qquad a_-^2 < 0 < a_+^2. \qquad \text{(4.219a–c)}$$

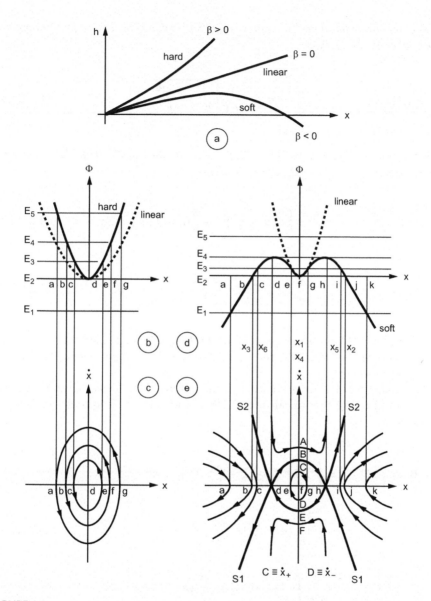

FIGURE 4.8

An example of non-linear restoring force (Figure 4.7) is a hard (soft) spring that (a) adds to (subtracts from) the linear restoring force of a harmonic oscillator, a cubic term leading to an anharmonic oscillator. The corresponding potential adds to (b) [subtracts from (d)] the quadratic term of a harmonic oscillator a quartic term corresponding to the anharmonic oscillator. For the hard spring the potential has a minimum at the origin (b) that is the center (Figure 4.1f) of closed periodic paths (c). For the soft spring the potential has (d) a minimum at the origin (two side maxima) corresponding (Figure 4.1f(e)) to a center (two saddle points) with (e) closed periodic paths in the potential well between the separatrices S_1 and S_2 and four families of open semi-infinite paths outside. The effect of cubic (quadratic) terms in the restoring force that is quartic (cubic) terms in the potential is shown in Figure 4.9.

A real amplitude has a positive square (4.219c) implying the choice (4.220b) for the range of motion (4.220a):

$$-a \leq x(t) \leq +a: \qquad a \equiv |a_+| \equiv \frac{1}{\sqrt{|\beta|}} \left| \sqrt{1 + \frac{4E|\beta|}{k}} - 1 \right|^{1/2}. \qquad (4.220a, b)$$

The integral (4.212b) with (4.215b; 4.218d) specifies the trajectory (4.221b) in inverse form t(x) with time zero at the origin (4.221a):

$$x(0) = 0: \qquad \omega_0 t = \int_0^x \left| (a^2 - \xi^2)\left[1 + \frac{\beta}{2}(a^2 + \xi^2)\right]\right|^{-1/2} d\xi; \qquad (4.221a, b)$$

$$\omega_0 \equiv \sqrt{\frac{k}{m}}: \qquad \frac{\tau}{4} = \int_0^a \left|\frac{2}{m}\left(E - \frac{k}{2}\xi^2 - \frac{k\beta}{4}\xi^4\right)\right|^{-1/2} d\xi, \qquad (4.221c, d)$$

the time to travel from the origin (4.221a) to the amplitude (4.220b) is (4.221b) one quarter of the period (4.221d), where (4.221c) the natural frequency for the linear or harmonic oscillator.

The trajectory (4.212b) in the case of the potential (4.215b) leading to (4.221b) is obtained using the amplitude (4.218d) and natural frequency (4.221c) as follows:

$$t_0 = 0 = x_0: \qquad t = \int_0^x \left|\frac{2}{m}\left(E - \frac{k}{2}\xi^2 - \frac{k\beta}{4}\xi^4\right)\right|^{-1/2} d\xi$$

$$= \int_0^x \left|\frac{k}{m}(a^2 - \xi^2) + \frac{k\beta}{2m}(a^4 - \xi^4)\right|^{-1/2} d\xi \qquad (4.222a-c)$$

$$= \frac{1}{\omega_0} \int_0^x \left|(a^2 - \xi^2)\left[1 + \frac{\beta}{2}(a^2 + \xi^2)\right]\right|^{-1/2} d\xi,$$

where: (i) the initial condition (4.212a) is (4.222a, b); (ii) a simple re-arrangement of (4.222c) leads to (4.221b). In the case of the linear oscillator (4.223a), the integral in (4.221b) ≡ (4.222c) is elementary (II.7.122a) ≡ (4.223b):

$$\beta = 0: \qquad \omega_0 t = \int_0^x \frac{d\xi}{\sqrt{a^2 - \xi^2}} = \arcsin\left(\frac{x}{a}\right), \qquad (4.223a, b)$$

leading to the displacement (4.223c) and velocity (4.223d):

$$x(t) = a\sin(\omega_0 t) = a\cos\left(\omega_0 t - \frac{\pi}{2}\right), \qquad \dot{x}(t) = a\cos(\omega_0 t) = a\sin\left(\omega_0 t + \frac{\pi}{2}\right).$$

$$(4.223\text{c, d})$$

The quarter period (4.221d) is given by (4.223a) with $x = a$ leading to (4.224a):

$$\tau = \frac{4}{\omega_0}\int_0^a \frac{dx}{\sqrt{a^2 - x^2}} = \frac{4}{\omega_0}arc\sin 1 = \frac{2\pi}{\omega_0}, \qquad x(t) = a\sin\left(\frac{2\pi t}{\tau}\right), \quad (4.224\text{a, b})$$

as confirmed by (4.224b). The result (4.223b) ≡ (4.223c) ≡ (4.224b) is equivalent to (2.56b) with the phase $\alpha = \pi/2$ such that at time $t = 0$ the harmonic oscillator is at the origin $x_0 = 0$. The present method thus: (i) leads to the same result as using the characteristic polynomial (section 2.2) for the harmonic oscillator that is specified by a linear second-order differential equation with constant coefficients; (ii) extends (subsections 4.4.6–4.4.15) to the anharmonic oscillator specified by a non-linear second-order differential equation (4.215a) to which the method of characteristic polynomials does not apply.

4.4.6 Closed (and Open) Trajectories for the Hard (Soft) Spring

The non-linear hard (soft) spring (Figure 4.8a) are compared next as regards the potential [Figure 4.8b(d)] and the paths in the phase plane [Figure 4.8c(e)]. The potential of the hard spring is larger than for the linear spring (Figure 4.7b), hence always positive, has an absolute minimum at the origin, so that only closed paths (Figure 4.7c) are possible: (i) no path is possible for a negative energy $E_1 < 0$; (ii) a zero energy $E_2 = 0$ corresponds to the position d of stable equilibrium; and (iii) increasing positive energies $0 < E_2 < E_3 < E_4$ lead to closed paths with larger amplitudes (c, e), (b, f) and (a, g). The soft spring changes the stable minimum at the origin from absolute to local (Figure 4.8d) and adds two symmetric absolute maxima that are unstable saddle points: (i) the energy $E_4 > 0$ corresponding to the saddle points d and h corresponds to separatrices $S1$ and $S2$ with closed (open) paths inside (outside); (ii) a larger positive energy $E_5 > E_4 > 0$ leads to open paths that pass through the origin with velocities A and B and do not stop anywhere; (iii) a smaller E_3 positive energy $E_4 > E_3 > 0$ leads to a closed path (e, g) inside the separatrices and two open paths c and i outside the separatrices; (iv) zero energy $E_2 = 0$ leads to stable equilibrium at the origin f and two open paths b and j outside the separatrices; and (v) a negative energy $E_1 < 0$, no matter how large in modulus, leads to two open paths with stopping points a and k progressively farther from the origin. Thus, whereas the hard spring implies always oscillatory motion, the soft spring allows both closed and open trajectories. The amplitude(s) is (are) specified by the roots of the biquadratic equation (4.218d) both for the hard (4.218a; 4.219b) [soft (4.225a; 4.225b)] spring and only one (4.219c) [both (4.225c)] root(s) is (are) real:

$$0 \le E \le \frac{k}{4|\beta|}: \qquad a_\pm^2 = \frac{1}{|\beta|}\left\{1 \mp \sqrt{1 - \frac{4E|\beta|}{k}}\right\}, \qquad a_\pm^2 > 0; \qquad (4.225a\text{–}c)$$

the condition (4.225a) corresponds to the energies (4.225d) in the Figure 4.8c, with the smaller (4.225e) [larger (4.225f)] amplitude corresponding to closed, periodic motion (open trajectory)

$$0 = E_2 \le E_3 \le E_4 = \frac{k}{4|\beta|}: \qquad |x(t)| \le a_-, \qquad |x(t)| \ge a_+; \qquad (4.225d\text{–}f)$$

for larger energies $E > E_4$ in Figure 4.8e there is no stopping point, so (4.219a) is impossible; for negative energies $E < E_2 = 0$ in Figure 4.8e, only the root a_+ in (4.225b) is real, and thus only the open trajectories (4.225f) exist.

4.4.7 Saddle Points for a Soft Spring

In the case (4.226a) of the soft spring (Figure 4.8d), the potential (4.215b) $\equiv$ (4.226b) vanishes (4.226c) at the points (4.226d):

$$0 > \beta = -|\beta|; \quad \Phi(x) = \frac{1}{2}kx^2\left(1 - \frac{|\beta|}{2}x^2\right), \quad \Phi(x_{1-3}) = 0: \qquad x_{1-3} = \left\{0, \pm\sqrt{\frac{2}{|\beta|}}\right\}.$$

$$(4.226a\text{–}d)$$

The restoring force (4.215a) $\equiv$ (4.227a) vanishes (4.227b) $\equiv$ (4.227c) at the equilibrium points (4.227d) $\equiv$ (4.227e):

$$0 = j(x) = -\Phi'(x) = -kx(1 - |\beta|x); \quad j(x_{4-6}) = 0: \quad x_{4-6} = \left\{0, \pm\frac{1}{\sqrt{|\beta|}}\right\} = \frac{x_{1-3}}{\sqrt{2}}.$$

$$(4.227a\text{–}e)$$

The second-order derivative (4.228a) of the potential shows that it is minimum (maximum) at x_4 (x_{5-6}); that is, the position of stable (4.228b) [unstable (4.228c)] equilibrium:

$$\Phi''(x) = k(1 - 3|\beta|x^2): \qquad \Phi''(x_4) = \Phi''(0) = k > 0,$$

$$(4.228a\text{–}c)$$

$$\Phi''(x_{5,6}) = k\left[1 - 3|\beta|(x_{5,6})^2\right] = -2k < 0.$$

The velocity is non-zero (4.229a) [zero (4.229b)] at the equilibrium (4.226d) [saddle (4.277e)] points:

$$\dot{x}_{1-3} = \sqrt{\frac{2E}{m}}, \qquad \dot{x}_{5,6} = 0; \qquad \dot{x}^2 + \omega_0^2 x^2 \left(1 - \frac{|\beta|}{2} x^2\right) = \frac{2E}{m} = \frac{k}{2m|\beta|} = \frac{\omega_0^2}{2|\beta|},$$

$$(4.229a\text{–}e)$$

the trajectories through the saddle points have energy (4.218b; 4.226a) ≡ (4.229c) leading to (4.229d, e), that is satisfied by (4.227d; 4.229b). Thus, *the separatrices (4.229c) passing through the saddle points (4.227d; 4.229b) of the undamped, unforced soft spring (4.226a) are (problem 108) two parabolas (4.229c) ≡ (4.230):*

$$\omega_0 \left(1 - |\beta| x^2\right) = \pm \dot{x} \sqrt{2|\beta|}, \qquad (4.230)$$

with opposite curvatures (S1 and S2 in Figure 4.7e).
   The **limit cycle oscillation** occurs if the initial condition (4.231a) lies between *the two parabolas; that is, both conditions (4.231b, c) are met:*

$$\{x_0, \dot{x}_0\} \equiv \{x(0), \dot{x}(0)\}: \qquad -\frac{1 - |\beta| x_0^2}{\sqrt{2|\beta|}} < \frac{\dot{x}_0}{\omega_0} < \frac{1 - |\beta| x_0^2}{\sqrt{2|\beta|}}. \qquad (4.231a\text{–}c)$$

In particular, if: (i) the initial velocity is zero (4.232a) the initial displacement must lie in the range (4.232b) between the saddle points (4.227d):

$$\dot{x}_0 = 0: \quad x_6 = -\frac{1}{\sqrt{|\beta|}} \le x_0 \le \frac{1}{\sqrt{|\beta|}} = x_5; \quad x_0 = 0: \quad \dot{x}_- \equiv -\frac{\omega_0}{\sqrt{2|\beta|}} \le \dot{x}_0 \le -\frac{\omega_0}{\sqrt{2|\beta|}} \equiv \dot{x}_+,$$

$$(4.232a\text{–}d)$$

(ii) if the initial displacement is zero (4.232c) the initial velocity must lie in the range (4.232d) between the points (C, D) in the Figure 4.8e. If one of the conditions (4.231b, c) is not met the path lies outside the separatrices and is open, corresponding to an unstable regime. Thus a limit cycle oscillation is possible only with amplitude not exceeding (4.233a):

$$a \equiv x_{\max} < |x_{5,6}| = \frac{1}{\sqrt{|\beta|}}: \qquad t\sqrt{\frac{2E}{m}} = \int_0^x \left|1 - \frac{k\xi^2}{2E} - \frac{k|\beta|\xi^4}{4E}\right|^{-1/2} d\xi, \qquad (4.233a, b)$$

and the trajectory is specified by (4.212b; 4.215b) ≡ *(4.233b) with initial time at the origin (4.221a).* The equation of motion of the anharmonic oscillator (4.215a) with bi-quadratic potential (4.215b) can be integrated [subsection 4.4.9 (4.4.11)] terms of elliptic functions (a series of Chebychev polynomials of the first kind); these exact solutions apply both to the closed (open) trajectories [subsections 4.4.5 (4.4.7)] corresponding to a limit cycle (unstable regime). The non-linearity affects the relation between the maximum velocity and maximum displacement or amplitude (subsection 4.4.8).

4.4.8 Non-Linearity Parameter and Maximum Velocity and Displacement

The times of zero displacement (4.234a, b) [velocity (4.234c, d)]:

$$\{x(t_1), \dot{x}(t_1)\} = \{0, b\}, \quad \{x(t_2), \dot{x}(t_2)\} = \{a, 0\}, \qquad (4.234a\text{--}d)$$

can be used (4.234f) [(4.234g)] to calculate total energy that is conserved (4.218b) ≡ (4.234e):

$$E = \frac{1}{2}m\dot{x}^2 + \frac{1}{2}kx^2 + \frac{1}{4}k\beta x^4 = \frac{1}{2}mb^2 = \frac{1}{2}ka^2\left(1 + \frac{\beta}{2}a^2\right). \quad (4.234e\text{--}g)$$

For a closed path or oscillatory motion the time of zero displacement (4.234a, b) [velocity (4.234c, d)] coincides with the time when the velocity (displacement) is maximum in modulus (4.235a, b) [(4.235c, d)]:

$$x(t_1) = 0: \ |\dot{x}(t)| \le b \equiv |\dot{x}(t_1)|; \qquad \dot{x}(t_2) = 0: \ |x(t)| \le a \equiv |x(t_2)|. \qquad (4.235a\text{--}d)$$

The parameter β in (4.234a) has the dimensions of the inverse square of length, and the **dimensionless non-linearity parameter** is defined by (4.236a):

$$\gamma \equiv \frac{\beta}{2}a^2: \qquad b = \sqrt{\frac{2E}{m}} = a\sqrt{\frac{k}{m}\left(1 + \frac{\beta}{2}a^2\right)} = \omega_0 a\sqrt{1+\gamma}, \qquad (4.236a\text{--}d)$$

and appears in the relation (4.236c) between the maximum velocity (4.236b) and amplitude. Thus *the ratio (4.236c)* ≡ *(4.236d) of the maximum velocity (4.235a, b) to the maximum displacement or amplitude (4.235c, d) increases (decreases) for the hard (soft) spring relative to the harmonic oscillator due to the effect of non-linearity.*

The total energy coincides with the potential energy where the velocity is zero, for example for the maximum displacement or amplitude (4.237a) of a periodic motion, and thus the equation of the trajectory (4.212b) becomes (4.237c) for any potential function:

$$E = \Phi(a); \qquad x(0) = 0: \qquad t = \int_0^x \left| \frac{2}{m} \left[\Phi(a) - \Phi(\xi) \right] \right|^{-1/2} d\xi, \qquad (4.237a\text{--}c)$$

with time zero at the origin (4.237b). In the case of the biquadratic potential (4.215b) the equation of motion (4.222b) ≡ (4.238):

$$t = \int_0^x \left| \frac{k}{m} \left[\left(a^2 - \xi^2 \right) + \frac{\beta}{2} \left(a^4 - \xi^4 \right) \right] \right|^{-1/2} d\xi, \qquad (4.238)$$

and can be put into a dimensionless form (4.239):

$$\omega_0 t = \int_0^x \left| a^2 - \xi^2 + \gamma \left(a^2 - \frac{\xi^4}{a^2} \right) \right|^{-1/2} d\xi, \qquad (4.239)$$

involving only the natural frequency (4.221a) and non-linearity parameter (4.236a). Since the radical in the integrand in (4.239) is a biquadratic polynomial, the integral can be evaluated (subsection 4.4.9) as a Jacobian elliptic function (section I.39.9).

4.4.9 Exact Solution in Terms of the Elliptic Sine

The Jacobian elliptic sine of variable ζ and modulus q is defined by (I.39.114c) ≡ (4.240a, b):

$$z = sn(\zeta; q): \qquad \zeta = \arg sn(z; q) \equiv \int_0^z \left| \left(1 - y^2 \right) \left(1 - q^2 y^2 \right) \right|^{-1/2} dy. \qquad (4.240a, b)$$

Rewriting (4.239) in the form (4.241):

$$\omega_0 t = \int_0^x \left| a^2 \left(1 + \gamma \right) - \xi^2 - \gamma \frac{\xi^4}{a^2} \right|^{-1/2} d\xi, \qquad (4.241)$$

suggests the change of variable (4.242a) leading to (4.242b):

$$y = \frac{\xi}{a}: \qquad \omega_0 t = \int_0^{x/a} \left| 1 + \gamma - y^2 - \gamma y^4 \right|^{-1/2} dy. \qquad (4.242a, b)$$

The integral (4.242b) can be re-written in the form:

$$\omega_0 t \sqrt{1+\gamma} = \int_0^{x/a} \left| 1 - \frac{y^2}{1+\gamma} - \frac{\gamma}{1+\gamma} y^4 \right|^{-1/2} dy = \int_0^{x/a} \left[(1-y^2) \left(1 + \frac{\gamma}{1+\gamma} y^2 \right) \right]^{-1/2} dy,$$

(4.243)

that coincides with the inverse elliptic since function (4.240a, b) with (4.244a–c):

$$z = \frac{x}{a}, \qquad \zeta = \omega_0 t \sqrt{1+\gamma} = \omega_0 t \sqrt{1 + \frac{\beta a^2}{2}}, \qquad q^2 = -\frac{\gamma}{1+\gamma} = -\frac{\beta a^2}{2 + \beta a^2}.$$

(4.244a–c)

It has been shown that *the trajectory of a **non-linear oscillator** (4.215a) with biquadratic potential (4.215b) is specified (problem 109) by the elliptic sine function (4.245):*

$$x = asn\left(\omega_0 t \sqrt{1+\gamma} ; \sqrt{-\gamma/(1+\gamma)} \right),$$

(4.245)

involving: (i) the amplitude (4.220b) that is a root of (4.218d); (ii) the natural frequency (4.221c) ≡ (2.54a) of the harmonic oscillator; (iii) the non-linearity parameter (4.236a) that specifies the modulus (4.244c) and modifies the variable (4.244a). In (4.245) the modulus from (2.444c) is imaginary (real) for the hard $\gamma > 0$ (soft $\gamma < 0$) spring. The exact oscillation frequency of the anharmonic oscillator is distinct from the natural frequency of the harmonic oscillator and depends on the amplitude (subsection 4.4.11). It is obtained from the expansion of the inverse elliptic sine as a series of circular sines (subsection 4.4.10). This leads to a series of Chebychev polynomials of the second kind (subsection 4.4.12) that specifies exactly the trajectory of the harmonic oscillator in inverse form with time as a function of position.

4.4.10 Inverse Elliptic Sine as a Series of Circular Sines

The changes of variable (4.246a, b) simplify the elliptic integral (4.240b; 4.244b) to (4.246c):

$$\sin\phi = z = \frac{x}{a}: \quad \sin\psi = y = \frac{\xi}{a}: \quad \omega_0 t \sqrt{1+\gamma} = \int_0^{\sin\phi} \left| 1 - q^2 \sin^2\psi \right|^{-1/2} d\psi$$

(4.246a–c)

$$\equiv \arg sn(\sin\phi; q).$$

The elliptic integral (4.246b) may be evaluated exactly in terms of elementary functions using two expansions: (i) first, the binomial series (I.25.37a–c) ≡ (4.247a, b) for the integrand of (4.246b):

$$\left|q^2\right| = \left|\frac{\gamma}{1+\gamma}\right| = \left|\frac{\beta a^2}{2+\beta a^2}\right| < 1: \quad \left|1-q^2\sin^2\psi\right|^{-1/2} = 1 + \sum_{n=1}^{\infty} a_n q^{2n}\sin^{2n}\psi, \quad (4.247a, b)$$

with coefficients (II.6.56a, b) ≡ (4.248a, b):

$$n \geq 1: \quad a_n \equiv (-)^n \binom{-1/2}{n} = \frac{(-)^n}{n!}\left(-\frac{1}{2}\right)\cdots\left(-\frac{1}{2}-n+1\right)$$

$$= \frac{2^{-n}}{n!}\, 1.3\ldots(2n-1) = \frac{(2n-1)!!}{n!\, 2^n} = \frac{(2n-1)!!}{(2n)!!} = \frac{(2n)!}{\left(n!\, 2^n\right)^2}, \quad (4.248a, b)$$

with the first seven values (II.6.57b–g); and (ii) the even powers of the circular sine in (4.247b) can be expanded as a finite sum of cosines of multiple angles (II.5.79a) ≡ (4.249):

$$\sin^{2n}\psi = 2^{-2n}\left\{\binom{2n}{n} + 2\sum_{m=0}^{n-1}(-)^{n-m}\binom{2n}{m}\cos\left[2(n-m)\psi\right]\right\}. \quad (4.249)$$

Substitution of (4.249) in (4.247b) leads to (4.250b):

$$\left|\gamma\right| < \left|1+\gamma\right|: \quad \left|1-q^2\sin^2\psi\right|^{-1/2} = 1 + \sum_{n=1}^{\infty}\left(\frac{q}{2}\right)^{2n}\frac{(2n-1)!!}{(2n)!!}$$

$$\times\left\{\frac{(2n)!}{(n!)^2} + 2\sum_{m=1}^{n-1}(-)^{n-m}\binom{2n}{m}\cos\left[2(n-m)\psi\right]\right\}, \quad (4.250a, b)$$

which is valid if the condition (4.247a; 4.244c) ≡ (4.250a) is met. Substitution of (4.250b) in (4.246b) allows immediate integration specifying the **inverse elliptic sine** *with variable* $\sin\phi$ *and modulus q as* (4.251) *a series of circular sines of even multiples of* ϕ:

$$\arg sn(\sin\phi; q) = A\phi + \sum_{n=1}^{\infty}\left(\frac{q}{2}\right)^{2n}\frac{(2n-1)!!}{(2n)!!}\sum_{m=0}^{n-1}(-)^{n-m}\binom{2n}{m}\frac{\sin\left[2(n-m)\phi\right]}{n-m}, \quad (4.251)$$

$$A(q) \equiv 1 + \sum_{n=1}^{\infty}\left(\frac{q}{2}\right)^{2n}\frac{(2n-1)!!}{(2n)!!}\frac{(2n)!}{(n!)^2} = 1 + \sum_{n=1}^{\infty}q^{2n}\left[\frac{(2n)!}{(n!)^2\, 2^{2n}}\right]^2, \quad (4.252a, b)$$

involving the **anharmonic factor** *(4.252a)* ≡ *(4.252b). In the passage from (4.252a) to (4.252b) was used:*

$$\frac{(2n-1)!!}{(2n)!!}\frac{(2n)!}{(n!)^2} = \frac{(2n)!}{[(2n)!!]^2}\frac{(2n)!}{(n!)^2} = \frac{[(2n)!]^2}{(n!\,2^n)^2(n!)^2} = \left[\frac{(2n)!}{(n!)^2\,2^n}\right]^2. \qquad (4.252c)$$

It is shown next (subsection 4.4.11) that the anharmonic factor (4.252a) ≡ (4.252b) relates the oscillation frequency of the anharmonic oscillator to the natural frequency of the harmonic oscillator.

4.4.11 Frequencies of the Anharmonic and Harmonic Oscillators

Using (4.253b), which is the inverse of (4.246a) and corresponds (4.253a) to the harmonic oscillator (4.223b), it follows that the amplitude corresponds to: (i) time equal (4.253c) to one-quarter of the period; and (ii) the value (4.253d) for ϕ:

$$\omega_0 t = \phi = arc\sin\left(\frac{x}{a}\right): \qquad\qquad x\left(\frac{\tau}{4}\right) = a \quad\Rightarrow\quad \phi = \frac{\pi}{2}. \qquad (4.253a\text{–}d)$$

Substituting (4.253a) in (4.251) and (4.246b) specifies the trajectory of the anharmonic oscillator in inverse form. Thus *the trajectory of the non-linear oscillator (4.215a) with bi-quadratic potential (4.215b) is specified (problem 110) exactly: (i) by (4.245) in terms of the elliptic sine in the direct form with position as a function of time; (ii) in elementary terms as a series (4.246b; 4.251) of circular sines in the inverse form with time as a function of position:*

$$\omega_0 t\sqrt{1+\gamma} = A\,arc\sin\left(\frac{x}{a}\right) + \sum_{n=1}^{\infty}\left(-\frac{\gamma}{\gamma+1}\right)^n\left[\frac{(2n)!}{(n!)^2\,2^{2n}}\right]^2$$

$$\times\sum_{n-m}^{n-1}\frac{(-)^{n-m}}{n-m}\frac{\sin\left[2(n-m)\,arc\sin(x/a)\right]}{m!\,(2n-m)!}. \qquad (4.254a)$$

In (4.254) the: (i) amplitude a is specified by the total energy (4.218d) that is conserved; (ii) the modulus (4.244c) accounts for the non-linearity (4.236a). In analogy with (4.252c) was used:

$$2^{-2n}\frac{(2n-1)!!}{(2n)!!}(2n)! = 2^{-2n}\frac{(2n)!}{[(2n)!!]^2}(2n)! = 2^{-2n}\frac{[(2n)!]^2}{(n!\,2^n)^2} = \left[\frac{(2n)!}{n!\,2^{2n}}\right]^2, \qquad (4.254b)$$

in the passage from (4.251) to (4.254a).

The amplitude (4.253b) corresponds to the displacement at one quarter period (4.253c) or the angle (4.253d), so that all terms of the series (4.254a) vanish for $x = a$ except the first:

$$\frac{\omega_0}{\omega} = \frac{\tau}{\tau_0} = \frac{\omega_0 \tau}{2\pi} = \frac{2\omega_0}{\pi} t(x = a) = \frac{2A}{\pi} \frac{\arcsin 1}{\sqrt{1+\gamma}} = \frac{A}{\sqrt{1+\gamma}}. \qquad (4.255)$$

The ratio of frequencies is specified by (4.252b) in terms of (4.244c) non-linearity parameter (4.236a) involving the anharmonic factor (4.256b) $\equiv$ (4.252a, b):

$$\frac{\omega_0}{\omega}\sqrt{1+\gamma} = 1 + \sum_{n=1}^{\infty}\left(-\frac{\gamma/4}{\gamma+1}\right)^n \frac{\left[(2n)!\right]^2}{(n!)^4\, 2^{2n}} = 1 + \sum_{n=1}^{\infty}\left(-\frac{\beta a^2}{2-\beta a^2}\right)^n \left[\frac{(2n)!}{(n!)^2\, 2^{2n}}\right]^2 \equiv A.$$

$$(4.256a, b)$$

The **oscillation frequency** ω of the non-linear oscillator is larger (4.257b) [smaller (4.257d)] than the natural frequency (4.221a) of the linear oscillator for the hard (4.257a) [soft (4.257c)] spring that is equivalent to an increased (decreased) resilience of the spring:

$$\gamma > 0: \qquad \frac{\omega_0}{\omega} < 1 - \frac{\gamma/4}{\gamma+1} < 1; \qquad \gamma < 0: \qquad \frac{\omega_0}{\omega} > 1. \qquad (4.257a\text{--}d)$$

The natural frequency (4.221c) [oscillation frequency (4.256a) $\equiv$ (4.256b)] of the harmonic (anhamonic) oscillator does not (does) depend on amplitude, because the amplitude is small (not small) for the linear (non-linear) oscillator. The proof of (4.257c, d) is made as follows. For the soft (4.257c) $\equiv$ (4.258a) spring: (i) the series (4.256b) has positive terms and starts with unity, so its sum exceeds unity (4.258b); (ii) since (4.258c) also exceeds unity follows (4.258d) $\equiv$ (4.275d):

$$\gamma < 0: \qquad A > 1 < \frac{1}{\sqrt{1+\gamma}} \quad \Rightarrow \quad \frac{\omega_0}{\omega} = \frac{A}{\sqrt{1+\gamma}} > 1; \qquad (4.258a\text{--}d)$$

$$\gamma > 0: \qquad A < 1 > \frac{1}{\sqrt{1+\gamma}} \quad \Rightarrow \quad \frac{\omega_0}{\omega} = \frac{A}{\sqrt{1+\gamma}} < 1, \qquad (4.258e\text{--}h)$$

the proof of (4.257a, b) for the hard (4.257a) $\equiv$ (4.258e) spring is made as follows: (i) the series (4.256b) has terms with alternating sign and decreasing modulus, and thus (example I.30.19) it converges to a sum that lies between successive terms, e.g., the first and second (4.258f); (ii) from (4.258g) follows (4.258h) $\equiv$ (4.257b). The exact solution (4.254a) for the trajectory of the non-linear oscillator in inverse form is equivalent to (4.251; 4.252a, b) and expresses the inverse elliptic sine as a series of Chebychev polynomials of the second kind (subsection 4.4.12).

4.4.12 Exact Solution as a Series of Chebychev Polynomials

The **Chebychev polynomials of the second kind** are defined by (II.5.129b) ≡ (4.259a) implying (4.259b, c):

$$U_N(\cos\phi) = 2^{1-N}\frac{\sin[(N+1)\phi]}{\sin\phi},\qquad (4.259a)$$

$$N = 2n - 2m - 1:\qquad \sin[2(n-m)\phi] = 2^{2n-2m-2}\sin\phi\, U_{2n-2m-1}(\cos\phi).$$

$$(4.259b, c)$$

Substitution of (4.259c) in (4.251) specifies *the inverse elliptic sine as a series of Chebychev polynomials of second kind:*

$$\arg sn(\phi;q) = A\phi + \sum_{n=1}^{\infty}\left(\frac{q}{2}\right)^{2n}\frac{(2n-1)!!}{(2n)!!}\sum_{m=0}^{n-1}\frac{(-)^{n-m}}{n-m}\binom{2n}{m}\qquad (4.260)$$

$$\times\, 2^{2n-2m-2}\sin\phi\, U_{2n-2m-1}(\cos\phi).$$

Using (4.246a; 4.259c) in (4.254a) leads to the equivalent result:

$$\omega_0 t\sqrt{1+\gamma} = A\,\arcsin\left(\frac{x}{a}\right) + \frac{x}{a}\sum_{n=1}^{\infty}\left(-\frac{\gamma}{\gamma+1}\right)^n\left[\frac{(2n)!}{(n!)^2\,2^{2n}}\right]^2$$

$$\times\sum_{m=0}^{n-1}\frac{(-)^{n-m}}{n-m}\frac{2^{2n-2m-2}}{m!\,(2n-m)!}U_{2n-2m-1}\left(\sqrt{1-\frac{x^2}{a^2}}\right),$$

$$(4.261)$$

that specifies (problem 111) in inverse form the trajectory of the anharmonic oscillator (4.215a) with biquadratic potential (4.215b) in terms of a Chebychev polynomial of the second kind (4.259a) using the amplitude a in (4.218b), natural frequency (4.221c), and non-linearity parameter (4.244c). In the linear case $\gamma = 0$, only the first term on the r.h.s. of (4.261) remains, leading to the harmonic oscillator (4.223b) since $A = 1$ in (4.256b). The lowest order $O(\gamma)$ non-linear perturbation is considered next as concerns: (i) the frequency (subsection 4.4.13); (ii) the amplitude (subsection 4.4.14); and (iii) the generation of harmonics (subsection 4.4.15).

4.4.13 Non-Linear Effects on the Frequency

The lowest order non-linear approximation (4.262a) to (4.246b) is (4.262b):

$$q^4 \ll 1: \qquad \omega_0 t \sqrt{1+\gamma} = \int_0^\phi \left[1 + \frac{q^2}{2} \sin^2 \phi \right] d\phi$$

$$= \int_0^\phi \left\{ 1 + \frac{q^2}{4} \left[1 - \cos(2\phi) \right] \right\} d\phi \qquad (4.262a, b)$$

$$= \left(1 + \frac{q^2}{4} \right) \phi - \frac{q^2}{8} \sin(2\phi).$$

The lowest order non-linear approximation (4.262a) corresponds by (4.244c) to (4.263a) and implies (4.263b, c):

$$\gamma^2 \ll 1: \qquad q^2 = -\frac{\gamma}{1+\gamma} = -\gamma + O(\gamma^2), \qquad 1 + \frac{q^2}{4} = 1 - \frac{\gamma}{4}. \qquad (4.263a\text{-}c)$$

The second term on the r.h.s. of (4.262b) has an upper bound (4.263d) that is small:

$$\left| -\frac{q^2}{8} \sin(2\phi) \right| \le \frac{|q|^2}{8} \sim \frac{\gamma}{8} = \frac{\beta a^2}{16}; \qquad (4.263d)$$

it is negligible compared with the r.h.s. of (4.262b) for sufficiently long time (4.264a), leading to the simplification (4.264b):

$$t \gg \frac{\gamma}{8\omega_0}: \qquad \phi = \omega_0 t \frac{\sqrt{1+\gamma}}{1 + q^2/4} = \omega_0 t \frac{1 + \gamma/2}{1 - \gamma/4} = \omega_0 t \left(1 + \frac{\gamma}{2} \right) \left(1 + \frac{\gamma}{4} \right) = \omega_0 t \left(1 + \frac{3\gamma}{4} \right).$$

$$(4.264a, b)$$

Setting (4.265a) in (4.246a) leads to (4.265b):

$$\phi \equiv \omega t: \qquad x(t) = a \sin(\omega t), \qquad \omega = \omega_0 \left(1 + \frac{3\beta a^2}{8} \right), \qquad (4.265a\text{-}c)$$

specifying the **oscillation frequency** (4.265c).

    Compared with the undamped unforced harmonic oscillator (2.54a) with natural frequency (2.54a) ≡ (4.221c), the non-linear oscillator (4.215a) with biquadratic potential (4.215b) to lowest order in the non-linearity parameter (4.236c) has higher ω > ω₀ *(lower $\omega < \omega_0$) frequency for the hard $\gamma > 0$ (soft $\gamma < 0$) spring that becomes stiffer (less stiff) as (problem 112) the displacement increases (Figure 4.8a).* The lowest order non-linear approximation (4.263a) ≡ (4.266a) to the ratio of oscillation and natural frequencies corresponds to the first two terms on the r.h.s. of (4.256b):

$$\gamma^2 \ll 1: \qquad \frac{\omega_0}{\omega} = \frac{1-(\gamma/4)/(\gamma+1)}{\sqrt{1+\gamma}} = \frac{1-\gamma/4}{1+\gamma/2} = 1 - \frac{\gamma}{4} - \frac{\gamma}{2} = 1 - \frac{3\gamma}{4} = 1 - \frac{3\beta a^2}{8},$$

$$(4.266a, b)$$

that is the inverse of (4.265c). The effect of non-linearity is considered to first-order not only on the frequency of oscillation (subsection 4.4.13) but also on the displacement as a function of time (subsection 4.4.14).

4.4.14 Effect of Non-Linearity on the Displacement

The path of the undamped, unforced non-linear oscillator (4.215a) with biquadratic potential (4.215b) is specified exactly: (i) by (4.245) in the direct form, with position as a function of time, in terms of the elliptic sine function; and (ii) by (4.254a) in inverse form, with time as a function of position, involving only elementary functions, namely circular sines. Next is obtained the trajectory in direct form (i) of position as a function of time in terms (ii) of circular functions, as an approximation to lowest order of non-linearity (4.262a) ≡ (4.263a) ≡ (4.267a). If the long time approximation (4.264a) is not made, for consistency to first-order of non-linearity (4.263b) the second term on the r.h.s. of (4.262b) should be included in the calculation of the path (4.267b) using the non-linearity parameter:

$$\gamma^2 \ll 1: \qquad \omega_0 t\left(1+\frac{\gamma}{2}\right) = \left(1-\frac{\gamma}{4}\right)\phi + \frac{\gamma}{8}\sin(2\phi); \qquad (4.267a, b)$$

from (4.267b) follows to the oscillation frequency (4.264b; 4.265a) ≡ (4.267c):

$$\omega t = \omega_0 t \frac{1+\gamma/2}{1-\gamma/4} = \phi + \frac{\gamma/8}{1-\gamma/4}\sin(2\phi) = \phi + \frac{\gamma}{4}\cos\phi\sin\phi. \qquad (4.267c)$$

Noting that (4.268a) implies (4.268b–d):

$$\chi \equiv \frac{\gamma}{4}\cos\phi\sin\phi: \qquad \chi^2 \ll 1; \qquad \cos\chi = 1, \qquad \sin\chi = \chi, \qquad (4.268a–d)$$

leads to:

$$\sin(\omega t) = \sin\left(\phi + \frac{\gamma}{8}\cos\phi\sin\phi\right) = \sin(\phi + \chi) = \sin\phi\cos\chi + \sin\chi\cos\phi$$

$$(4.269)$$

$$= \sin\phi + \chi\cos\phi,$$

that is equivalent to:

$$\sin(\omega t) = \sin\phi\left(1 + \frac{\gamma}{4}\cos^2\phi\right),\tag{4.270}$$

using (4.268a). Substituting (4.246a) in (4.270) leads to the same first-order of non-linearity to:

$$\frac{x}{a} = \sin\phi = \frac{\sin(\omega t)}{1 + (\gamma/4)\cos^2(\omega t)} = \sin(\omega t)\left[1 - \frac{\gamma}{4}\frac{1+\cos(2\omega t)}{2}\right]$$

$$= \left(1 - \frac{\gamma}{8}\right)\sin(\omega t) - \frac{\gamma}{8}\frac{\sin(3\omega t) - \sin(\omega t)}{2} = \left(1 - \frac{\gamma}{16}\right)\sin(\omega t) - \frac{\gamma}{16}\sin(3\omega t),$$

$$\tag{4.271}$$

where (II.5.82a; II.5.88c) were used. It follows from (4.271) that the non-linear effect changes the amplitude (subsection 4.4.14) and generates harmonics (subsection 4.4.15).

4.4.15 Generation of Harmonics in a Non-Linear Oscillator

Writing (4.271) in the form:

$$x(t) = a\left(1 - \frac{\beta a^2}{32}\right)\sin(\omega t) - \frac{\beta a^3}{32}\sin(3\omega t) = -x(-t),\tag{4.272a, b}$$

shows that *an undamped unforced anharmonic oscillator (4.215a) with (problem 113) biquadratic potential (4.215b) has, relative to a harmonic oscillator (2.54a, c; 4.223c) to lowest order in the non-linear parameter (4.236a) in the case of the hard $\beta > 0$ (soft $\beta < 0$) spring: (i) an increase (decrease) in frequency (4.265c) and a decrease (increase) of the amplitude of the oscillation at the fundamental frequency, corresponding to the first term on the r.h.s. of (4.272a); (iii) corresponding to the second term on the r.h.s. of (4.272a) a second harmonic with triple frequency is generated, with small amplitude, and out of (in) phase with the fundamental. The cubic non-linearity in the restoring force (4.215a) leads to lowest-order in the non-linearity parameter (4.236a) to the generation of the second harmonic rather than the first harmonic, since the second term on the r.h.s. of (4.272a) has frequency 3ω instead of 2ω; this implies that the displacement remains an odd function of time (4.272b) implying a zero displacement at time zero.* Thus a non-linear restoring force leads to the generation of harmonics of the natural frequency of the linear oscillator; this is confirmed next (section 4.5) using the approximate method of perturbation expansions, applied to a more general non-linear restoring force containing a quadratic in addition to the cubic term.

4.5 Perturbation Expansions and Generation of Harmonics

It has been shown (section 4.4) that a non-linear restoring force: (i) changes the frequency of oscillation relative to the natural frequency of the linear system (subsection 4.4.13); and (ii) generates harmonics of the fundamental mode (subsection 4.4.15). The most direct way to obtain the frequency changes and amplitudes of the harmonics is to use a perturbation expansion (section 4.5). The perturbation expansion is an approximate method (section 4.5) alternative to the exact solution (section 4.4). It is applied to a more general restoring force including linear, quadratic, and cubic terms (subsection 4.5.1), to obtain the first (second) order non-linear corrections [subsection 4.5.2 (4.5.3)] to the harmonic oscillator, that specify (subsection 4.5.4) the change in fundamental frequency and the generation of harmonics.

4.5.1 Restoring Force with Quadratic and Cubic Terms

The unforced, undamped, unharmonic oscillator (4.216b) is reconsidered with a restoring force containing linear, quadratic, and cubic terms (4.273) corresponding to the potential (4.274) with quadratic, cubic, and quartic terms:

$$m\ddot{x} = -j(x) = -kx\left(1 + \varepsilon\alpha x + \varepsilon^2\beta x^2\right) = -\frac{d\Phi}{dx}, \tag{4.273}$$

$$\Phi(x) = \frac{1}{2}kx^2\left(1 + \frac{2}{3}\varepsilon\alpha x + \frac{1}{2}\varepsilon^2\beta x^2\right), \tag{4.274}$$

where the small parameter ε indicates that quadratic (cubic) term in the restoring force (4.273a; that is, the cubic (quartic) term in the potential (4.273b) is of first (second) order of smallness. The potential (4.274) satisfies (4.214b, c) and thus the origin is a position of stable equilibrium, and its neighborhood is a potential well corresponding to oscillatory motion. Introducing the natural frequency (4.221c) $\equiv$ (4.275a) of the harmonic oscillator (4.54c) leads to (4.275b):

$$\omega_0 = \sqrt{\frac{k}{m}}: \qquad \ddot{x} + \omega_0^2 x = -\omega_0^2\left(\varepsilon\alpha x^2 + \varepsilon^2\beta x^3\right), \tag{4.275a, b}$$

involving the quadratic (cubic) non-linearity parameters $\alpha(\beta)$. The displacement (4.276a) and frequency (4.276b) are specified by **parametric expansions**:

$$x(t) = x_1(t) + \varepsilon x_2(t) + \varepsilon^2 x_3(t) + ..., \quad \omega = \omega_0 + \varepsilon\omega_1 + \varepsilon^2\omega_2 + ... \tag{4.276a, b}$$

A rigorous approach would require proof existence of the parametric expansions and of their convergence with n terms as $n \to \infty$. In most problems only the first $O(\varepsilon)$ or second $O(\varepsilon^2)$ terms can be obtained and this is considered as an acceptable approximation for small ε. Substituting (4.276a, b) in (4.273a) gives:

$$\ddot{x}_1 + \varepsilon \ddot{x}_2 + \varepsilon^2 \ddot{x}_3 + \left(\omega - \varepsilon \omega_1 - \varepsilon^2 \omega_2\right)^2 \left(x_1 + \varepsilon x_2 + \varepsilon^2 x_3\right)$$
$$= -\varepsilon \alpha \omega_0^2 \left(x_1 + \varepsilon x_2 + \varepsilon x_3\right)^2 - \varepsilon^2 \beta \omega_0^2 \left(x_1 + \varepsilon x_2 + \varepsilon^2 x_3\right)^3 , \tag{4.277}$$

valid to $O(\varepsilon^2)$, for example in the second term of (4.277):

$$\varepsilon^3 \ll 1: \qquad \omega_0^2 = \left(\omega - \varepsilon \omega_1 - \varepsilon^2 \omega_2\right)^2 = \omega^2 - \varepsilon 2\omega_1 \omega + \varepsilon^2 \left(\omega_1^2 - 2\omega_2 \omega\right). \tag{4.278a, b}$$

Equating the coefficients of successive powers of ε gives (4.279a; 4.280; 4.281):

$$\varepsilon^0: \qquad\qquad \ddot{x}_1 + \omega^2 x_1 = 0, \qquad x_1(t) = a \sin(\omega t), \tag{4.279a, b}$$

$$\varepsilon^1: \qquad\qquad \ddot{x}_2 + \omega^2 x_2 = 2\omega\omega_1 x_1 - \omega_0^2 \alpha x_1^2 , \tag{4.280}$$

$$\varepsilon^2: \qquad \ddot{x}_3 + \omega^2 x_3 = 2\omega\omega_1 x_2 + \left(2\omega\omega_2 - \omega_1^2\right) x_1 - 2\omega_0^2 \alpha x_1 x_2 - \omega_0^2 \beta x_1^3. \tag{4.281}$$

The method of parametric expansions (4.276a, b) transforms (4.277) a non-linear problem (4.273) into (problem 114) an infinite recursive sequence of linear problems (4.279a; 4.280; 4.281;...) with each order: (i) satisfying the same linear differential equation; (ii) forced by linear and non-linear combinations of the preceding orders. In the case of the anharmonic oscillator with restoring force (4.273a): (i) the lowest order (4.279a) is an oscillation (4.279b) at a frequency ω to be determined; (ii) the order n involves forcing terms of lower order (1, 2, . . . , n − 1) generating harmonics with frequencies up to $n\omega$. This method of parametric expansions is demonstrated next with the first (second) order approximation [subsections (4.5.2 (4.5.3–4.5.4)].

4.5.2 First-Order Perturbation and Secular Condition

Substituting the zero order solution (4.279b) in the first-order approximation (4.280) leads to:

$$\ddot{x}_2 + \omega^2 x_2 = 2\omega\omega_1 a \sin(\omega t) - \frac{\alpha}{2} \omega_0^2 a^2 \left[1 - \cos(2\omega t)\right]. \tag{4.282}$$

The first term on the r.h.s. of (4.282) is a forcing at the frequency ω, and would lead to an undamped resonance (subsection 2.7.3) with amplitude growing linearly with time (2.176b); this would imply that the energy of the oscillator would grow quadratically with time (2.178c), which is not possible because there is no external force acting on the system and the total energy is conserved (4.210c). Thus this term must be excluded, imposing the **secular condition** that its amplitude is zero; the secular condition implies that there is no frequency change to first order (4.283a), and (4.282) simplifies to (4.283b):

$$\omega_1 = 0: \qquad\qquad \ddot{x}_2 + \omega^2 x_2 = -\frac{\alpha}{2}\,\omega_0^2 a^2 \left[1 - \cos(2\omega t)\right]. \qquad\qquad (4.283a, b)$$

The solution of (4.283b) consists (4.284a) of a constant part (4.284b) plus an oscillation of frequency 2ω with amplitude (4.284c):

$$x_2(t) = x_{20} + a_2 \cos(2\omega t), \qquad \omega^2 x_{20} = -\frac{\alpha a^2 \omega_0^2}{2}, \qquad -3\omega^2 a_2 = \frac{\alpha}{2} a^2 \omega_0^2. \quad (4.284a\text{–}c)$$

Bearing in mind (4.283a) that the frequency (4.276b) is unchanged (4.285a) to lowest order in the non-linearity and the coefficients (4.284b, c) simplify to (4.285b, c) in (4.284a) $\equiv$ (4.285d):

$$\omega = \omega_0: \quad x_{20} = -\frac{\alpha a^2}{2}, \quad a_2 = -\frac{\alpha a^2}{6}, \quad x_2(t) = -\frac{\alpha a^2}{6}\left[3 + \cos(2\omega t)\right]. \quad (4.285a\text{–}d)$$

The first order perturbation (4.285d) may be added to the fundamental (4.279b) to the first order in (4.276a, b) $\equiv$ (4.286a, b):

$$\omega = \omega_0: \qquad x(t) = x_1(t) + x_2(t) = -\frac{\alpha a^2}{2} + a\sin(\omega_0 t) - \frac{\alpha a^2}{6}\cos(2\omega_0 t), \quad (4.286a, b)$$

showing that *to first-order of non-linearity $O(\varepsilon)$ the motion of undamped unforced anhamonic oscillator is (problem 115) perturbed only by the quadratic term in the restoring force (4.273a) and cubic term in the potential (4.274), implying (Figure 4.9a) that: (i) the potential has an inflexion point at the origin, so that the mean position is shifted to the side of lowest potential, as shown by the first term on the r.h.s. of (4.286a); (ii) the cubic potential does not contribute to the potential well, so the frequency is unchanged (4.286a) from the natural frequency (4.275a) of a harmonic oscillator; (iii) to the oscillation at the natural frequency is added at the next order the first harmonic or oscillation at twice the natural frequency because the restoring force has a quadratic term proportional to the square of the displacement:*

$$\left[x_1(t)\right]^2 = a^2 \sin(\omega_0 t) = \frac{a^2}{2}\left[1 - \cos(2\omega_0 t)\right]; \qquad\qquad (4.287)$$

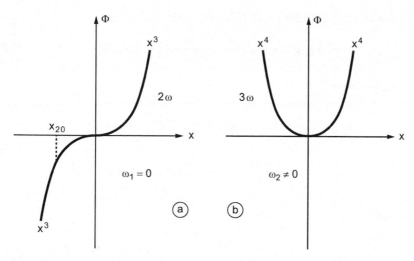

FIGURE 4.9
The quadratic (cubic) term in the restoring force of a non-linear, undamped, unforced oscillator corresponds to a cubic (quartic) term in the potential [Figure 4.9a (b)] that: (i) has a inflexion point (a minimum) at the origin, so that the average position is shifted to the side of lower potential (remains at the origin); (ii) does not (does) contribute to the potential well at the origin created by the linear restoring force or quadratic potential, and thus does not (does) change the oscillation frequency of the non-linear oscillator relative to the natural frequency of the harmonic oscillator; and (iii) creates harmonics starting at the double (triple) of natural frequency, that is, the first (second) harmonic.

and (iv) the first harmonic is out of phase by $-\pi/4$ relative to the fundamental:

$$\cos(2\omega_0 t) = \sin\left[2(\omega_0 t + \pi/4)\right].\tag{4.288}$$

The next second order of non-linearity $O(\varepsilon^2)$ will be affected both by the quadratic and cubic terms in the restoring force (subsections 4.5.3–4.5.4).

4.5.3 Effect of Quadratic and Cubic Terms on the Second-Order Perturbation

Substituting (4.283a) simplifies the second-order approximation (4.281) to:

$$\ddot{x}_3 + \omega^2 x_3 = \left(2\omega_2\omega - 2\omega_0^2\alpha x_2 - \omega_0^2\beta x_1^2\right)x_1.\tag{4.289}$$

Substituting the zero (4.279b) and first (4.285d) order approximations in (4.289) leads to:

$$\ddot{x}_3 + \omega^2 x_3 = \left\{2\omega_2\omega + \omega_0^2\alpha^2 a^2 + \frac{\omega_0^2\alpha^2 a^2}{3}\cos(2\omega t) - \frac{\omega_0^2\beta a^2}{2}\left[1 - \cos(2\omega t)\right]\right\}a\sin(\omega t).$$
$$\tag{4.290}$$

Using the identity (II.5.88c) $\equiv$ (4.117b) $\equiv$ (4.291):

$$2\cos(2\omega t)\sin(\omega t) = \sin(3\omega t) - \sin(\omega t), \tag{4.291}$$

in (4.290) leads to:

$$\ddot{x}_3 + \omega^2 x_3 = \left(2\omega_2\omega + \omega_0^2\alpha^2 a^2 - \frac{\omega_0^2\beta a^2}{2}\right)a\sin(\omega t)$$
$$+ \frac{\omega_0^2 a^3}{2}\left(\frac{\alpha^2}{3} + \frac{\beta}{2}\right)\left[\sin(3\omega t) - \sin(\omega t)\right], \tag{4.292}$$

that simplifies to:

$$\ddot{x}_3 + \omega^2 x_3 = \left[2\omega_2\omega + \omega_0^2 a^2\left(\frac{5\alpha^2}{6} - \frac{3\beta}{4}\right)\right]a\sin(\omega t) + \frac{\omega_0^2 a^3}{2}\left(\frac{\alpha^2}{3} + \frac{\beta}{2}\right)\sin(3\omega t), \tag{4.293}$$

consisting of a resonant (non-resonant) term, namely the first (second) on the r.h.s. of (4.293).

4.5.4 Frequency Change and Generation of Harmonics

The resonant term first on the r.h.s. of (4.293) is eliminated by the secular condition setting the coefficient equal to zero (4.294a), and the first order (4.283a) approximation $\omega \sim \omega_0$ can be substituted in the second order leading to (4.294b):

$$\omega - \omega_0 = \omega_2 = \frac{\omega_0^2 a^2}{2\omega}\left(\frac{3\beta}{4} - \frac{5\alpha^2}{6}\right) \sim \frac{\omega_0 a^2}{2}\left(\frac{3\beta}{4} - \frac{5\alpha^2}{6}\right). \tag{4.294a, b}$$

The solution of the differential equation (4.293) is forced by the second term on the r.h.s. alone:

$$x_3(t) = -\frac{\omega_0^2 a^3}{16\omega^2}\left(\frac{\alpha^2}{3} + \frac{\beta}{2}\right)\sin(3\omega t). \tag{4.295}$$

The first-order approximation (4.285a) may be used in the second-order approximation (4.294a, b), leading to the frequency:

$$\omega = \omega_0\left[1 + \frac{a^2}{4}\left(\frac{3\beta}{2} - \frac{5\alpha^2}{3}\right)\right], \tag{4.296}$$

that appears in the total displacement:

$$\varepsilon^3 \ll 1: \qquad\qquad x(t) = x_1(t) + x_2(t) + x_3(t), \qquad\qquad (4.297a, b)$$

consisting of the sum of (4.279b), (4.285d) and (4.295):

$$x(t) = a\sin(\omega t) - \frac{\alpha a^2}{6}\left[3 + \cos(2\omega t)\right] - \frac{a^3}{16}\left(\frac{\alpha^2}{3} + \frac{\beta}{2}\right)\sin(3\omega t). \qquad (4.298)$$

The undamped, unforced, anharmonic oscillation with (problem 116) linear, quadratic, and cubic terms in the restoring force (4.273) has: (i) oscillation frequency (4.296) modified at the second-order because the cubic restoring force (4.273) and the quartic potential (4.274) affects the potential well (Figure 4.9b); (ii) the quadratic term in the restoring force (4.273) and cubic term in the potential (4.274) adding the first (second) harmonic at the first $O(\varepsilon)$ [second $O(\varepsilon^2)$] order of non-linearity; (iii) the cubic term in the restoring force (4.273) or quadric term in the potential (4.274) appearing first at the second-order of non-linearity $O(\varepsilon^2)$ and generating the second harmonic (II.5.86b) $\equiv$ (4.299) with triple frequency:

$$\left[x_1(t)\right]^3 = a^3\sin^3(\omega t) = \frac{a^3}{4}\left[3\sin(\omega t) - \sin(3\omega t)\right]. \qquad (4.299)$$

In the absence of the quadratic term in the restoring force $\alpha = 0$: (i) the frequency (4.296) simplifies to (4.265c); (ii) there is no first harmonic in (4.298) and the second harmonic coincides with that in (4.272b): (iii) the ratio of amplitudes of the second harmonic and fundamental in (4.272b) is (4.300c), which coincides with the ratio (4.300d) in (4.298), bearing in mind that (4.300a) is small (4.300b):

$$\delta \equiv \frac{\beta a^2}{32}, \quad \delta^2 \ll 1: \qquad -\frac{a\delta}{a(1-\delta)} = -\frac{\delta}{1-\delta} = -\delta + 0(\delta^2). \qquad (4.300a\text{–}d)$$

The (i) approximate method of the perturbation expansion for a non-linear restoring force with quadratic and cubic terms (section 4.5) is (ii) simpler than the exact solution and its approximation with cubic term alone (section 4.4); thus method (i) will be applied in the next (section 4.6) to non-linear resonance.

4.6 Non-Linear Resonance, Amplitude Jumps, and Hysterisis

The resonance of an oscillator forced at the natural frequency applies both to the linear or harmonic (non-linear or anharmonic) oscillator sections [2.7–2.8 (4.6)]. Since a non-linear restoring force generates harmonics (sections 4.4–4.5), the anharmonic oscillator forced at multiples (sub-multiples) of the natural frequency [subsections 4.6.5–4.6.8 (4.6.4)], also resonates at the natural frequency, although with smaller amplitude than for direct forcing at the natural frequency (subsections 4.6.1–4.6.3). The non-linear resonance of a damped harmonic oscillator exhibits a number of phenomena (subsection 4.6.9) not possible for a harmonic oscillator; for example, the non-linear resonance due to forcing at: (i) the natural frequency (subsection 4.6.1) leads to amplitude jumps (subsections 4.6.2) and a hysteresis loop (subsection 4.6.3); (ii) also at the first sub-harmonic (subsection 4.6.4), with smaller amplitude; (iii) at the first harmonic (subsection 4.6.6) leads to amplitude jumps and suppression of oscillations (subsection 4.6.7); and (iv) at the second harmonic (subsection 4.6.7) leads to no oscillation unless a threshold initial amplitude is satisfied (subsection 4.6.9).

4.6.1 Non-Linear Resonance by Forcing at the Natural Frequency

Consider a linearly damped anharmonic oscillator with mass m, restoring force with linear, quadratic, and cubic terms (4.273), linear (2.15) damping μ, and a sinusoidal external applied force with applied frequency (2.165b):

$$m\ddot{x} + \mu\dot{x} + kx = -kx^2(\alpha + \beta x) + F_a \cos(\omega_a t). \qquad (4.301)$$

The applied frequency (4.302c) is close (4.302a) to the natural frequency (4.221a) ≡ (4.302b):

$$\varepsilon^2 \ll \omega_0^2: \qquad \omega_0 = \sqrt{\frac{k}{m}}, \qquad \omega_a = \omega_0 + \varepsilon, \qquad \lambda \equiv \frac{\mu}{2m}, \qquad f_a \equiv \frac{F_a}{m}, \qquad (4.302\text{a–e})$$

and using the damping (2.23b) ≡ (4.302d) and force (2.23c) ≡ (2.302e) per unit mass leads to:

$$\ddot{x} + 2\lambda\dot{x} + \omega_0^2 x = -\omega_0^2 x^2(\alpha + \beta x) + f_a \cos(\omega_a t). \qquad (4.303)$$

The non-linear effects change the oscillation frequency (4.296) relative to the natural frequency in proportion to the square of amplitude (4.304a) with a factor (4.304b):

$$\omega_0 \to \omega = \omega_0(1 + b^2\psi), \qquad\qquad \psi \equiv \frac{3\beta}{8} - \frac{5\alpha^2}{12}, \qquad (4.304\text{a, b})$$

where: (i) the amplitude a of the free oscillation (4.295) is replaced by the amplitude b of the forced oscillation (4.304a); (ii) the dimensionless frequency shift, that is, the second term in the curved bracket in (4.304a), is proportional to the square of the amplitude, and thus negligible in the linear case; (iii) in the non-linear case the coefficient is the **non-linearity factor** (4.304b) combining the coefficients $\alpha(\beta)$ of the quadratic (cubic) of the restoring force first on the r.h.s. of (4.301). Thus, instead of (4.303) it is sufficient to consider the linear damped oscillator (2.192a) with natural frequency replaced by (4.304a, b) and the same forcing:

$$\ddot{x} + 2\lambda\dot{x} + \omega_0^2 \left(1 + b^2\psi\right)^2 x = \mathrm{Re}\left\{f_a \exp\left[i(\omega_0 + \varepsilon)t\right]\right\} \equiv \mathrm{Re}\left\{f_a \exp(i\omega_a t)\right\}. \quad (4.305)$$

Excluding free oscillations, the forced oscillation (4.306a) at the applied frequency in (4.305) leads to (4.306b):

$$x(t) = \mathrm{Re}\left\{b\exp(i\omega_a t)\right\}: \qquad f_a = b\left[\omega_0^2\left(1 + b^2\psi\right)^2 + 2i\lambda\omega_a - \omega_a^2\right], \quad (4.306a, b)$$

where (4.302c) may be substituted leading to (4.306c):

$$f_a = b\left[2\omega_0^2 b^2\psi + \omega_0^2 b^4\psi^2 + 2i\lambda\omega_0 + 2i\lambda\varepsilon - 2\varepsilon\omega_0 - \varepsilon^2\right]. \quad (4.306c)$$

Neglecting the higher order terms (4.307a–c), the relation (4.306c) between the amplitudes of forcing f_a and oscillation b simplifies to (4.307d):

$$\varepsilon^2, \lambda\varepsilon \ll \omega_0^2 b^2\psi; \qquad b^4\psi^2 \ll 1: \qquad f_a = 2b\omega_0\left(\omega_0 b^2\psi - \varepsilon + i\lambda\right). \quad (4.307a–d)$$

Taking the modulus of (4.307d) leads to (4.308a):

$$\left(\frac{|f_a|}{2\,\omega_0}\right)^2 = b^2\left[\left(\omega_0 b^2\psi - \varepsilon\right)^2 + \lambda^2\right]: \qquad \omega_\pm - \omega_0 = \varepsilon_\pm = \omega_0 b^2\psi \pm \sqrt{\left(\frac{|f_a|}{2\omega_0 b}\right)^2 - \lambda^2},$$

$$(4.308a, b)$$

implying that for each pair of amplitudes $\left(f_a, b\right)$ there are two oscillation frequencies $\omega_\pm$ shifted from the natural frequency by (4.308b).

The roots of (4.308a) specify the amplitude of oscillation b as a function of the frequency difference ε from the natural frequency (Figures 4.10a–d). Differentiation of (4.308a) with regard to ε leads to:

$$0 = \frac{db}{d\varepsilon}\left[\left(\omega_0 b^2\psi - \varepsilon\right)^2 + \lambda^2\right] + b\left(\omega_0 b^2\psi - \varepsilon\right)\left(2\omega_0\psi b\frac{db}{d\varepsilon} - 1\right)$$

$$(4.309a)$$

$$= \frac{db}{d\varepsilon}\left[\left(\omega_0 b^2\psi - \varepsilon\right)\left(3\omega_0\psi b^2 - \varepsilon\right) + \lambda^2\right] - b\left(\omega_0 b^2\psi - \varepsilon\right),$$

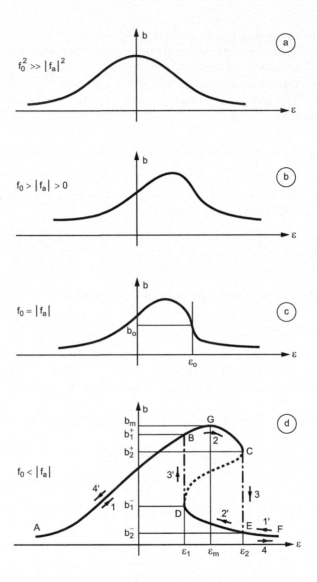

FIGURE 4.10

The resonance of a non-linear oscillator with the potential a quartic function of the position leads to an amplitude b that is a symmetric function of the difference $\varepsilon = \omega_a - \omega_0$ between the applied and natural frequency when the excitation amplitude is small (a). As the excitation amplitude increases (b) the curve becomes unsymmetric with larger amplitude of oscillation at applied larger than natural frequency, until (c) a vertical slope occurs at a threshold amplitude f_0. Beyond (d) the threshold amplitude: (i) there is one amplitude for each frequency as before (a–c) only outside the range $(\varepsilon_1, \varepsilon_2)$; (ii) inside this range of frequencies there are three amplitudes, of which only the highest and lowest are stable. This leads to three situations: (i/ii) starting at $A(F)$ at low (high) frequency, and increasing (decreasing) the frequency there is a downward CE (upward DB) **amplitude jump** at the higher ε_2 (lower ε_1) frequency in the hysteresis range; (iii) combining (i) and (ii) it is possible to stay within the hysteresis range $\varepsilon_1 \leq \varepsilon \leq \varepsilon_2$ by described the **hysteresis loop** with jumps at CE and DB.

that specifies the slope:

$$\frac{db}{d\varepsilon} = \frac{b\left(\omega_0 b^2 \psi - \varepsilon\right)}{\varepsilon^2 - 4\omega_0 b^2 \psi \varepsilon + 3\omega_0^2 \psi^2 b^4 + \lambda^2}. \tag{4.309b}$$

The cubic equation (4.308a) in b^2 with real coefficients shows that for each forcing amplitude $|f_a|$ and frequency shift ε there are three roots for the square of the amplitude of oscillation b^2, of which: (i) one must be real (ii-a, b, c); the other two are either (ii-a) real and distinct, (ii-b) a double real root, or (ii-c) a complex conjugate pair. The three cases (a, b, c) are considered next (subsection 4.6.2).

4.6.2 Multiple Amplitudes for the Same Frequency

In the case of the linear (4.310a) damped oscillator the amplitude (4.308a) is symmetric (Figure 4.10a) relative to the natural frequency (4.310b), where it is maximum (4.310c):

$$\psi = 0: \qquad b(\varepsilon) = \frac{f_a}{2\omega_0 \sqrt{\varepsilon^2 + \lambda^2}}, \qquad b_m = b(0) = \frac{f_a}{2\omega_0 \lambda}, \tag{4.310a–c}$$

in agreement with (2.201d) $\equiv$ (4.310c). If the non-linearity parameter is non-zero the amplitude curve becomes unsymmetric (Figure 4.10b) and is shifted toward frequencies above the natural frequency, as follows from (4.309b) $\equiv$ (4.311a) where (4.311b) are the roots of the denominator:

$$\frac{db}{d\varepsilon} = \frac{b\left(\omega_0 b^2 \psi - \varepsilon\right)}{(\varepsilon - \varepsilon_+)(\varepsilon - \varepsilon_-)}, \qquad \varepsilon_\pm = 2\omega_0 b^2 \psi \pm \sqrt{\omega_0^2 b^4 \psi^2 - \lambda^2}. \tag{4.311a, b}$$

The amplitude curve has one (two) vertical tangent(s) [Figure 4.10c(d)] when $db/d\varepsilon = \infty$ in (4.311a), implying that the denominator vanishes, and the frequency difference (4.308b) equals one of the roots (4.311b) that may coincide (be different). Thus three cases arise. The intermediate case II of a single (4.312a) vertical tangent (Figure 4.10c) in (4.311b):

$$\frac{db}{d\varepsilon} = 0: \qquad \left[b(\varepsilon_0)\right]^2 = \frac{\lambda}{\omega_0 \psi} \equiv b_0^2, \qquad \varepsilon_0 \equiv \varepsilon_+ = \varepsilon_- = 2\omega_0 b_0^2 \psi = 2\lambda, \tag{4.312a–c}$$

$$|f_a| = 2\sqrt{2}\psi\omega_0^2 b_0^3 = 2\sqrt{2}\omega_0 \lambda b_0 \equiv f_0, \tag{4.312d}$$

leading (4.308a) to a forcing amplitude (4.312d).

In the case I of forcing amplitude (4.313a) less than (4.312d), the oscillation amplitude (4.313b) is less than (4.312a), the roots (4.311b) are complex conjugate (4.313c), and there is no vertical tangent anywhere:

$$|f_a| < f_0: \qquad b^2 < \frac{\lambda}{\omega_0 \psi}, \qquad \varepsilon_- = (\varepsilon_+)^*, \qquad \frac{db}{d\varepsilon} \neq \infty, \qquad (4.313\text{a–d})$$

so there is only one amplitude for each frequency (Figure 4.10b). In the case III of forcing amplitude (4.314a) larger than (4.312d): (i) the oscillation amplitude (4.314b) exceeds (4.312b); (ii) there are two (4.314d, e) distinct (4.314c) vertical tangents:

$$|f_a| > f_0: \qquad b^2 > \frac{\lambda}{\omega_0 \psi}, \qquad \varepsilon_+ > \varepsilon_- > 0, \qquad \frac{db}{d\varepsilon_+} = 0 = \frac{db}{d\varepsilon_-}, \qquad (4.314\text{a–e})$$

$$\omega_0 + \varepsilon_- < \omega < \omega_0 + \varepsilon_+: \qquad \Delta\omega = \varepsilon_+ - \varepsilon_- = 2\sqrt{\omega_0^2 b^4 \psi^2 - \lambda^2} \leq 2\omega_0 b^4 \psi, \qquad (4.314\text{f–h})$$

(iii) outside (inside) the frequency range (4.314f) of width (4.314g) between the vertical tangents there is one (are three) amplitudes for each frequency (Figure 4.10d); (iv) the frequency range (4.314g) is largest in the absence of damping (4.314h), reduces in the presence of damping, and disappears if damping exceeds the value (4.312b); and (v) in the frequency range (4.314f) are possible amplitude jumps and a hysteresis loop (subsection 4.6.3), corresponding to the case III above (subsection 4.6.2).

4.6.3 Amplitude Jumps and Hysteresis Loop

The theorem on stability boundaries (subsection 4.8.2) states that the parts of the curve facing downward (upward), in a dashed (solid) line in Figure 4.10d are unstable (stable); thus, in the interval $(\varepsilon_1, \varepsilon_2)$ only the highest and lowest amplitude of the motion are stable (Figure 4.10d) implying that: (i) starting from a high frequency the amplitude follows the curve *FED*, then jumps *DB* up at ε_1 from b_1^- to b_1^+, and then follows *BA*; (ii) starting at low-frequency the amplitude follows *ABC*, then jumps down *CE* at ε_2 from b_2^+ to b_2^-, and then follows *EF*; (iii) starting in the interval $(\varepsilon_1, \varepsilon_2)$ the amplitude goes around the **hysteresis loop** *BGCEDB*, thus performing (extracting) work due to the applied external force, depending on the direction clockwise (counterclockwise) in the Figure 4.10d. The maximum amplitude of oscillation corresponds to zero slope (4.315a) in (4.309b) ≡ (4.311a), leading to (4.315b) and the frequency (4.315c) ≡ (4.304a):

$$\frac{db}{d\varepsilon_m} = 0: \quad \varepsilon_m = \omega_0 b^2 \psi; \quad \omega_0 + \varepsilon_m = \omega, \quad |f_a| = 2\omega_0 \lambda b_m, \quad b_m = \frac{|f_a|}{2\omega_0 \lambda}, \qquad (4.315\text{a–e})$$

this corresponds (4.308a) to the forcing amplitude (4.315d) that is related to the maximum amplitude (4.315e) as in the linear case (2.201d) $\equiv$ (4.310c) = (4.315e). Thus *the forcing of the linearly damped anharmonic oscillator (4.301) close (problem 117) to the natural frequency (4.303; 4.302a–e) leads to an oscillation with a single amplitude (4.308a) $\equiv$ (4.308b) at each frequency (Figure 4.10a–c) if the forcing amplitude does not exceed the value (4.312d). If it does exceed (4.314a) the value (4.312d) there is a single amplitude (Figure 4.10d) only outside the frequency range (4.314f) of the hysteresis loop (4.311a, b). The peak amplitude is the same in the linear (4.310a–c) and non-linear (4.315a–e) cases.*

4.6.4 Non-Linear Resonance by Forcing at a Sub-Harmonic Frequency

Forcing a linear oscillator (4.316b) at an applied frequency close (4.302a) to one half of the natural frequency (4.316a) leads to the non-resonant response (4.316c):

$$\omega_a = \frac{\omega_0 + \varepsilon}{2}: \qquad \ddot{x}_1 + \omega_0^2 x_1 = f\cos(\omega_a t), \qquad x_1(t) = \frac{4f_a}{3\omega_0^2}\cos\left[\frac{\omega_0 + \varepsilon}{2}t\right]. \qquad (4.316a\text{–}c)$$

The first-order non-linear approximation for the damped anharmonic oscillator (4.303) is (4.317a):

$$\ddot{x}_2 + 2\lambda\dot{x}_2 + \omega_0^2\left(x_2 + \alpha x_2^2 + \beta x_2^3\right) = -\alpha\omega_0^2 x_1^2 + \dots$$

$$= -\frac{16f_a^2\alpha}{9\omega_0^2}\cos^2\left(\frac{\omega_0 + \varepsilon}{2}\right) + \dots \qquad (4.317a)$$

$$= -\frac{8f_a^2\alpha}{9\omega_0^2}\cos\left[(\omega_0 + \varepsilon)t\right] + \dots,$$

where on the r.h.s. was retained only the quadratic term of the restoring force leading to forcing at the natural frequency (4.316a). The non-linear terms (4.303) are taken into account replacing (4.304a, b) the natural by the oscillation frequency (4.317b):

$$\ddot{x}_2 + 2\lambda\dot{x}_2 + \omega^2 x_2 = -\frac{8f_a^2\alpha}{9\omega_0^2}\cos\left[(\omega + \varepsilon)t\right]. \qquad (4.317b)$$

The oscillation (4.317b) is similar to (4.303) $\equiv$ (4.305): (i) with the substitution of (4.318a) for $|f_a|$ in (4.308a), leading to the relation (4.318b) between the amplitude of forcing $|f_a|$ and oscillation:

$$-f_1 = \frac{8\alpha|f_a|^2}{9\omega_0^2}: \qquad b^2\left[\left(\omega_0 b^2\psi - \varepsilon\right)^2 + \lambda^2\right] = \left(\frac{f_1}{2\omega_0}\right)^2 = \left[\frac{4\alpha|f_a|^2}{9\omega_0^3}\right]^2; \qquad (4.318a, b)$$

(ii) the same substitution (4.318a) in (4.312d) leads (4.319a) to the minimum forcing amplitude (4.319b) for the existence of an hysteresis loop:

$$\frac{8\alpha|f_a|^2}{9\omega_0^2} > 2\sqrt{2}\omega_0\lambda b_0: \qquad\qquad |f_a|^2 > \frac{9\sqrt{2}}{4\alpha}\omega_0^3\lambda b_0 \equiv f_{10}; \qquad\qquad (4.319a, b)$$

(iii) the frequency range for hysteresis is the same (4.314f–h). Thus, *the forcing of a linearly damped anharmonic oscillator (4.303) by a quadratic restoring force (4.317a) at an applied frequency (4.316a) close (4.302a) to one-half of the natural frequency (4.302b) leads (problem 118) to a* **first-subharmonic resonance**, *similar to the resonance at the natural frequency (amplitude jumps and hysteresis loop) in Figures 4.10a–c(d) with reduced amplitude both of oscillation (4.318a, b) and of triggering the limit cycle (4.319a, b). The hysteresis loop leads to a growing (decaying) oscillation if it is described in the clockwise (counterclockwise) direction, so that the external forcing provides (extracts) work, increasing (decreasing) the energy in each cycle.* The amplitude jumps and hysteresis loop correspond to the fold or cusp bifurcation (subsection 4.8.8) of which real examples or analogues include flutter, wake breaking, and shock waves (subsection 4.6.5).

4.6.5 Flutter, Wave Breaking, and Shock Formation

The examples of divergent or unstable hysteresis loops include the flutter (Figure 4.11a) of the control surfaces of an airplane with aerodynamic forcing. This occurs at large angles of attack and or sideslip far from equilibrium conditions. The large aerodynamic forces require large deflections of the control surfaces. If the hysteresis loop is entered as a **limit cycle** with energy input the amplitude of the oscillations increases; eventually the control surfaces reach **saturation**; that is, attain the maximum mechanically possible deflections and hit the **bump stops**. The violent oscillations and repeated collisions with the bump stops may lead to break-up or disintegration. Flutter can be avoided, ensuring that the applied force does not exceed the limit (4.312d); a smaller aerodynamic force may imply keeping the velocity below a **flutter speed**. Thus, an airplane may have a **flutter envelope** that is smaller than its **flight envelope** (Figure 4.11b); that is, there are speeds and altitudes at which it could fly if there was no risk of flutter. The more desirable situation is the inverse, of a flutter envelope larger than the flight envelope.

The evolution of the curves in the Figures 4.10a–d correspond to cuts in a folded sheet (subsection 4.8.8) and look like a wave breaking on a beach. The analogy is justified because a wave on the surface of water is linear (non-linear) if the amplitude is (is not) small compared with the depth (Figure 4.12a); as the linear ocean wave approaches the sloping beach its amplitude ceases to be small compared with the depth, and the **water wave** becomes (Figure 4.12b) a

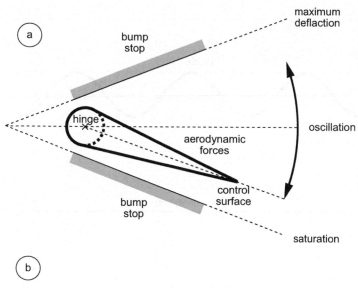

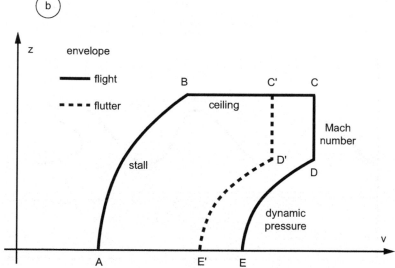

FIGURE 4.11

A control (elevator, rudder, aileron, canard, taileron) or high-lift (flap, slat) surface of an aircraft rotates around an hinge with saturation limits imposed by bump stops that dictate the maximum deflection (Figure 4.11a). When flutter occurs, the aerodynamic forces do work in a hysteresis loop or limit cycle and the oscillation amplitude increases until repeated collision with the bump stops and possible beak-up of the hinge or the control or high-lift surface lead to loss of control. The flight envelope (Figure 4.11b) of an aircraft in terms of airspeed v versus altitude z has typically four boundaries (i) stall due to lack of lift at low speed (*AB*); (ii) maximum altitude (*BC*) for zero climb rate; (iii) maximum speed at high altitude (*CD*) due to drag or shock waves; and (iv) maximum speed at low altitude (*DE*) due to high air density and dynamic pressure. Flutter may reduce the flight envelope by limiting the maximum speed (*C'D'E'*) below the aerodynamic and propulsive capabilities of the aircraft (*CDE*).

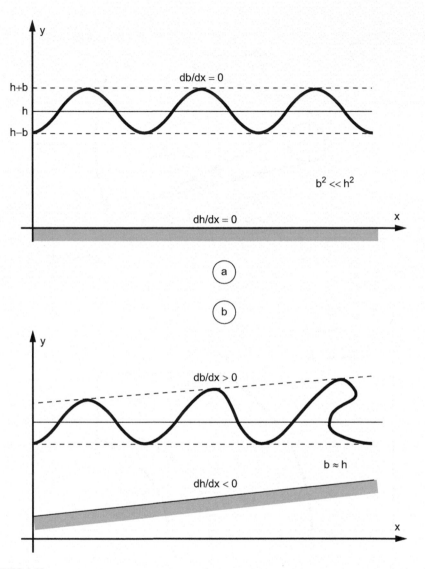

FIGURE 4.12
A water wave on the surface of the ocean (Figure 4.12a) is linear because the amplitude is small $b^2 \ll h^2$ compared with the depth. When approaching the coastline on a sloping beach, $dh/dx < 0$ and the wave steepens (Figure 4.12b) $db/dx > 0$ as in Figure 4.10a–d and eventually breaks.

non-linear **breaking wave**. Another analogy (Figures 4.13a, b) is an **acoustic wave** that is linear (non-linear) if the pressure perturbation is (is not) small compared the mean pressure of the ambient medium. Thus a non-linear sound wave increases (decreases) the total ambient plus wave pressure in the **compression (depression) phase** (Figure 4.13a); the compression (rarefaction) front travels faster (slower) that the average wave speed, and the relative lead

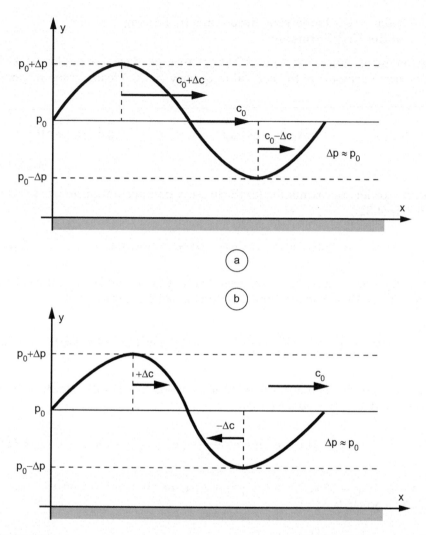

FIGURE 4.13
A non-linear sound wave (Figure 4.13a) causes a large perturbation Δp of the ambient pressure p_0 so that: (i) the total pressure $p_0 + \Delta p \,(p_0 - \Delta p)$ is increased (decreased) in the compression (depression) phase; (ii) the compression front (depression tail) moves ahead Δc (lags behind-Δc) the average speed c_0, and the wave steepens (Figure 4.13b) as in Figures 4.10a–d until it forms a shock wave.

(lag) causes the steepening of the wave front (Figure 4.13b) that eventually forms a **shock wave**, like the **sonic boom** of a supersonic aircraft. Thus the non-linear resonance of an anharmonic oscillator is a prototype problem for other non-linear wave phenomena. Having considered [subsection(s) 4.6.4 (4.6.2–4.6.3)] the non-linear resonance of an anharmonic oscillator at the first subharmonic (fundamental frequency), next are considered the first (second) harmonics [subsections 4.6.6–4.6.7 (4.6.8–4.6.9)].

4.6.6 Non-Linear Parametric Resonance by Forcing at the First Harmonic

The forcing of a linear oscillator (4.320b) at an applied frequency close (4.302b) to the first harmonic of the natural frequency (4.320a) leads to the response (4.320b):

$$\omega_a = 2(\omega_0 + \varepsilon): \quad \ddot{x}_1 + \omega_0^2 x_1 = f_a \cos(\omega_a t), \quad x_1(t) = -\frac{f_a}{3\omega_0^2} \cos\left[2(\omega_0 + \varepsilon)t\right].$$

$$(4.320a\text{--}c)$$

The first-order approximation for the linearly damped anharmonic oscillator (4.303) is (4.321):

$$\ddot{x}_2 + 2\lambda\dot{x}_2 + \omega_0^2\left(x_2 + \alpha x_2^2 + \beta x_2^3\right) = -2\alpha\omega_0^2 x_1 x_2 + \dots \tag{4.321}$$

where the r.h.s. retained only the term leading to forcing at twice the natural frequency (4.320a). Substitution of (4.320c) in (4.321) leads to:

$$\ddot{x}_2 + 2\lambda x_2 + \omega_0^2 x_2 \left\{1 - \frac{2\alpha f_a}{3\omega_0^4}\cos\left[2(\omega_0 + \varepsilon)t\right]\right\} + \omega_0^2\left(\alpha x_2^2 + \beta x_2^3\right) = 0, \tag{4.322a}$$

that corresponds (section 4.3) to parametric resonance (4.181) in terms (4.322b) of the oscillation frequency (4.304a):

$$\ddot{x}_2 + 2\lambda x_2 + \omega^2 x_2 = \omega^2 \frac{2\alpha f_a}{3\omega_0^4} x_2 \cos\left[2(\omega_0 + \varepsilon)t\right]. \tag{4.322b}$$

Substituting (4.323a, b) in the parametric resonance term on the r.h.s. of (4.322b) leads to (4.323c):

$$\omega \sim \omega_0, \quad x_2(t) \sim b\cos\left[(\omega_0 + \varepsilon)t\right]:$$

$$\frac{2\alpha f_a b}{3\omega_0^2}\frac{\omega^2}{\omega_0^2} x_2 \cos\left[2(\omega_0 + \varepsilon)t\right] \sim \frac{2\alpha f_a b}{3\omega_0^2}\cos\left[(\omega_0 + \varepsilon)t\right]\cos\left[2(\omega_0 + \varepsilon)t\right]$$

$$= \frac{\alpha f_a b}{3\omega_0^2}\left\{\cos\left[(\omega_0 + \varepsilon)t\right] + \cos\left[3(\omega_0 + \varepsilon)t\right]\right\};$$

$$(4.323a\text{--}c)$$

the term of frequency $3(\omega_0 + \varepsilon)$ is non-resonant, and is omitted from (4.323c) when substituting in the r.h.s. of (4.322b), leading to (4.323d):

$$\ddot{x}_2 + 2\lambda\dot{x}_2 + \omega^2 x_2 = \frac{\alpha f_a b}{3\omega_0^2}\cos\left[(\omega_0 + \varepsilon)t\right]. \tag{4.323d}$$

The oscillation (4.323d) is similar to (4.303) ≡ (4.305) with the substitution (4.324a):

$$f_2 = \frac{\alpha |f_a| b}{3 \, \omega_0^2}: \qquad b^2 \left[\left(\omega_0 b^2 \psi - \varepsilon \right)^2 + \lambda^2 \right] = \left(\frac{f_2}{2\omega_0} \right)^2 = \left[\frac{\alpha |f_a| b}{6\omega_0^3} \right]^2, \qquad (4.324a, b)$$

the substitution of $|f_a|$ in (4.308a) by (4.324a) leads to the relation (4.324b) between the amplitudes of forcing $|f_a|$ and oscillation b. This relation can lead to: (i) amplitude jumps as before (subsections 4.6.1–4.6.4); also to (ii) suppression of oscillations as shown next (subsection 4.6.7).

4.6.7 Amplitude Jumps and Suppression of Oscillations

The amplitude equation (4.324b) with the approximation (4.325a) can be factorized (4.325b):

$$\varepsilon^2 \ll \omega_0^2 b^4 \psi^2: \qquad 0 = b^2 \left[\omega_0^2 b^4 \psi^2 - 2\omega_0 \varepsilon b^2 \psi + \lambda^2 - \left(\frac{\alpha |f_a|}{6 \, \omega_0^3} \right)^2 \right], \qquad (4.325a, b)$$

that has three roots for b^2: (i) the zero root (4.326a); (ii–iii) the roots (4.326b, c):

$$b_0 = 0, \qquad \left(b_{\pm} \right)^2 = \frac{1}{\omega_0 \psi} \left\{ \varepsilon \pm \sqrt{\left(\frac{\alpha |f_a|}{6 \, \omega_0^3} \right)^2 - \lambda^2} \right\}. \qquad (4.326a\text{–}c)$$

If the damping is not too strong (4.327a), the roots (4.326b, c) are real (4.327a, b):

$$\lambda < \frac{\alpha |f_a|}{6\omega_0^3}: \qquad \left(b_{\pm} \right)^2 = \frac{\varepsilon \pm \varepsilon_0}{\omega_0 \psi}, \qquad \varepsilon_0 \equiv \left| \left(\frac{\alpha |f_a|}{6\omega_0^3} \right)^2 - \lambda^2 \right|^{1/2}. \qquad (4.327a\text{–}c)$$

The roots (4.327b, c) vanish (4.324a, b) at the points $\pm\varepsilon_0$:

$$b_+ \left(-\varepsilon_0 \right) = 0 = b_- \left(\varepsilon_0 \right), \qquad b_* \equiv b_+ \left(\varepsilon_0 \right) = \frac{2\varepsilon_0}{\omega_0 \psi}, \qquad (4.328a\text{–}c)$$

and at ε_0 the second root has the value (4.328c). As shown in the Figure 4.14: (i) the root b_+ is real beyond the point $-\varepsilon_0$ and is stable (by the theorem of

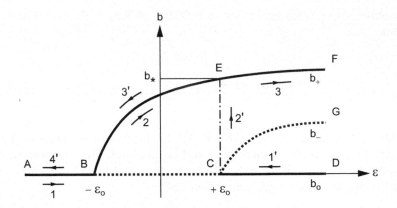

FIGURE 4.14
In the non-linear resonance of an anharmonic oscillator with a potential that is a quartic function of position with forcing at the first harmonic; that is, applied frequency equal to twice the natural frequency, there: (i) is one zero amplitude in the lower frequency range $\varepsilon < -\varepsilon_0$; (ii) three amplitudes of which only the highest and lowest (that is zero) are stable in the higher frequency range $\varepsilon > \varepsilon_0$; and (iii) two amplitudes of which only the higher is stable in the intermediate frequency range $-\varepsilon_0 < \varepsilon < \varepsilon_0$. This leads to three situations: (i) starting in the lower frequency range $\varepsilon < -\varepsilon_0$, there is no oscillation AB until $\varepsilon = -\varepsilon_0$ and then the amplitude increases with frequency BEF; (ii) starting with non-zero amplitude at F in the high-frequency range, the reverse of (i) happens with the amplitude reducing FEB until the oscillations are suppressed AB at $\varepsilon = -\varepsilon_0$; and (iii) starting with zero amplitude at D in the high-frequency range $\varepsilon > \varepsilon_0$, the oscillation starts at $\varepsilon = \varepsilon_0$ with an amplitude jump CE, then the amplitude decreases EB until the oscillation stops again AB at $\varepsilon = -\varepsilon_0$.

subsection 4.8.2) because it corresponds to the segment BEF facing upward; (ii) the root b_- is real beyond $+\varepsilon_0$, but is always unstable, because it corresponds to the segment CG that lies between the other roots $b_+ > b_- > 0 = b_0$ and so faces downwards; and (iii) the root $b_0 = 0$ corresponds to the real axis, and is unstable (stable) between (outside) $-\varepsilon_0$ and $+\varepsilon_0$, where it faces downwards (upwards). The motion may be described as follows: (i) starting from high-frequency there is no oscillation along DC up to $+\varepsilon_0$, where there is a jump to E, with amplitude (4.328c), followed by an oscillation with decreasing amplitude along EB, that is suppressed at $-\varepsilon_0$ along BA; and (ii) starting at low frequency there is no oscillation along AB, and the oscillation starts smoothly at B with $-\varepsilon_0$ with increasing amplitude along BEF. Figure 4.14 applies if $\psi > 0$, and in the opposite case $\psi < 0$ it is inverted relative to the ε − axis. Thus a linearly *damped anharmonic oscillator (4.303) forced at an applied frequency (4.320a) close (4.302a) to the first harmonic of the natural frequency, has* (problem 119) *non-linear resonance at the natural frequency (Figure 4.14) with: (i) oscillation always in the range* $(-\varepsilon_0, +\varepsilon_0)$ *in (4.327c); (ii)* **suppression of oscillations** *always at lower frequencies* $\varepsilon < -\varepsilon_0$; *and (iii) at higher frequencies the oscillations are suppressed if coming from above* $\varepsilon > \varepsilon_0$, *until* $\varepsilon = \varepsilon_0$ *when there is an amplitude jump (4.328c), joining the oscillations whose amplitude grows smoothly if coming from below. This corresponds to the Hopf bifurcation (subsection 4.8.5).*

4.6.8 Non-Linear Parametric Resonance by Forcing at the Second Harmonic

The forcing of a linear oscillator (4.329b) at an applied frequency (4.329a) close (4.302a) to the second harmonic of the natural frequency leads to the response (4.329c):

$$\omega_a = 3(\omega_0 + \varepsilon): \qquad \ddot{x}_1 + \omega_0^2 x_1 = f_a \cos(\omega_a t), \quad x_1(t) = -\frac{f_a}{8\omega_0^2} \cos\left[3(\omega_0 + \varepsilon)t\right].$$

$$(4.329a-c)$$

The first-order approximation for the damped, non-linear oscillator (4.303) is (4.330):

$$\ddot{x}_2 + 2\lambda\dot{x}_2 + \omega_0^2 x_2 + \alpha x_2^2 + \beta x_2^3 = -3\beta x_1 x_2^2 + ..., \qquad (4.330)$$

where only the term linear in the first order and quadratic in the second order was retained. The reason is that for an oscillation (4.331b) close to the natural frequency (4.331a):

$$\omega = \omega_0 + \varepsilon: \qquad x_2(t) = b\cos\left[(\omega_0 + \varepsilon)t\right] = b\cos(\omega t), \qquad (4.331a, b)$$

the forcing term on the r.h.s. of (4.330):

$$-3\beta x_1(t)\left[x_2(t)\right]^2 = \frac{3\beta f_a b^2}{8\,\omega_0^2} \cos^2\left[(\omega_0 + \varepsilon)t\right]\cos\left[3(\omega_0 + \varepsilon)t\right]$$

$$= \frac{3\beta f_a b^2}{16\,\omega_0^2}\left\{1 + \cos\left[2(\omega_0 + \varepsilon)t\right]\right\}\cos\left[3(\omega_0 + \varepsilon)t\right] \qquad (4.331c)$$

$$= \frac{3\beta f_a b^2}{32\,\omega_0^2}\left\{\cos\left[(\omega_0 + \varepsilon)t\right] + ... \right\},$$

(i) contains a triple frequency multiplied by the square of the frequency; (ii) the latter introduces a double frequency; (iii) the product of the double and triple frequency reintroduces the oscillation frequency that causes resonance; and (iv) the resonant term on the r.h.s. of (4.330) $\equiv$ (4.331c) and the substitution of the l.h.s. of (4.330) as in (4.303) by the oscillation frequency (4.304a) as in (4.305) leads to:

$$\ddot{x}_2 + 2\lambda\dot{x}_2 + \omega^2 x_2 = \frac{3\beta f_a b^2}{32\omega_0^2}\cos\left[(\omega_0 + \varepsilon)t\right]. \qquad (4.332)$$

Thus *forcing of an anharmonic, linearly damped oscillator (4.303) near (4.302a) at the second harmonic (4.329a) leads (4.332) to (problem 120) resonance at the natural frequency through the cubic term of the non-linear restoring force (4.330), with a double parametric resonance (section 4.3) involving a product of three cosines (4.331c).* The factor (4.333a) on the r.h.s. of (4.332) replaces $|f_a|$ in the relation (4.308a) between the amplitude of forcing $|f_a|$ and oscillation leading to (4.333b):

$$f_3 = \frac{3\beta |f_a| b^2}{32\omega_0^2}: \qquad b^2 \left[\left(\omega_0 b^2 \psi - \varepsilon \right)^2 + \lambda^2 \right] = \left(\frac{f_3}{2\omega_0} \right)^2 = \left[\frac{3\beta |f_a| b^2}{64\omega_0^3} \right]^2. \qquad (4.333a, b)$$

Whereas forcing a linearly damped anharmonic oscillator at the natural frequency (first harmonic) causes [subsections 4.6.1–4.6.4 (4.6.6)] amplitude jumps and hysteresis loop (suppression of oscillations) [subsection 4.6.5 (4.6.7)], forcing at the third harmonic (subsection 4.6.8) leads to (subsection 4.6.9) a threshold amplitude and frequency shift.

4.6.9 Threshold Amplitude and Frequency Shift

The amplitude equation (4.333b) $\equiv$ (4.334b) involves the coefficient (4.333a):

$$B \equiv \frac{9\beta^2}{8192\omega_0^6}: \qquad 0 = b^2 \left[\left(\omega_0 b^2 \psi - \varepsilon \right)^2 + \lambda^2 - 2B |f_a|^2 b^2 \right]$$

$$= b^2 \left[\omega_0^2 \psi^2 b^4 - 2\left(\omega_0 \varepsilon \psi + B |f_a|^2 \right) b^2 + \varepsilon^2 + \lambda^2 \right]; \qquad (4.334a\text{–}c)$$

the cubic (4.334b) $\equiv$ (4.334c) in b^2 has the roots (4.335a–c) for (4.325a):

$$b_0 = 0: \qquad \omega_0^2 \psi^2 \left(b_\pm \right)^2 = \omega_0 \varepsilon \psi + B |f_a|^2 \pm \sqrt{2B |f_a|^2 \omega_0 \varepsilon \psi + B^2 |f_a|^4 - \omega_0^2 \psi^2 \lambda^2}.$$

$$(4.335a\text{–}c)$$

There is no oscillation (4.335a) unless the roots (4.335b, c) are real, leading to (4.336a):

$$\varepsilon > \frac{\omega_0 \psi \lambda^2}{2B |f_a|^2} - \frac{B |f_a|^2}{2\omega_0 \psi} \equiv \varepsilon_0; \qquad \omega_0^2 \psi^2 \left(b_\pm \right)^2 = \omega_0 \varepsilon \psi + B |f_a|^2 \pm |f_a| \sqrt{2B\omega_0 \psi (\varepsilon - \varepsilon_0)} > 0;$$

$$(4.336a\text{–}c)$$

the roots (4.335b, c) $\equiv$ (4.336b, c) coincide for $\varepsilon = \varepsilon_0$, leading to (4.337a–c):

$$(b_*)^2 \equiv \left[b_\pm (\varepsilon_0) \right]^2 = \frac{\varepsilon_0}{\omega_0 \psi} + \frac{B |f_a|^2}{\omega_0^2 \psi^2} = \frac{\lambda^2}{2B |f_a|^2} + \frac{B |f_a|^2}{2\omega_0^2 \psi^2} > 0. \qquad (4.337a\text{–}c)$$

Thus no oscillation is possible for $\varepsilon < \varepsilon_0$ because the only root (4.335a) implies zero amplitude. For $\varepsilon \geq \varepsilon_0$ in (4.336a) the root b_+ (b_-) is stable (unstable) because (Figure 4.15) it faces upwards (downwards). Thus the motion may be described as follows: (i) starting at a low frequency no oscillation is possible along DE, jumping at ε_0 to A, and along AB an oscillation is possible with amplitude $b_+ > b_*$; (ii) starting at a high frequency with zero amplitude no oscillation is possible; (iii) starting at high frequency with amplitude b_+ the oscillation is possible until ε_0 with amplitude b_* and is suppressed for lower frequencies. Thus b_* is the **threshold amplitude** below which no oscillations can occur. It has been shown that *the linearly damped, anharmonic oscillator (4.303) forced (problem 120) at a frequency (4.329a) close (4.302b) to the second harmonic of the natural frequency, (i) has non-linear resonance only above the **frequency shift** (4.336a) if it attains a minimum amplitude b_+ in (4.335b) $\equiv$ (4.335c) $\equiv$ (4.336b) $\equiv$ (4.336c) that at least equals the threshold b_* in (4.337a) $\equiv$ (4.337b) $\equiv$ (4.337c); (ii) there is suppression of oscillations beyond ε_0 for amplitudes below the threshold; and (iii) no oscillation is possible in any case below ε_0.*

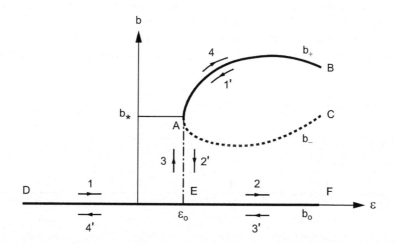

FIGURE 4.15
In the non-linear resonance of an anharmonic oscillator with the potential a quartic function of the position, with forcing at the second harmonic, that is applied frequency the triple of the natural frequency: (i) there is no oscillation in the low frequency range $\varepsilon < \varepsilon_0$; (ii) in the high frequency range $\varepsilon > \varepsilon_0$ there are three amplitudes of which are stable the highest and lowest with the latter being zero. Thus there are two reversible possibilities: (i) a continuous amplitude is possible over the whose frequency range only if it is zero, that is, there no oscillation for increasing DEF or decreasing FED frequency; (ii) starting in the lower frequency range $\varepsilon < \varepsilon_0$, the oscillation can start only at $\varepsilon = \varepsilon_0$ with an upward amplitude jump EA; (iii) starting at B in the higher-frequency range $\varepsilon > \varepsilon_0$ with non-zero amplitude with decreasing frequency the oscillations are suppressed at $\varepsilon = \varepsilon_0$ with a downward amplitude jump AE.

4.6.10 Non-Linear Resonance by Forcing at Harmonics and Sub-Harmonics

The forcing of a damped, anharmonic oscillator at an applied frequency (4.338c) close (4.338d) to any rational multiple (4.338a, b) of the natural frequency:

$$p, q \in | N: \qquad \omega_a = \frac{p}{q}\omega_0 + \varepsilon, \qquad \varepsilon^2 \ll \omega_0^2, \qquad (4.338a\text{–}d)$$

can cause (problem 121) resonance at the natural frequency, with smaller amplitude for larger positive integer (p, q). These resonances do not introduce new phenomena beyond (Table 4.4) those already described: (i) amplitude jumps, for example at the natural frequency and first subharmonic (Figure 4.10d) and at the first (second) harmonic [Figure 4.14 (4.15)]; (ii) hysteresis loops, for example subsections [4.6.1– 4.6.3 (4.6.4–4.6.5)] at the fundamental (first subharmonic); (iii) amplitude thresholds, for example at the second harmonic (subsections 4.6.8–4.6.9); (iv) suppression of oscillations, for example at the first (second) harmonic [subsections 4.6.6–4.6.7 (4.6.8–4.6.9)]. The non-linear resonance was considered for restoring force with: (i–ii) quadratic (cubic) non-linearity for the first (second) harmonic [subsections 4.6.6–4.6.7 (4.6.8–4.6.9)]; (iii–iv) both quadratic and cubic non-linearities for [subsections 4.6.1–4.6.3 (4.6.4–4.6.5)] the fundamental (first sub-harmonic). Next is considered non-linear damping and amplification (section 4.7).

TABLE 4.4

Resonances of an Anharmonic Oscillator

Case	I	II	III	IV
subsection	4.6.1–4.6.3	4.6.4–4.6.5	4.6.6–4.6.7	4.6.8–4.6.9
figure	4.10–4.13	4.10–4.13	4.14	4.15
forcing at	natural frequency	first sub-harmonic	first harmonic	second harmonic
amplitude jumps	X	X	X	X
hysteresis loop	X	X	-	–
suppression of oscillations	–	–	X	X
amplitude threshold	–	–	–	X

Note: The phenomena associated with non-linear resonance include (i) amplitude jumps, (ii) hysteresis loops, (iii) suppression of oscillations, and (iv) amplitude thresholds.

4.7 Non-Linear Damping and Self-Excited Oscillations

Following non-linear restoring forces (sections 4.4–4.6), non-linear damping or amplification is considered (section 4.7). If there is damping at all times, a limit cycle is not possible (subsection 4.7.1) and the system spirals down to rest. Self-excited oscillations are possible (subsection 4.7.2) if there is damping for large and amplification for small amplitudes or velocities so that the system can neither diverge nor decay to rest. An example is the homopolar dynamo (subsections 4.7.3–4.7.5); the dynamo effect may be related to magnetic field generation in rotating bodies like the earth, sun, and other planets and stars (subsections 4.7.6–4.7.8). The reverse of non-linear restoring force with linear damping (section 4.6) is linear restoring force with non-linear damping (subsection 4.7.15). An example of restoring and friction forces both non-linear is the large amplitude motion of a pendulum. The undamped, non-linear motion of a pendulum includes (subsections 4.7.9–4.7.14) three cases: (i) if the total energy exceeds the potential energy at the highest point, that is, circulatory motion; (ii) if the total energy equals the potential energy at the highest point, that is, a position of unstable equilibrium, the pendulum takes an infinite time to reach the position of rest; and (iii) if the total energy is less than the potential energy at the highest point, the latter cannot be reached, and there are large amplitude oscillations that include linear oscillations for small amplitude around the lowest position of stable equilibrium. In the presence of damping, the pendular motion ultimately decays to the equilibrium position regardless of the initial condition, whether it is circulatory or a large or small oscillation. This is shown for small oscillations with quadratic damping (subsection 4.7.15). The quadratic damping is also considered for large amplitude pendulum motion (subsection 4.7.16); if the initial angular velocity is high enough the pendulum performs a finite number of revolutions, after which it settles to a damped oscillation leading to rest (subsection 4.7.17).

4.7.1 Dissipation Function, Damping, and Amplification

The unforced (4.339a) general second-order system with separable non-linear restoring and friction forces (4.208c) is (4.339b):

$$F(t) = 0: \qquad m\ddot{x} - j(\dot{x}) = h(x) = -\frac{d\Phi}{dx}. \qquad (4.339a, b)$$

The rate of change with time of the total (4.210b), that is, kinetic plus potential energy (4.340a) is specified:

$$E \equiv \frac{m}{2}\dot{x}^2 + \Phi(x): \qquad \frac{dE}{dt} = m\dot{x}\ddot{x} + \frac{d\Phi}{dx}\frac{dx}{dt} = \dot{x}\left[m\ddot{x} - h(x)\right] = \dot{x}j(\dot{x}) \equiv -\Psi(\dot{x}),$$

$$(4.340a, b)$$

*by a **dissipation function** (4.340b) that equals is the work performed by the friction force per unit time:*

$$dW = j(\dot{x})dt = j(\dot{x})\dot{x}dt = -\Psi(\dot{x})dt. \tag{4.341}$$

If the system has a limit cycle, the energy returns to the same value after a period and the integral of the dissipation function must be zero:

$$0 = \oint dE = \oint (dE/dt)dt = -\oint \Psi(\dot{x})dt. \tag{4.342}$$

*Thus two cases are possible: (i) if there is dissipation (amplification) at least some of the time $\Psi > 0$, $\Psi < 0$, then (4.342) cannot be met, and a limit cycle cannot exist, that is, there is **braking (growth)** of the system down to rest (away from rest); (ii) a limit cycle is possible if dissipation $\Psi > 0$ and amplification $\Psi < 0$ balance over a period (4.342). An example is a **self-excited oscillator**: (i) damping dominates for large velocity, so the path is bounded; (ii) amplification dominates for small velocity so it cannot come to rest. As an example with both first and second-order cases is the homopolar disk dynamo considered next (subsection 4.7.2).*

4.7.2 Homopolar Disk Dynamo Under Uniform Rotation

The electric current J due an homopolar disk dynamo corresponds (4.343a) to an electrical circuit with total induction L and resistance R:

$$L\dot{J} + RJ = F_e, \qquad F_e = \pm \frac{\Phi_m \Omega}{2\pi}, \qquad \Phi_m = ZJ, \tag{4.343a–c}$$

where the electromotive force is due to the magnetic flux Φ_m through the rim of the disk (4.343b) and Ω is the angular velocity; the magnetic flux (4.343c) equals the current J times the mutual induction Z. The upper $+$ (*lower* $-$) sign in (4.343b) corresponds to corotation (counterrotation); that is, the external wire carrying the current (Figure 4.16) in the same (opposite) direction as that of the rotation. Substituting (4.343c) in (4.343b) specifies the electromotive force (4.344b) driving the circuit (4.343a) $\equiv$ (4.344a):

$$L\dot{J} + RJ = F_e = \pm \frac{Z\Omega}{2\pi}J; \qquad J_\pm(t) = J_\pm(0)\exp\left[-\left(R \mp \frac{Z\Omega}{2\pi}\right)\frac{t}{L}\right], \tag{4.344a, b}$$

the integral (4.344b) of (4.344a) shows that an *homopolar electric disk dynamo for (problem 122) counterrotation, that is the + sign below in (4.344b) acts as a* **brake** *in (4.345a) since the mutual inductance and angular velocity act (4.345c)*

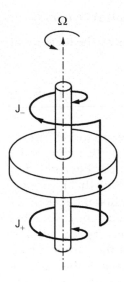

FIGURE 4.16
A homopolar disk dynamo with electric current in the direction opposite J_- to that of rotation acts (i) as a brake. If the current J_+ is in the direction of rotation, the angular velocity is (ii) constant for a critical resistance; higher (lower) resistance leads to (iii) a reduced braking relative to (i) [an exponential increase in current with time (iv) without bound in the absence of rotational inertia and driving torque (see Figure 4.17)].

to increase the effective resistance (4.345b), causing a faster exponential decay of the current:

$$J_+(t) = J_+(0)\exp\left(-\frac{\tilde{R}}{L}t\right): \qquad \tilde{R} = R + \frac{Z\Omega}{2\pi} = R + \bar{R}, \qquad \bar{R} = \frac{Z\Omega}{2\pi}. \qquad (4.345a, b)$$

In the case (problem 123) of corotation; that is, the − sign above in (4.344b), the mutual induction and rotation oppose the resistance leading to three cases:

$$\lim_{t\to\infty} J_-(t) = J_-(0) \times \begin{cases} \exp\left[-|R - \bar{R}|t/L\right] = 0 & \text{if} \quad R > \bar{R}, & (4.346a) \\ 1 & \text{if} \quad R = \bar{R}, & (4.346b) \\ \exp\left[|R - \bar{R}|t/L\right] = \infty & \text{if} \quad \bar{R} > R, & (4.346c) \end{cases}$$

(i) for the critical resistance a constant current; (ii) for larger resistance a (4.346b) reduced braking effect $R - \bar{R}$; (iii) for smaller resistance (4.346c) the current grows exponentially with time. The preceding results assume that the angular velocity is kept constant by providing whatever driving torque is needed; the torque is related to the angular acceleration through the moment of inertia. Thus the preceding **electrical dynamo** (subsection 4.7.2) should for consistency be complemented by the mechanical effects of rotational inertia in the **electromechanical dynamo** (subsection 4.7.3).

4.7.3 Effects of Rotational Inertia in the Electromechanical Dynamo

The product of the angular acceleration $\dot{\Omega}$ around the axis by the moment of inertia I relative to the axis is the inertia torque (4.347a):

$$I\dot{\Omega} = M_e = aF_e = \pm\frac{aZ}{2\pi}\Omega J, \qquad (4.347a\text{–}d)$$

that is balanced by the driving torque (4.347b), equal to the product (4.347c) of the electromotive force (4.344a) by the moment arm a; that is, the radius of the disk (Figure 4.16) leading to (4.347d). The result (4.347d) can be interpreted in terms of the Lorentz force (subsection I.28.3.2 and note III.8.10) that is proportional to the velocity and magnetic field; these are proportional respectively to the angular velocity Ω and magnetic flux or (4.343c) current J. Thus the torque should scale on the product of the electric current by the angular velocity (4.347d) like the electromotive force (4.344a), and they are in a constant ratio a in (4.348b) corresponding to the moment arm a. If the resistive term in (4.344a) is small compared with that due to the mutual induction (4.348a), the current equation simplifies to (4.346b):

$$2\pi R \ll Z\Omega: \qquad j = \frac{F_e}{L} = \frac{M_e}{aL} = \frac{I}{aL}\dot{\Omega}; \qquad \frac{J}{\Omega} = \frac{I}{aL}. \qquad (4.348a\text{–}e)$$

The current equation (4.348b) is similar (4.348c) to the equation (4.347c) for the angular velocity, showing (4.348d) that it is in a constant ratio to the angular velocity (4.348e) if they are both zero at some initial time. Substitution (4.348e) in (4.344a) leads to the differential equation for the electric current (4.349c):

$$S \equiv \frac{Z\Omega}{2\pi J} = \frac{ZaL}{2\pi I}: \qquad L\dot{j} = -RJ \pm SJ^2, \qquad (4.349a\text{–}c)$$

that involves the constant **dynamo parameter** (4.349a, b). Thus, *the electromechanical homopolar dynamo (Figure 4.16) in the case of small resistance compared with the mutual induction (4.348a), has electric current and angular velocity in a constant ratio (4.348e) and satisfies the same non-linear equation (4.349c) with coefficient (4.349a) $\equiv$ (4.349b).* The equation (4.349c) is solved next (subsection 4.7.4).

4.7.4 Case of Small Resistance Compared with the Induction

The non-linear differential equation of the first order (4.349c) is separable (4.350a) and can be integrated (4.350b) from the initial current $J_{\pm 0}$ at time $t = 0$ to the current $J_\pm$ at time t:

$$-\frac{R}{L}t = \int_{J_0}^{J_\pm} \frac{dJ}{J \mp SJ^2/R} = \int_{J_0}^{J_\pm}\left(\frac{1}{J} \pm \frac{1}{R/S \mp J}\right)dJ = \left[\log\left(\frac{J}{R/S \mp J}\right)\right]_{J_0}^{J_\pm}, \qquad (4.350a, b)$$

using a the partial fraction decomposition (section I.31.8; subsection 1.5.3) to obtain (4.350b) ≡ (4.351a, b):

$$J_0 \equiv J_{\pm}(0): \qquad \exp\left(-\frac{R}{L}t\right) = \frac{J_{\pm}}{J_0} \frac{R \mp J_0 S}{R \mp J_{\pm} S}. \qquad (4.351a, b)$$

The equation (4.351b) specifies the electric current and angular velocity (4.348e) as a function of time:

$$\frac{I}{aL}\Omega_{\pm}(t) = J_{\pm}(t) = \frac{J_0 R}{(R \mp J_0 S)\exp(Rt/L) \pm J_0 S}. \qquad (4.352a, b)$$

If it can be checked that (4.352b) satisfies the initial condition (4.351a) at time $t = 0$; the current and angular velocity also vanish in (4.352b) as $t \to \infty$, showing that the retarding torque due to the Lorentz force will stop the motion unless there is a driving torque. In the preceding solution the approximation of small resistance (4.348a) was made in (4.348c). If the approximation had also been made in (4.349b) it would become (4.353a):

$$L\dot{J} = \pm SJ^2; \quad \pm\frac{S}{L}t = \int_{J_0}^{J_{\pm}}\frac{dJ}{J^2} = \frac{1}{J_0} - \frac{1}{J_{\pm}}, \quad \frac{I}{aL}\Omega_{\pm}(t) = J_{\pm}(t) = \frac{J_0}{1 \mp J_0 St/L}; \quad (4.353a\text{--}c)$$

the solution (4.353b) also confirms that J_0 is the initial value and that $J \to 0$ as $t \to \infty$ algebraically in (4.353c) instead of exponentially in (4.352b). Thus, *the electromechanical homopolar dynamo (Figure 4.16) with small resistance compared with the mutual induction (4.348a) has electric current and angular velocity (4.348e) both decaying to zero due to the braking torque associated with the Lorentz magnetic force. The decay is exponential (4.342a, b) [algebraic (4.353a, b)] if [problem 124 (125)] the approximation (4.348a) is not (is also) made in the differential equation (4.349c) [(4.353a)].* The case when the approximation (4.348a) is not used at all, and there is no restriction of the ratio of the resistance to the mutual induction, is considered next (subsection 4.7.5).

4.7.5 Electromechanical Dynamo in the Phase Plane

It is possible to eliminate between the electrical (4.344a) and mechanical (4.347d) relations without making the approximation (4.348a) or any other; that is, the exact solution of the electromechanical dynamo problem satisfies:

$$\frac{d}{dt}\left(\frac{1}{J}\frac{dJ}{dt}\right) = \frac{d}{dt}\left(-\frac{R}{L} \pm \frac{Z\Omega}{2\pi L}\right) = \pm\frac{Z}{2\pi L}\frac{d\Omega}{dt}$$

$$= \pm\frac{Z}{2\pi L}\left(\pm\frac{aZ}{2\pi I}\Omega J\right) = \pm\frac{aZ}{2\pi IL}\left(L\frac{dJ}{dt} + RJ\right). \qquad (4.354)$$

This is a second-order non-linear differential equation (4.355b) involving the constant coefficient (4.355a):

$$b \equiv \frac{aZ}{2\pi I}: \qquad\qquad \ddot{J} - \left(\frac{\dot{J}}{J} \pm bJ\right)\dot{J} \mp b\frac{R}{L}J^2 = 0. \qquad (4.355a, b)$$

The corresponding autonomous system of two first-order differential equations (4.355b) ≡ (4.356a, b) is:

$$\dot{J} \equiv G, \qquad\qquad \dot{G} = \frac{G^2}{J} \pm bJ\left(G + \frac{R}{L}J\right); \qquad (4.356a, b)$$

using (4.357a) leads to the path (4.357b) in the phase plane $(J, \dot{J}) \equiv (J, G)$:

$$\dot{G} \equiv \frac{dG}{dt} = \frac{dG}{dJ}\frac{dJ}{dt} = G\frac{dG}{dJ}: \qquad\qquad \frac{dG}{dJ} = \frac{G}{J} \pm bJ\left(1 + \frac{RJ}{LG}\right). \qquad (4.357a, b)$$

The latter (4.357b) becomes (section 3.7) a separable equation (4.358b) using the homogeneous change of variable (4.358a):

$$H \equiv \frac{G}{J} = \frac{\dot{J}}{J} = \frac{d}{dt}(\log J): \qquad \pm bJ\left(1 + \frac{R}{LH}\right) = -H + \frac{d}{dJ}(HJ) = J\frac{dH}{dJ}. \qquad (4.358a, b)$$

The integration of (4.358b) leads to:

$$\pm bJ = \int \frac{LHdH}{R + LH} = \int\left(1 - \frac{R}{R + LH}\right)dH = H - \frac{R}{L}\log(R + LH) + C, \quad (4.359)$$

where C is a constant; substituting (4.356a, 4.358a) in (4.359) yields:

$$\pm bJ^2 = \dot{J} - J\frac{R}{L}\log\left(\frac{L}{J}\dot{J} + R\right) + CJ. \qquad (4.360)$$

When $J \to 0$ it follows: (i) from (4.360) that $\dot{J} \to 0$; (ii) from (4.359) that $H = \dot{J}/J \to$ const; (iii) from (4.355b) that $\ddot{J} \to 0$. Since all $(J, \dot{J}, \ddot{J})$ vanish simultaneously there is a steady state of zero current. Thus *the electromechanical homopolar dynamo (Figure 4.16), whatever the relative importance of the resistance and mutual induction in (4.355a, b), always leads (4.359; 4.360) to a zero electric current if (problem 126) the braking torque of the Lorentz force is not balanced by an external driving torque.* The electrochemical homopolar dynamo is reconsidered next (subsection 4.7.6) in the presence of a driving torque balancing the torque of the Lorentz magnetic force.

4.7.6 Dynamo with Driving Torque Balancing the Joule Dissipation

In the presence of a driving torque the equation (4.347d) for the angular velocity becomes (4.361a):

$$I\dot{\Omega} = M \pm \frac{aZ}{2\pi}\Omega J; \qquad M\Omega = RJ^2. \qquad (4.361a, b)$$

For a steady motion, the work performed by the driving torque should balance electrical dissipation by the Joule effect (4.361b). For a general driving torque, elimination between (4.361a) and (4.344b), as in (4.354), leads to:

$$\frac{d}{dt}\left(\frac{1}{J}\frac{dJ}{dt}\right) = \pm\frac{Z}{2\pi L}\frac{d\Omega}{dt} = \pm\frac{Z}{2\pi L}\left(\pm\frac{a}{2\pi I}\frac{Z}{\Omega}\Omega J + \frac{M}{I}\right) = \pm\frac{aZ}{2\pi IL}\left(L\frac{dJ}{dt} + RJ + \frac{M}{a}\right),$$

$$(4.362)$$

that has an extra term compared with (4.354). Thus the driving torque modifies the differential equation, specifying the variation with time of the electric current of the electromechanical homopolar dynamo from (4.354) ≡ (4.355a, b) to (4.362) ≡ (4.363):

$$\ddot{J} - \left(\frac{\dot{J}}{J} \pm bJ\right)\dot{J} \mp bJ\left(\frac{R}{L}J + \frac{M}{aL}\right) = 0. \qquad (4.363)$$

The solution of (4.363) depends on the relation $M(\Omega, J)$ between the driving torque, angular velocity, and electric current; if the driving torque is constant the last term in (4.363) includes a linear amplification (oscillation) for the upper $-$ (lower $+$) sign. The simplest case is the driving torque (4.361b) balancing the torque due to the Lorentz force in (4.361a), leading to steady rotation (4.364a); the latter implies by (4.361b) a steady electric current (4.364b). In this case (4.344a) specifies the constant angular velocity (4.364c):

$$\dot{\Omega} = 0: \qquad \dot{J} = 0, \qquad \Omega = \pm\frac{2\pi R}{Z}; \qquad J = \frac{M\Omega}{RJ} = \mp\frac{aZ\Omega^2}{2\pi R} = \mp\frac{2\pi aR}{Z}, \qquad (4.364a-d)$$

then (4.361b) specifies the steady electric current (4.364d) from (4.361a). Thus *an electromechanical homopolar dynamo (Figure 4.16) with total induction L, resistance R, and mutual induction Z, rotating at constant angular velocity Ω in (4.344a) acts (problem 122) as a brake (4.345a–c) if the winding is opposite to the direction of rotation (J_ in the Figure 4.16); if (problem 123) the winding is in the direction of rotation, the current is constant (4.346b) ≡ (4.364b) for the angular velocity $|\bar{\Omega}|$ in (4.364c), and decreases (increases) exponentially with time (4.346a) [(4.346c)] for smaller $\Omega < |\bar{\Omega}|$ (larger $\Omega > |\bar{\Omega}|$) values. Taking into account the retarding torque* *due to the Lorentz force (4.347a–d) and the rotational moment of inertia I leads to (problem 125) an algebraic decay of the current (4.353c) if the effect of resistance is small compared with that of the mutual inductance; in this case the electric current is proportional to the angular velocity, and (problem 124) including the electrical resistance only in the current equation (4.349a–c) leads to an exponential decay of the electric current (4.352a, b). If the approximation (4.348a) is not made the equations (4.344b; 4.347d) for the electromechanical dynamo lead to (4.355a, b), whose (problem 126) paths in the phase plane are specified by (4.360) and include a steady state of zero current. The steady rotation (4.364a, c) is compatible (4.344b) with a steady electric current (4.364b, d) if (problem 127) there is a driving torque balancing the torque due to the Lorentz magnetic force (4.361a) such that the mechanical work of driving torque equals electrical dissipation by the Joule effect (4.361b).*

4.7.7 The Self-Excited Dynamo and Magnetic Field Generation

A simple example of a first-order equation with steady solution is (4.365c) that balances the inertia force against: (i) a quadratic damping (second term on the r.h.s.); (ii) a linear amplification (damping) if the upper + (lower –) sign is chosen in the first term on the r.h.s.:

$$a > 0 < b: \qquad\qquad m\dot{v}_\pm = \pm b v_\pm - a v_\pm^2. \qquad\qquad (4.365\text{a–c})$$

The r.h.s. of (4.365c) with (4.365a, b) is the reverse of (4.349c) that involves linear damping and quadratic damping or amplification. The two signs in (4.365c) lead (Figure 4.17) to two cases: (i) in case I of the lower sign in (4.365c), the steady solution (4.366a) in the presence of linear and non-linear damping (4.366b) can only be an asymptotic decay to zero (4.366c):

$$\dot{v}_- = 0: \qquad\qquad 0 = v_-(b + av_-) \quad\Rightarrow\quad \lim_{t\to\infty} v_-(t) = 0; \qquad (4.366\text{a–c})$$

$$\dot{v}_+ = 0: \qquad\qquad 0 = v_+(-b + av_+) \quad\Rightarrow\quad \lim_{t\to\infty} v_+(t) = \frac{b}{a}, \qquad (4.367\text{a–c})$$

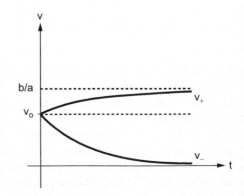

FIGURE 4.17
An electromechanical homopolar disk dynamo (Figure 4.16), including rotational inertia and driving torque can lead either to a zero v_- or a constant v_+ asymptotic velocity or current. In the presence of both linear and quadratic damping there is a decay v_- to zero. In the presence of a linear amplification and non-linear damping, both decay and growth are limited, and there is an asymptotic steady state of equilibrium v_+ with a non-zero value.

(ii) in the case II of the upper sign in (4.365c), the steady solution (4.367c) corresponds to a balance between linear amplification and quadratic damping. These expectations can be checked by solving the first order differential equation (4.365c) that is separable, and can be integrated:

$$
\pm \frac{b}{m} t = \int_{v_0}^{v_\pm} \frac{dv}{v \mp a v^2/b} = \int_{v_0}^{v_\pm} \left(\frac{1}{v} \pm \frac{1}{b/a \mp v} \right) dv
$$

$$
= \left[\log \left(\frac{v}{b/a \mp v} \right) \right]_{v_0}^{v_\pm} = \log \left(\frac{v_\pm}{v_0} \frac{b \mp a v_0}{b \mp a v_\pm} \right),
$$

(4.368)

using a partial fraction decomposition (section I.31.9). The integrations (4.350a, b; 4.351a, b) and (4.368) are similar with the substitutions:

$$
\frac{R}{L} \leftrightarrow \mp \frac{b}{m}, \qquad v \leftrightarrow J, \qquad \frac{S}{R} \leftrightarrow \frac{a}{b}, \qquad \text{(4.369a–c)}
$$

that also apply to the solution (4.352a, b) ↔ (4.370a, b) for the velocity:

$$
v_\pm(0) = v_0: \qquad v_\pm(t) = \frac{v_0 b}{(b \mp v_0 a) \exp(\mp b t/m) \pm v_0 a}.
$$

(4.370a, b)

The solution (4.370a, b) of (Figure 4.17) the equation of motion (4.365c) with quadratic damping (4.365a) plus linear damping or amplification (4.365b): (i) confirms that (4.370a) in (4.370b) is the initial velocity; (ii/iii) the lower (upper) sign in (4.370b) corresponding [problem 128 (129)] to linear damping (amplification), combined with non-linear damping, leads asymptotically for a long time to a zero (non-zero) steady solution (4.366a–c) [(4.367a–c)].

4.7.8 Magnetic Field Generation in Planets and Stars

The earth and other planets and the sun and other stars have a steady magnetic field (Figure 4.18). The possibility of a **primordial magnetic field**, dating from the origins of the celestial bodies, can be excluded because the electrical conductivity is too high, and such a magnetic field would already have been dissipated, given the age of the planet or star. The remaining possibility is a permanently generated steady magnetic field associated with an analogue of the **dynamo effect**: (i) a cloud of particles (gas) collapses under mutual gravitational attraction to form a planet (star); (ii) the cloud is initially rotating, and the conservation of the angular momentum implies that as it collapses the angular velocity increases; (iii) the gravitational collapse increases the mass density, pressure and temperature sufficiently to separate electrons from atoms that become ions; and (iv) the ionized matter subject

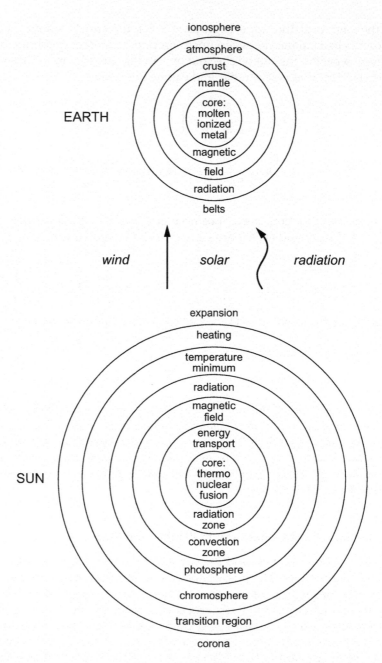

FIGURE 4.18
The dynamo effect generates the magnetic fields of the earth (top) [sun (bottom)] that are typical of planets (stars). Two of the interactions in the solar-terrestrial system (middle) are: (i) the earth's magnetic field captures some of the solar wind (due to expansion of the solar atmosphere), into radiation belts that form the ionosphere of the earth; (ii) the atmosphere and ionosphere of the earth protect the surface from some of the strongest solar radiation.

to rotation acts as the homopolar dynamo and generates a steady magnetic field. The role of the magnetic field is best understood considering the successive layers of a planet or star from the center outward; the internal structure of planets (stars) is best known for the earth (sun).

In the case of the sun (bottom of the Figure 4.18), the central temperature in the **core** $T \sim 10^7 K$ is sufficiently high to cause **thermonuclear fusion** of hydrogen and its isotopes (deuterium and tritium) into helium, providing the main source of energy. The energy is transmitted through a **radiation zone**, then interacts with matter in the **convection zone**, where interaction with rotation generates the magnetic field through the dynamo effect. The magnetic field permeates the solar atmosphere, starting with the deepest layer, the **photosphere,** so named because it is the source of most solar radiation. The temperature decreases with altitude towards a temperature minimum in the **chromosphere,** so named because it gives the sun its yellow colour. The waves generated by hydromagnetic turbulence in the photosphere propagate upward through the chromosphere and are dissipated in the **transition region** to the **corona**. The steep temperature gradients in the transition region lead to high coronal temperature $T \sim 1.8x10^6 K$, leading to its expansion into the **solar wind**. The corona owes its name to its irregular shape, which is due to flows directed by the magnetic field. The solar wind then reaches the earth and other planets.

4.7.9 Solar-Terrestrial System and Solar Cycle

The **Earth** (top of the Figure 4.18) has its own magnetic field generated in the **outer core** made of molten metal by the dynamo effect associated with rotation. The **inner core** is solid, as is the **mantle** separating the core from the **crust,** which is the solid layer underlying the oceans and the land masses. The **atmosphere** supports life on the earth. Above the earth its magnetic field traps the solar wind in **radiation belts** forming the **ionosphere**. The ionosphere and atmosphere protect the surface of the earth from the most intense solar radiation. The combination of the sun and earth and their interactions forms the **solar-terrestrial system** (Figure 4.18). The remaining planets in the **solar system** have their own magnetic fields interacting with the sun in a way that is significantly affected by the distance; for example, the **inner planets** (Mercury and Venus) are much hotter than the **outer planets** (Mars, Jupiter, Saturn, Neptune and Uranus), with the earth as an intermediate case. The magnetic fields of planets are quite stable, varying slowly in direction and magnitude over time; this is not the case for the sun.

The solar magnetic field changes polarity during a **solar cycle** of 11–14 years. The **change of polarity** or inversion of direction of the solar magnetic field implies that it is weak at the start and end of a cycle and is strongest in mid-cycle. The evolution of the solar magnetic field during the solar cycle is observed in **sunspots** that are dark regions of strong magnetic field in the photosphere. The number and size of sunspots and their motion change during the solar cycle from a maximum at mid cycle to a minimum at the start; the solar cycles are not identical with the observed **Maunder minimum** occurring in 1645–1715. Conversely, in periods of **high solar activity** the stronger magnetic field affects the whole, including the atmosphere up to the corona, leading to the observation of (i) **solar flares** that liberate large amounts of energy and hence strong radiation; and (ii) **coronal mass ejections** that lead to particle streams more intense than the average solar wind. Both (i) and (ii) can be a hazard to space travel or habitation outside the earth's atmosphere and ionosphere.

The dynamo effect and its consequences are still largely open research topics. Thus *the combination of linear amplification (damping), that is the upper+ (lower –) sign in (4.365b, c) with quadratic damping in (4.365a, c) leads to (Figure 4.17) an asymptotic steady state (4.367a–c) [asymptotic decay (4.366a–c)]. A similar* **dynamo effect** *is thought to generate the magnetic field of the earth, sun, and other celestial bodies through the non-linear interaction between rotation and the ionized flow. The electrical conductivity of the earth is such that any original magnetic field would have decayed at its present age, so a steady mechanism to generate the existing magnetic field must be present.* Having considered non-linear restoring forces (sections 4.4–4.6) and non-linear damping or amplification (subsections 4.7.1–4.7.9), non-linear restoring force (subsections 4.7.10–4.7.11) with non-linear damping (subsections 4.7.12–4.7.15) for the large amplitude motion of a pendulum is considered next.

4.7.10 Small and Large Amplitude Motion of a Pendulum

Consider a circular pendulum (Figure 4.19a) consisting of a mass m linked to the rotation point O by an inextensible rod of length L and negligible mass compared to m. In the uniform gravity field with constant acceleration g the equation of motion (4.371a) balances: (i) the inertia force, equal to mass times the linear acceleration that in turn is the product of the angular acceleration $\ddot{\theta}$ by the radius L of the circular trajectory; (ii) against the weight directed downward $-mg$ and projected along the tangent to the trajectory:

$$mL\ddot{\theta} = -mg\sin\theta = -\frac{\Phi'(\theta)}{L}, \qquad \Phi(\theta) = -mgL\cos\theta, \qquad \Phi(\pi/2) = 0, \qquad (4.371a\text{–}c)$$

where (4.371b) is the potential energy measured (4.371c) from the height of the center of rotation. The equation (4.371a) $\equiv$ (I.8.2a) corresponds to the balance of tangential forces (section I.8). The potential has extrema (4.372b, c) where the tangential force vanishes (4.372a):

$$F(\theta) = -L\Phi'(\theta) = -mg\sin\theta = 0: \qquad \theta_1 = 0 \quad or \quad \theta_1 = \pi; \qquad (4.372a\text{–}c)$$

$$\Phi''(\theta) = mgL\cos\theta: \qquad \Phi''(\theta_1) = mgL > 0, \qquad \Phi''(\pi) = -mgL < 0, \qquad (4.373a\text{–}c)$$

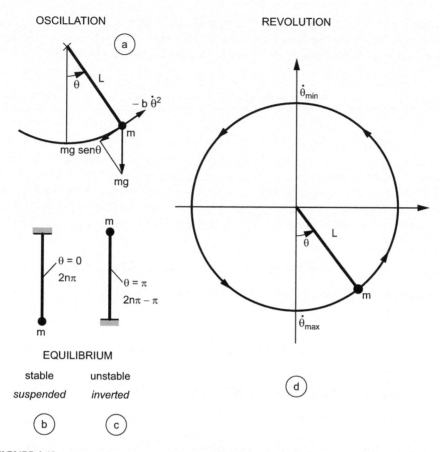

OSCILLATION   REVOLUTION

EQUILIBRIUM

stable   unstable

suspended   *inverted*

FIGURE 4.19
In the absence of damping there are three possible cases of motion of a circular pendulum: (i) if the angular velocity never vanishes, the pendulum circulates (d) across the lowest (b) and highest (c) positions, and is a pendulum of revolution; (iii) if the angular velocity vanishes before the highest position (b), that is, never reached, the pendulum passes back and forth through the lowest position (c) in an oscillatory motion, and is an oscillating pendulum. The intermediate case (ii) between the (i) circulating and (iii) oscillating pendulum is stoppage at the top position after an infinite time, in the absence of perturbations. Since the highest position corresponds to unstable equilibrium, any perturbation that increases (decreases) the energy leads to (i) circulatory [(iii) oscillatory] motion. The lowest position of stable equilibrium is where the pendulum comes to rest in the presence of damping, for any initial condition (Figure 4.21).

the equilibrium positions lead to (4.373a) a minimum (4.373b) [maximum (4.373c)] of the gravity potential at the lowest (4.372b) [highest (4.372c)] position that is stable (unstable) and corresponds to the suspended (inverted) pendulum [Figure 4.19b(c)]; the linear (4.374a) motion in the vicinity of the position of stable (4.374d) [unstable 4.374d)] equilibrium is specified by (4.374c) [(4.374e)]:

$$\theta^2 \ll 1: \quad \sin\theta \sim \theta, \quad L\ddot{\theta} + g\theta = 0, \quad \theta \to \theta + \pi: \quad L\ddot{\theta} - g\theta = 0, \quad (4.374a\text{–}e)$$

in agreement with the lower (upper) sign in (2.71a, b). The leads (subsection 2.2.3) to an oscillation (divergence) with natural frequency (characteristic exponent) specified by (4.375a) in the exact equation of motion (4.371a) ≡ (4.375b):

$$\omega_0 = \sqrt{\frac{g}{L}}: \qquad \ddot{\theta} + \omega_0^2 \sin\theta = 0, \qquad 0 \le \theta \le 2\pi. \qquad (4.375\text{a–c})$$

The equation of motion (4.375b) will be solved exactly in the sequel (subsections 4.7.10–4.7.15) over the full range (4.375c) of angles, rather than just near the equilibrium points, so that the separation of suspended or inverted pendulum is no longer relevant: in a circulatory motion or oscillation with large amplitude $\theta_{max} > \pi/2$, the pendulum is suspended (for $|\theta| \le \pi/2$) part of the time and inverted (for $\pi/2 \le |\theta| < \pi$) the rest of the time (subsection 4.7.11).

4.7.11 Conservation of Energy and Classification of Pendular Motions

Multiplying the equation of motion of the pendulum (4.371a) by the tangential velocity:

$$0 = mL^2\ddot{\theta}\dot{\theta} + mgL\dot{\theta}\sin\theta = \frac{d}{dt}\left(\frac{1}{2}mL^2\dot{\theta}^2 - mgL\cos\theta\right) = \frac{dE}{dt}, \qquad (4.376)$$

follows the (4.377a) **conservation of energy** (4.377b), kinetic plus potential (4.371b):

$$const = E = \frac{1}{2}mL^2\dot{\theta}^2 - mgL\cos\theta = \frac{1}{2}mL^2\left(\dot{\theta}^2 - 2\omega_0^2\cos\theta\right); \quad (4.377\text{a–c})$$

thus the term in curved brackets in (4.377c) is a constant that may be evaluated (4.378c) from any initial state $\left(\theta_0, \dot{\theta}_0\right)$:

$$\dot{\theta}^2 - 2\omega_0^2\cos\theta = \dot{\theta}_0^2 - 2\omega_0^2\cos\theta_0. \qquad (4.378)$$

For any pendular motion to exist, whether linear or non-linear; that is, with small or large amplitude, the velocity cannot be zero (4.379a) at the lowest position (4.379b) leading to (4.378) ≡ (4.379c):

$$\theta_0 = 0 \ne \dot{\theta}_0: \qquad \dot{\theta}^2 = \dot{\theta}_0^2 + 2\omega_0^2(\cos\theta - 1) = \dot{\theta}_0^2 - 4\omega_0^2\sin^2\left(\frac{\theta}{2}\right); \qquad (4.379\text{a–d})$$

the passage from (4.379c) to (4.379d) used (II.5.61c).

It has been shown that *the large amplitude motion of a pendulum (4.371a–c) satisfies the conservation of the total, kinetic plus potential, energy (4.377a–c) and thus depends (4.379d) only on two parameters: (i) the natural frequency (4.375a) that is a physical property of the pendulum and coincides with the oscillation frequency only in the linear case of small oscillations near the lowest position of stable equilibrium; and (ii) the velocity (4.379b) at the lowest point (4.379a), which implies that there can be no rest at the position of stable equilibrium (Figure 4.19b) provided it be non-zero. This leads to three possible cases of motion. The intermediate case II is for an angular velocity at the bottom equal to twice the natural frequency (4.380a) implying (4.379d) ≡ (4.380b):*

$$\left|\dot{\theta}_0\right| = 2\omega_0: \qquad \dot{\theta} = \dot{\theta}_0 \sqrt{1 - \sin^2\left(\frac{\theta}{2}\right)} = \dot{\theta}_0 \cos\left(\frac{\theta}{2}\right); \qquad \text{(4.380a, b)}$$

*it follows (problem 130) that there is a **stopping point** (4.380c) at the highest position (4.380d). The stoppage at the highest position (Figure 4.19c) occurs because the total energy (4.377c):*

$$\dot{\theta} = 0: \qquad \theta = \pi, \qquad E = mgL = \Phi(\pi), \qquad \text{(4.380c–e)}$$

equals (4.380e) the potential energy (4.371b) measured from the height of the center of rotation.

In case II, when the angular velocity at the lowest point is larger than twice the natural frequency (4.381a), the total energy (4.377c) is larger than the potential energy at the highest position (4.381b):

$$\left|\dot{\theta}_0\right| > 2\omega_0: \qquad E > \frac{1}{2}mL^2\left(\dot{\theta}_0^2 - 2\omega_0^2\right) > mL^2\omega_0^2 = mgL; \qquad \text{(4.381a, b)}$$

it follows that the pendulum passes through the highest position with a non-zero minimum angular velocity (4.382a):

$$\dot{\theta}_{min} = \dot{\theta}(\pi) = \sqrt{\dot{\theta}_0^2 - 4\omega_0^2} \le \dot{\theta} \le \dot{\theta}_0 \equiv \dot{\theta}(0) = \dot{\theta}_{max}. \qquad \text{(4.381c, d)}$$

*The maximum angular velocity at the lowest position (4.381d) ≡ (4.379a, b) specifies the range of angular velocities for (problem 131) the **circulatory motion** of the pendulum (Figure 4.19d). In the remaining case III of angular velocity at the lowest position smaller than twice the angular velocity (4.382a), the total energy (4.377c) is less than the potential energy at the highest position (4.382b) that cannot be reached:*

$$\left|\dot{\theta}_0\right| < 2\omega_0: \qquad E < \frac{1}{2}mL^2\left(\dot{\theta}_0^2 - 2\omega_0^2\right) = mgL; \qquad \text{(4.382a, b)}$$

*it follows (4.379d) that the angular velocity is zero (4.382c) at the angle (4.382d) corresponding to the amplitude (4.382e) of (problem 132) the **oscillatory motion** of the pendulum (Figure 4.19a):*

$$\dot{\theta} = 0: \qquad \sin\left(\frac{\theta_{max}}{2}\right) = \frac{|\dot{\theta}_0|}{2\omega_0} < 1, \qquad -\theta_{max} \leq \theta \leq \theta_{max}. \qquad (4.382\text{c–e})$$

The case III includes large amplitude oscillations, and (problem 133) small amplitude oscillations (4.383a) are the sub-case (4.383b) of small angular velocity at the lowest position:

$$(\theta_{max})^2 \ll 1: \qquad \dot{\theta}_0^2 \ll 4\omega_0^2 = \frac{4g}{L}. \qquad (4.383\text{a, b})$$

Having classified (subsection 4.7.11) the three cases (four sub-cases) of motion of a pendulum (Table 4.5) next are considered the corresponding equations of motion (subsection 4.7.12).

4.7.12 Pendular Motion Near Stable and Unstable Equilibrium Positions

The general equation (4.379d) for the pendular motion with angular velocity (4.379b) at the lowest position can be integrated:

$$t = \int_0^\theta \frac{d\theta}{\dot{\theta}} = \int_0^\theta \left| \dot{\theta}_0^2 - 4\omega_0^2 \sin^2\left(\frac{\theta}{2}\right) \right|^{-1/2} d\theta. \qquad (4.384)$$

The simplest case (problem 133) of small amplitude oscillations (chapter I.8; subsections 2.2.4–2.2.5) is reviewed briefly before passing to the non-linear cases. The linear case (4.385a, b) simplifies (4.384) to (4.385c):

$$\theta^2 \ll 1: \qquad \sin\left(\frac{\theta}{2}\right) \sim \frac{\theta}{2}: \qquad t = \int_0^\theta |\dot{\theta}_0^2 - \omega_0^2\theta^2|^{-1/2} d\theta; \qquad (4.385\text{a–c})$$

the integration of (4.385c) ≡ (4.386b) is performed via the change of variable (4.386a):

$$\phi \equiv \frac{\omega_0}{\dot{\theta}_0}\theta: \qquad \omega_0 t = \int_0^\phi \frac{d\phi}{\sqrt{1-\phi^2}} = arc\sin\phi = arc\sin\left(\frac{\omega_0}{\dot{\theta}_0}\theta\right). \qquad (4.386\text{a, b})$$

TABLE 4.5

Motions of a Simple Pendulum

Parameter	Oscillation		Unstable		Circulation	
	Small Amplitude	Large Amplitude	Divergence	Stoppage	Without Damping	With Damping
differential equation	(4.374a–c)	(4.371a–c) ≡ (4.375a–c)	(4.374a, d, e)	(4.380a, b)	(4.371a–c) ≡ (4.375a–c)	(4.418a, b)
initial conditions	(4.387a, b)	(4.387a, b)	(4.387a, b)	(4.380c–e)	(4.382a–e)	(4.425a)
equation of motion	(4.387c, d)	(4.392a, b)	(2.73a–c)	(4.390a, b)	(4.396a, b)	(4.419a–c)
stability	stable	stable	unstable	unstable	stable	–
period	(4.407a)	(4.404a, b)	–	(4.413a–c; 4.414a–c)	(4.399a–d; 4.400a, b; 4.403a, b)	–
section	4.7.12	4.7.10–4.7.15	2.2.4	4.7.10–4.7.15	4.7.12–4.7.15	4.7.17–4.7.18
figure	2.7a	4.19a	2.7b	4.19c	4.19d	4.20

Note: Comparison of the small amplitude (stable or unstable) and large amplitude (oscillation, stoppage, or circulation) of a circular pendulum with or without damping.

Thus *(problem 133) the small oscillations of a pendulum with zero initial displacement (4.387a) and non-zero initial velocity (4.387b) are specified by the displacement (4.387c) and velocity (4.387d) as a function of time:*

$$\theta_0 = 0 \neq \dot\theta_0: \qquad \theta(t) = \frac{\dot\theta_0}{\omega_0}\sin(\omega_0 t), \qquad \dot\theta(t) = \dot\theta_0\cos(\omega_0 t). \qquad (4.387a–d)$$

In the opposite case of the initial velocity $\dot\theta_0 = 0$ and non-zero initial displacement $\theta_0 = a = 0$, the displacement (velocity) is given by (4.223c) [(4.223d)] as a function of time. Both cases are included in the displacement (2.56b) [velocity (2.56c)] as a function of time for arbitrary initial conditions (subsection 2.2.1).

The motion near (problem 130) the upper unstable equilibrium point (4.380a) ≡ (4.388a) is specified (4.384) by (4.388b) ≡ (4.380b):

$$|\dot\theta_0| = 2\omega_0: \qquad \omega_0 t = \frac{1}{2}\int_0^\theta \frac{d\theta}{\cos(\theta/2)}. \qquad (4.388a, b)$$

The change of variable (4.389a) leads to (4.389b):

$$\zeta \equiv \sin\left(\frac{\theta}{2}\right): \qquad d\theta = \frac{2d\zeta}{\cos(\theta/2)} = \frac{2d\zeta}{\sqrt{1-\zeta^2}}, \qquad (4.389a, b)$$

and substitution in (4.389b) performs the integration:

$$\omega_0 t = \int_0^\zeta \frac{d\zeta}{1-\zeta^2} = arc \tanh \zeta = arc \tanh\left[\sin\left(\frac{\theta}{2}\right)\right]. \qquad (4.389c)$$

It follows (4.389c) ≡ (4.390a) that *(problem 130) the pendulum with energy (4.380e) corresponding (4.380a)* ≡ *(4.388a) to stoppage at the top (4.380c, d) has equation of motion:*

$$\sin\left(\frac{\theta}{2}\right) = \tanh(\omega_0 t): \qquad \lim_{\theta \to \pi} \sin\left(\frac{\theta}{2}\right) = 1 = \lim_{t \to 0} \tanh(\omega_0 t), \qquad (4.390a, b)$$

that implies (4.390b) that it takes an infinite time to reach (or move away from) the upper position of equilibrium; since it is unstable any disturbance that increases (decreases) energy will lead to the circulatory (oscillatory) motion [problem 131 (132)] considered next (subsection 4.7.13).

4.7.13 Two Equivalent Exact Solutions in Term of Elliptic Sines

The change of variable (4.391a) in the general equation of energy for a pendulum (4.384) leads to (4.391b):

$$\psi \equiv \frac{\theta}{2}: \qquad \frac{\dot{\theta}_0 t}{2} = \int_0^\psi \left|1 - \left(\frac{2\omega_0}{\dot{\theta}_0}\right)^2 \sin^2 \psi\right|^{-1/2} d\psi = arg\, sn\left(\sin \psi; \frac{1}{q}\right), \qquad (4.391a, b)$$

that is (4.246b) an elliptic sine integral (4.392b) with modulus $1/q$ given by the **non-linearity parameter** (4.392a):

$$q \equiv \frac{\dot{\theta}_0}{2\omega_0}: \qquad \sin\left(\frac{\theta}{2}\right) = sn\left(\frac{\dot{\theta}_0 t}{2}; \frac{2\omega_0}{\dot{\theta}_0}\right). \qquad (4.392a, b)$$

Thus (4.392b) is an exact solution for the large amplitude pendular motion. An alternative form of this exact solution can be obtained rewriting (4.391b) ≡ (4.393a) as (4.393b):

$$\frac{\dot{\theta}_0 t}{2} = \int_0^\psi \left|1 - q^{-2} \sin^2 \psi\right|^{-1/2} d\psi \quad \Leftrightarrow \quad \omega_0 t = \frac{\dot{\theta}_0 t}{2q} = \int_0^\psi \left|q^2 - \sin^2 \psi\right|^{-1/2} d\psi.$$

$$(4.393a, b)$$

In (4.393b) may be made the change of variable (4.394a):

$$\sin\psi = q\sin\varphi: \qquad \frac{d\psi}{d\varphi} = q\frac{\cos\varphi}{\cos\psi} = q\left|\frac{1-\sin^2\varphi}{1-\sin^2\psi}\right|^2 = \left|\frac{q^2-\sin^2\psi}{1-q^2\sin^2\varphi}\right|^{1/2}, \qquad \text{(4.394a, b)}$$

leading from (4.393b) to (4.395):

$$\omega_0 t = \int_0^\varphi \left|1-q^{-2}\sin^2\varphi\right|^{-1/2} d\varphi = \arg sn(\sin\varphi;q); \qquad \text{(4.395)}$$

this is equivalent (4.395) ≡ (4.396a) to:

$$sn\left(\omega_0 t; \frac{\dot\theta_0}{2\omega_0}\right) = \sin\varphi = \frac{1}{q}\sin\psi = \frac{2\omega_0}{\dot\theta_0}\sin\left(\frac{\theta}{2}\right), \qquad \text{(4.396a, b)}$$

using (4.392a) and (4.394a; 4.391a) in (4.396b).

It has been shown that *the large amplitude motion of a pendulum is specified exactly and equivalently by (4.392b) ≡ (4.396b) involving only two parameters: (i) the maximum angular velocity (4.379b) at the lowest point (4.379a); and (ii) the natural frequency (4.375a) of linear oscillations (4.387a–d) that depends on the length of the pendulum and acceleration of gravity. There are four cases:*

$$q^2 \equiv \left(\frac{\dot\theta_0}{2\omega_0}\right)^2 \begin{cases} > 1 & \textit{circulation motion,} & \text{(4.397a)} \\ = 1 & \textit{stopage at the top,} & \text{(4.397b)} \\ < 1 & \textit{large oscillations,} & \text{(4.397c)} \\ \ll 1 & \textit{small oscillations,} & \text{(4.397d)} \end{cases}$$

namely: (I) circulatory motion (4.397a) passing (Figure 4.19d) at the top (4.381a–d) with (problem 131) minimum angular velocity (4.381c); (II) stoppage (Figure 4.19c) at top (4.397b) unstable (problem 130) equilibrium position (4.380a–e), that takes (4.390a) an infinite time to reach (4.390b) or deviate from; (III) oscillations (4.397c) with (Figure 4.19a) large (problem 132) amplitude (4.382a–e); (iv) in the linear case (4.385a) ≡ (4.398a) the (problem 133) amplitude of oscillations (4.382b) ≡ (4.398b) is small (4.398c) and (4.396b) simplifies to (4.398d):

$$\theta^2 \ll 1: \qquad a \equiv \theta_{max} = \frac{\dot\theta_0}{\omega_0}, \qquad a^2 \ll 1, \qquad \sin(\omega_0 t) = \frac{\omega_0}{\dot\theta_0}\theta(t), \qquad \text{(4.398a–d)}$$

in agreement with (4.387c) ≡ (4.398d). The amplitude (4.398b) and the natural frequency (4.375a) apply only in the linear case IV of small oscillations; the circulatory (oscillatory) motion in the case I (III) are also periodic, and their periods are calculated next (subsection 4.7.13).

4.7.14 Periods of the Circulatory and Oscillatory Motions

The period of the circulatory motion is (4.393a) twice the time taken to travel from $\theta = 0$ to $\theta = \pi$, or (4.391a) from $\psi = 0$ to $\psi = \pi/2$ leading to (4.399a, b):

$$\tau_1 = 2t(\pi) = \frac{4}{\dot{\theta}_0} \int_0^{\pi/2} \left|1 - \frac{1}{q^2}\sin^2\psi\right|^{-1/2} d\psi$$

$$= \frac{4}{\dot{\theta}_0}\frac{\pi}{2}\left[1 + \sum_{n=1}^{\infty}\left(\frac{1}{2q}\right)^2 \frac{(2n-1)!!}{(2n)!!}\frac{(2n)!}{(n!)^2}\right] = \frac{2\pi}{\dot{\theta}_0}A\left(\frac{1}{q}\right),$$

(4.399a–d)

where (4.392a) [(4.252b)] were used in (4.399c) [(4.399d)]. The function $A(q^{-1})$ has power series (4.252b) $\equiv$ (4.400a) with coefficients (4.400b):

$$A\left(\frac{1}{q}\right) = 1 + \sum_{n=1}^{\infty}\left(a_n q^{-n}\right)^2, \qquad a_n \equiv \frac{(2n)!}{(n!)^2\, 2^{2n}}. \qquad (4.400a, b)$$

The coefficients are given by (4.400b) $\equiv$ (4.400d) $\equiv$ (4.400e):

$$a_n = \frac{2n(2n-1)(2n-2)....3.2.1}{\left[2n(2n-2)(2n-4)...6.4.2\right]^2}$$

$$= \frac{(2n-1)(2n-3)....3.1}{2n(2n-2)....4.2} = \frac{(2n-1)!!}{(2n)!!},$$

(4.400c–e)

in agreement with (4.400b–e) $\equiv$ (4.248a, b) $\equiv$ (II.6.56a, b) $\equiv$ (II.6.57a–d). The oscillatory motion has amplitude (4.382b; 4.392a) $\equiv$ (4.401a) corresponding (4.394a) to (4.401b) and hence (4.401c):

$$\sin\theta_{max} = \frac{|\dot{\theta}_0|}{2\omega_0} = q: \qquad \sin\varphi_{max} = \frac{1}{q}\sin\theta_{max} = 1, \qquad \varphi_{max} = \pi/2. \qquad (4.401a–c)$$

The period of oscillation is four times the time taken to travel from $\theta = 0$ to $\theta = \theta_{max}$; that is, from $\varphi = 0$ to $\varphi = \varphi_{max} = \pi/2$ in (4.395):

$$\tau_2 = \frac{4}{\omega_0}\int_0^{\pi/2}\left|1 - q^2\sin^2\varphi\right|^{-1/2}d\varphi = \frac{4}{\omega_0}A(q)\frac{\pi}{2} = \frac{2\pi}{\omega_0}A(q), \qquad (4.402a–c)$$

in terms of the same function (4.400a, b) with q replacing $1/q$.

It has been shown that the period of *the circulatory (oscillatory) motion [Figure 4.19d (a)] is [problem 131 (132)] given by (4.399d)* ≡ (4.403a, b) [(4.402c) ≡ (4.404a, b)]:

$$2\omega_0 < |\dot{\theta}_0|: \qquad \tau_1 = \frac{2\pi}{|\dot{\theta}_0|}\left[1 + \sum_{n=1}^{\infty}(a_n)^2\left(\frac{2\omega_0}{\dot{\theta}_0}\right)^{2n}\right], \qquad (4.403a, b)$$

$$|\dot{\theta}_0| < 2\omega_0: \qquad \tau_2 = \frac{2\pi}{\omega_0}\left[1 + \sum_{n=1}^{\infty}(a_n)^2\left(\frac{\dot{\theta}_0}{2\omega_0}\right)^{2n}\right], \qquad (4.404a, b)$$

where: (i) the leading term involves the inverse of the maximum angular velocity of the lowest point (the natural frequency of the linear oscillator); (ii) the correcting factor is a power series (4.403b) [(4.404b)] of even powers of the inverse (of) the non-linearity parameter (4.392a); (iii) the series have the same coefficients (4.400b–e) ≡ *(4.248a, b)* ≡ *(II.6.57a–g); and (iv) in both cases the powers in (4.403b) [(4.404b)] apply to a variable less than unity (4.403a) [(4.404a)] ensuring convergence of the series. Thus the two lowest-order non-linear terms in the period of the circulatory (4.403b) [oscillatory (4.404b)] motion:*

$$\tau_1 = \frac{2\pi}{|\dot{\theta}_0|}\left(1 + \frac{\omega_0^2}{\dot{\theta}_0^2} + \frac{9\omega_0^4}{4\,\dot{\theta}_0^4} + ...\right) = \frac{2\pi}{|\dot{\theta}_0|} + \frac{2\pi\omega_0^2}{|\dot{\theta}_0|^3} + \frac{9\pi\omega_0^4}{2|\dot{\theta}_0|^5} + ..., \qquad (4.405)$$

$$\tau_2 = \frac{2\pi}{\omega_0}\left(1 + \frac{\dot{\theta}_0^2}{16\omega_0^2} + \frac{9\dot{\theta}_0^4}{1024\omega_0^4} + ...\right) = \frac{2\pi}{\omega_0} + \frac{\pi\dot{\theta}_0^2}{8\omega_0^3} + \frac{9\pi\dot{\theta}_0^4}{512\omega_0^5} + ... \qquad (4.406)$$

are (4.405) [(4.406)].

The lowest-order non-linear approximation to the period of oscillation (4.407b) gives a correction factor for the period of the linear oscillator (4.407a):

$$\tau_0 = \frac{2\pi}{\omega_0}: \qquad \tau_2 = \frac{2\pi}{\omega_0}\left(1 + \frac{\dot{\theta}_0^2}{16\omega_0^2}\right) = \tau_0\left(1 + \frac{\dot{\theta}_0^2\tau_0^2}{64\pi^2}\right),$$

$$\omega_2 = \frac{2\pi}{\tau_2} = \frac{2\pi}{\tau_0}\left(1 - \frac{\dot{\theta}_0^2\tau_0^2}{64\pi^2}\right) = \omega_0\left(1 - \frac{\dot{\theta}_0^2}{16\omega_0^2}\right),$$

$$(4.407a–c)$$

and corresponds to the frequency (4.307c). The frequency of the non-linear oscillation (4.407b) has, relative to the natural frequency (4.375a) of linear oscillation, a correction (4.408a) involving the amplitude (4.398b) whose linear approximation may be substituted in the lowest-order non-linear correction:

$$\omega_2 = \omega_0\left(1 - \frac{a^2}{16}\right) = \omega_0^2\left(1 + \frac{3\beta a^2}{8}\right): \qquad\qquad \beta = -\frac{1}{6}; \qquad\qquad (4.408a\text{--}c)$$

comparing with the oscillation frequency of an anharmonic oscillator with cubic restoring force (4.265c) ≡ (4.408b) it follows that the cubic coefficient is (4.408c). This can be confirmed from the restoring force (4.372a) ≡ (4.409a) for the pendulum using the first two terms of the power series for the circular sine (II.7.13b) ≡ (4.409b).

$$F(\theta) = -mg\sin\theta = -mg\theta\left[1 - \frac{\theta^2}{6} + O(\theta^4)\right] = -mg\theta\left[1 + \beta\theta^2 + O(\theta^4)\right]. \quad (4.409a\text{--}c)$$

Thus, *to the lowest order of non-linearity the circular pendulum (4.371a) is equivalent to a soft spring (4.215a) with cubic restoring force (4.409a–c) with cubic coefficient (4.408c). The oscillation frequency for the anharmonic oscillator (4.408a–c) [for the circular pendulum (4.404a, b)] is approximate (exact) result to the lowest order (all orders) of non-linearity.* The exact periods of the circulatory (oscillatory) pendular motion are compared next (subsection 4.7.15) using their series expansions.

4.7.15 Convergence of the Series Specifying the Exact Period

The periods of the circulatory (4.403b) [oscillatory (4.404b)] circular pendular motion are specified exactly by series whose convergence is specified by the ratio (4.410) of successive coefficients (4.400e):

$$\left(\frac{a_{n+1}}{a_n}\right)^2 = \left(\frac{2n+1}{2n+2}\right)^2 = \left[\frac{1 + 1/(2n)}{1 + 1/n}\right]^2 = \left[\left(1 + \frac{1}{2n}\right)\left(1 - \frac{1}{n}\right)\right]^2$$

$$= \left(1 - \frac{1}{2n}\right)^2 = 1 - \frac{1}{n} + O\left(\frac{1}{n^2}\right). \qquad\qquad (4.410)$$

The combined convergence test (subsection I.29.1.1) specifies the convergence of the series (4.403b; 4.404b) for all values of the non-linearity parameter (4.392a). The logarithmic series (II.3.48b) ≡ (4.411a) has coefficients (4.411b) like the arithmetic series (4.411c) that have the same ratio (4.411d) ≡ (4.410) to order n^{-2}:

$$-\log(1-z) = \sum_{n=1}^{\infty}\frac{z^n}{n} = \sum_{n=1}^{\infty}d_n z^n, \quad d_n \equiv \frac{1}{n}, \quad \frac{d_{n+1}}{d_n} = \frac{n}{n+1} = 1 - \frac{1}{n} + O\left(\frac{1}{n^2}\right).$$

$$(4.411a\text{--}d)$$

The combined convergence test (I.29.2c) ≡ (4.412a):

$$\left(\frac{a_{n+1}}{a_{n+1}}\right)^2 = 1 - \frac{g}{n} + O\left(\frac{1}{n^2}\right); \qquad g = 1, \qquad (4.412a, b)$$

implies (4.412b) and thus: (i) the series have radius of convergence unity; (ii/iii) they converge absolutely (uniformly) for variable $|z| < 1 (|z| \le 1 - \varepsilon$ *with* $0 < \varepsilon < 1)$; (iv) they diverge (I.29.5b) at $z = 1$. The variable is $z = q^2$ in (4.403b) and $z = 1/q^2$ in (4.404b).

Thus *the period of the circulatory (oscillatory) motion [problem 131 (132)] of a circular pendulum is specified exactly by the series (4.403b) [(4.404b)] that: (i) converges absolutely (sections I.21.3–I.21.4), that is, the series of moduli converges if* $q^2 > 1 (q^2 < 1)$ *corresponding to (4.403a) [(4.404a)]; and (ii) converges uniformly (section I.21.5–I.21.6), that is, independently of the value of* q^2, *if* $q^2 > 1 + \delta$ *with* $\delta > 0$ $(q^2 < 1 - \varepsilon$ *with* $0 < \varepsilon < 1)$. *The two series coincide on the boundary of convergence (4.413a) where they diverge (4.413b) because this corresponds to the stoppage at the top (4.390a) that takes an infinite time (4.390b) corresponding to an infinite period (4.413c):*

$$q = 1: \qquad |\dot{\theta}_0| = 2\omega_0, \qquad D \equiv A(1) = 1 + \sum_{n=1}^{\infty} (a_n)^2 = \infty. \qquad (4.413a–c)$$

In the stoppage limit (4.413a–c) the period of the oscillation (4.414a) is twice (4.414c) the period of circulation (4.414b):

$$\tau_2 = \frac{2\pi D}{\omega_0} = \frac{4\pi D}{|\dot{\theta}_0|}, \qquad \tau_1 = \frac{4\pi D}{|\dot{\theta}_0|} = \frac{2\pi D}{\omega_0}: \qquad \tau_2 = 2\tau_1, \qquad (4.414a–c)$$

because a complete oscillation (circulation) goes twice (once) around the circle. The period of the circulatory (4.403a, b; 4.405) [oscillatory (4.404a, b; 4.406)] circular pendular motion increases for lower (higher) angular velocity at the bottom, and they coincide (4.413a–c) in the case of stoppage at the top, when both are infinite (4.414a, b) in the ratio (4.414c). The non-linear motion of the circular pendulum has been considered in the absence of damping (subsections 4.7.9–4.7.14) and is extended next (subsections 4.7.15–4.7.17) to quadratic damping.

4.7.16 Harmonic Oscillator with Non-Linear Damping

In the case of the homopolar dynamo (subsection 4.7.2), there is no oscillation because the differential equation (4.343a) is of first order with real coefficients. An oscillation will occur for a second-order system with a linear restoring force and non-linear damping (4.315a):

$$\ddot{x} + h(\dot{x}) + \omega_0^2 x = 0: \qquad v = \dot{x}, \qquad \dot{v} = -h(v) - \omega_0^2 x, \qquad (4.415a–c)$$

that is equivalent to the autonomous differential system (4.415b, c). The paths in the phase plane are (4.415c) ≡ (4.416b):

$$\psi = \phi + \frac{\pi}{2}: \qquad \tan\phi = \frac{dv}{dx} = \frac{dv/dt}{dx/dt} = \frac{\dot{v}}{v} = -\frac{h(v) + \omega_0^2 x}{v} \equiv -\frac{H(v,x)}{v} = -\cot\psi.$$

$$(4.416a, b)$$

This implies (4.416a) that paths are orthogonal (Figure 4.20a) to the straight line obtained as follows: (i) draw $x = h(v)$, which is the dashed line passing through the origin called the **characteristic curve**; (ii) take any point A in the phase plane and join it to the point B on the characteristic curve by an horizontal line; (iii) project B vertically on the real axis at C. Then the straight line CA has slope ψ in (4.416a); (iv) the path is orthogonal (4.416b) to CA at the point A; that is, the curve with tangent of slope ϕ in (4.416b). The particular case of **quadratic friction force** is proportional to the square of the velocity with opposite direction:

$$h(v) = v|v| = v^2 \operatorname{sgn}(v) = \begin{cases} v^2 & \text{if} \quad v \le 0 \quad \text{and} \quad x = a, \\ -v^2 & \text{if} \quad v \ge 0 \quad \text{and} \quad x = -a, \end{cases} \qquad (4.417a, b)$$

and the characteristic consists of two semi-infinite straight lines orthogonal to the x-axis at $\pm a$ in Figure 4.20b. The path in the lower (upper) half plane

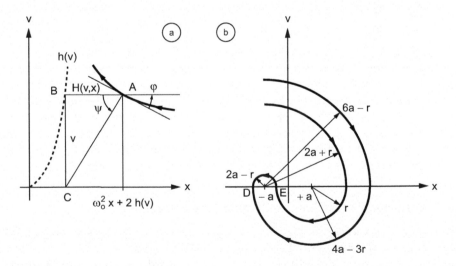

FIGURE 4.20
In the case of a harmonic oscillator with non-linear damping its path in the phase plane (a) is such that it must come to rest on the real axis (b) at the points A or B. In the case of quadratic damping, the paths in the phase plane are circles with centers at $(\pm a, 0)$ and radii r, $2a \pm r$, $4a - 3r$ and $6a - r$.

are circles with center at $+a(-a)$, a with positive $v > 0$ (negative $v < 0$) velocity, and the motion stops on the real axis at D and E.

4.7.17 Non-Linear Pendulum with Quadratic Damping

The linear damping with non-linear restoring force (section 4.6), and linear restoring force with non-linear damping (subsections 4.7.1–4.7.7), were considered in connection with the self-excited dynamo that may have amplification, attenuation, or a steady value. The non-linear pendular motion (subsections 4.7.9–4.7.15), apart from the stoppage at the top, is an oscillatory or circulatory motion in the absence of damping. In the presence of damping: (i) the oscillations come to rest as all energy is dissipated; (ii) the circulatory motion becomes an oscillation when the total energy decays below the potential energy at the highest point. Thus the circulatory motion of a pendulum continues indefinitely (is limited to a finite number of circulations) in the absence (presence) of damping. The transition from circulatory to oscillatory motion of the pendulum is analyzed next for quadratic damping, that adds to the equation of motion (4.371a) the second term in (4.418b):

$$\mu > 0: \qquad\qquad mL\ddot{\theta} + \mu\dot{\theta}|\dot{\theta}| + mg\sin\theta = 0, \qquad\qquad \text{(4.418a, b)}$$

where m is the mass L, the length, g the acceleration of gravity and (4.418a) the friction coefficient. These quantities appear in the equation of motion (4.419c) only in two combinations, namely the natural frequency of small amplitude oscillations (4.419a) and the damping (4.419b):

$$\omega_0 \equiv \sqrt{\frac{g}{L}}, \quad \lambda \equiv \frac{\mu}{mL}: \qquad\qquad \ddot{\theta} + \lambda\dot{\theta}|\dot{\theta}| + \omega_0^2\sin\theta = 0. \qquad\qquad \text{(4.419a–c)}$$

The second-order differential equation (4.419c) is equivalent to the autonomous system:

$$\dot{\theta} = \Omega, \qquad\qquad \dot{\Omega} = -\lambda\Omega|\Omega| - \omega_0^2\sin\theta, \qquad\qquad \text{(4.420a, b)}$$

where Ω is the angular velocity. The path in the phase plane is (4.421b) in terms of the angular velocity:

$$\Theta \equiv \Omega^2: \qquad -2\omega_0^2\sin\theta = 2\lambda|\Omega|\Omega + 2\Omega\frac{d\Omega}{d\theta} = \pm2\lambda\Theta + \frac{d\Theta}{d\theta}; \qquad \text{(4.421a–c)}$$

using the square of the angular velocity as variable (4.421a) transforms the non-linear (4.421b) into a linear forced differential equation of first order

(4.421c), where the upper (lower) sign corresponds to $\Omega > 0(\Omega < 0)$. The solution of the corresponding unforced equation is (4.422a):

$$\Theta(\theta) = C_\pm(\theta)\exp(\mp 2\lambda\theta); \qquad \frac{dC_\pm}{d\theta} = -2\omega_0^2\exp(\pm 2\lambda\theta)\sin\theta, \qquad \text{(4.422a, b)}$$

using the method of variation of parameters (notes 1.2–1.4) the constant $C_\pm$ is replaced by a function $C_\pm(\theta)$, and substitution of (4.422a) in (4.421c) leads to (4.422b) as follows from:

$$-2\omega_0^2\sin\theta = \pm 2\lambda C_\pm(\theta)\exp(\mp 2\lambda\theta) + \frac{d}{d\theta}\Big[C_\pm(\theta)\exp(\mp 2\lambda\theta)\Big] = \exp(\mp 2\lambda\theta)\frac{dC_\pm}{d\theta}.$$

$$\text{(4.422c)}$$

The integration of (4.222c) is elementary:

$$C_\pm(\theta) = i\omega_0^2\int\Big[\exp(\pm 2\lambda\theta + i\theta) - \exp(\pm 2\lambda\theta - i\theta)\Big]d\theta$$

$$= i\omega_0^2\exp(\pm 2\lambda\theta)\left(\frac{e^{i\theta}}{\pm 2\lambda + i} - \frac{e^{-i\theta}}{\pm 2\lambda - i}\right) + D_\pm, \qquad \text{(4.422d)}$$

$$= \frac{2\omega_0^2}{1 + 4\lambda^2}\exp(\pm 2\lambda\theta)(\cos\theta \mp 2\lambda\sin\theta) + D_\pm,$$

using (II.5.3a, b) to within an added constant $D_\pm$.

4.7.18 Number of Revolutions and Damped Oscillation into Rest

Substituting (4.422d) with a distinct added constant $D_+(D_-)$ for $\Omega \geq 0(\Omega \leq 0)$ and upper (lower) sign in (4.422a; 4.421a) leads to:

$$\omega_1^2 = \frac{\omega_0^2}{1 + 4\lambda^2}: \qquad \big[\Omega(\theta)\big]^2 = \begin{cases} D_+e^{-2\lambda\theta} + 2\omega_1^2(\cos\theta - 2\lambda\sin\theta) & \text{if} \quad \Omega \geq 0, \\ D_-e^{2\lambda\theta} + 2\omega_1^2(\cos\theta + 2\lambda\sin\theta) & \text{if} \quad \Omega \leq 0. \end{cases}$$

$$\text{(4.423a–c)}$$

Thus *the non-linear pendulum (Figure 4.19 a) with quadratic friction (4.418a, b) has (problem 134) paths (4.423b, c) in the phase plane where (4.423a) is the oscillation frequency for small amplitude and (4.419b) is the damping coefficient. The large amplitude motion (Figure 4.21) curls around (moves away from) the points (4.424a) in the phase plane that are positions of stable (unstable) equilibrium (4.424b) [(4.424c)] where the pendulum is suspended (inverted) in the Figure 4.19b(c):*

$$\dot{\theta} = 0, \qquad \theta_n = n\pi = \begin{cases} 2n\pi & \text{stable equilibrium}, \\ 2n\pi - \pi & \text{unstable equilibrium}. \end{cases} \qquad \text{(4.424a–c)}$$

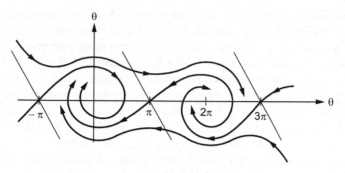

FIGURE 4.21
A circular pendulum (Figure 4.19a) with total energy larger than the potential energy at the top has non-zero angular velocity and performs a circulatory motion (Figure 4.19d) that would go on indefinitely in the absence of damping. In the presence of damping the phase diagram (Figure 4.21) for the angle θ versus angular velocity $\dot{\theta}$ shows that: (i) the pendulum moves away from the positions of unstable equilibrium $\theta = \mp\pi, \pm 3\pi, \ldots$; (ii) the paths curl around the positions of stable equilibrium $\theta = 0, \pm 2\pi, \pm 4\pi, \ldots$. Thus the damped pendulum starting with circulatory (Figure 4.19d) motion: (i) performs a finite number of circulations until the angular velocity decays sufficiently that it cannot reach the top position; (ii) it then performs oscillations (Figure 4.19a) that are damped to rest at the lowest position of stable equilibrium (Figure 4.19b).

The path starting (problem 135) at the highest point at rest (4.425a, b) has angular velocity (4.425c) at the lowest point:

$$\{\dot{\theta}, \theta\} = \{0, 2n\pi - \pi\}, \quad D_- = 2\omega_1^2: \quad \{\Omega_{2n-1}(0)\}^2 = 2\omega_1^2 \{\exp[2\lambda(2n-1)\pi] - 1\}.$$

$$(4.425a\text{–}c)$$

If (problem 136) the pendulum starts at $\theta_0 = 0$ with initial angular velocity $\Omega_{2n-1} < \dot{\theta}_0 \leq \Omega_{2n+1}$ it will (i) make n revolutions (Figure 4.19d), then (ii) go into a damped oscillation (Figure 4.19a) ending (iii) at rest in the position of stable equilibrium (Figure 4.19b).

4.8 Bifurcations, Iterated Maps, and Chaos

The dynamical systems may be represented by: (i) the trajectory in physical space; (ii) the path in phase space; or (iii) the bifurcations in parameter space (subsection 4.8.1). The parameter space is used to prove the theorem on stability boundaries (subsection 4.8.2) used before in connection with non-linear resonance (section 4.6). The bifurcations apply both to linear (nonlinear) dynamical systems [subsection(s) 4.8.4 (4.8.5–4.8.11)] that include gradient and conservative dynamical systems (subsection 4.8.3). The perturbation of a linear system can lead to non-linear systems, for example the Hopf supercritical or subcritical bifurcation (subsection 4.8.5) of a spiral flow (subsection 4.8.4). In addition to the previous bifurcations for linear and non-linear oscillators, the van der Pol oscillator (subsection 4.8.7) is a non-linear example of non-separable displace-ment and velocity terms (subsection 4.8.6); it has a Hopf subcritical bifurcation with a hysteresis loop (subsection 4.8.7). Two of the most common or important bifurcations are the fold (cusp) catastrophe [subsection 4.8.8 (4.8.9)] corresponding to a one (two)- parameter potential with a cubic (quartic) term. The nature of the orbits of dynamical systems can be investigated with the help of iterated maps (subsection 4.8.10); these can indicate the emergence of chaos (subsection 4.8.11), for example the transition from laminar to turbulent flow (subsection 4.8.12). The transition from laminar to turbulent flow is associated with a loss of lift that can cause an aircraft to stall (subsection 4.8.13); stall is a departure from stable flight that can lead to a spin (subsection 4.8.14); that is, a hazardous falling spiral motion that needs recovery within the available altitude to avoid an accident.

4.8.1 Physical, Phase, and Parameter Spaces

A second-order dynamical system has several equivalent representations in different spaces, and may include one (or more) parameter(s) q. The trajectory in the physical space (x, y) is specified by (4.1a, b) ≡ (4.426a, b) the two components of the velocity as a system of two coupled first-order differential equations:

$$\frac{dx}{dt} \equiv \dot{x}(x,y;t;q), \qquad \frac{dx}{dt} \equiv \dot{y}(x,y;t;q); \qquad \text{(4.426a, b)}$$

choosing an initial position (4.426c, d) at time t_0 leads to the trajectory (4.426e, f) for all time:

$$x_0 \equiv x(t_0), y_0 \equiv y(t_0): \quad x = x(t;x_0,y_0;q), \qquad y = y(t;x_0,y_0;q). \qquad \text{(4.426c–f)}$$

Solving (4.426a) for y leads to (4.427a) and substitution in (4.426b) leads to a single second-order differential equation (4.427b) for the dependent variable x:

$$y = y(x,\dot{x};t;q): \qquad F(x,\dot{x},\ddot{x};t,q) = 0, \qquad \text{(4.427a, b)}$$

and likewise for y; choosing an initial position (4.427c) and velocity (4.427d) at time t_0 leads from (4.427b) to the trajectory (4.427e) for all time:

$$x_0 \equiv x(t_0), \dot{x}_0 \equiv \dot{x}(t_0): \qquad x = x(t;x_0,\dot{x}_0;q). \qquad \text{(4.427c–e)}$$

From (4.227e) follows the velocity (4.428a) and eliminating time with (4.427e) leads to the path (4.428b) in the **phase plane**:

$$\dot{x} = \dot{x}(t;x_0,\dot{x}_0;q): \qquad G(x,\dot{x};\dot{x}_0;q) = 0, \qquad \text{(4.428a, b)}$$

whose coordinates are the position and velocity.

In the case of a force depending only on position it: (i) specifies the acceleration (4.429a); and (ii) derives from a potential (4.429b) specified to within an added constant (4.429c):

$$\ddot{x} = f(x;q) = -\frac{\partial \Phi(x;q)}{dx}, \qquad \Phi(x;q) = \int^x f(\xi;q)d\xi + C. \qquad (4.429a\text{–}c)$$

The **equilibrium position(s)** correspond to zero force (4.430a) [stationary potential (4.430b)] and are **stable (unstable)** if: (i) the force has negative (positive) gradient (4.430c) [(4.430e)]; and (ii) the potential is minimum (4.430d) [maximum (4.430f)]:

$$0 = f(\bar{x};q) = -\frac{\partial \Phi(\bar{x};q)}{\partial \bar{x}}: \qquad \begin{cases} \text{stable:} & -\frac{\partial f}{\partial x} = \frac{\partial^2 \Phi}{\partial x^2} > 0, \\ \\ \text{unstable:} & -\frac{\partial f}{\partial \bar{x}} = \frac{\partial^2 \Phi}{\partial \bar{x}^2} < 0. \end{cases} \qquad (4.430a\text{–}d)$$

Next to be considered is the dependence of the equilibrium positions $\bar{x}$ on the parameter(s) q that leads to bifurcations and stability boundaries (subsection 4.8.2).

4.8.2 Equilibria, Bifurcations, and Stability Boundaries (Poincaré 1881)

The equilibrium position (4.430a) ≡ (4.431a) leads to (4.431b) by differentiation with regard to the parameter:

$$0 = f(\bar{x};q): \qquad 0 = \frac{df}{dq} = \frac{\partial f}{\partial q} + \frac{\partial f}{\partial \bar{x}}\frac{d\bar{x}}{dq}, \qquad (4.431a, b)$$

where the equilibrium position depends on the parameter (Figure 4.22). If the force is stationary (4.432a); that is, the potential has an inflection point (4.432b):

$$0 = \frac{\partial f}{\partial \bar{x}} = -\frac{\partial^2 \Phi}{\partial \bar{x}^2}: \qquad \frac{\partial f}{\partial q}\begin{cases} = 0 & \text{singular bifurcation,} \\ \neq 0 & \text{vertical bifurcation,} \end{cases} \qquad (4.432a\text{–}d)$$

two cases arise: (i) in the case (4.432c) $d\bar{x}/dq$ is indeterminate, corresponding to a singular point, where no curve or several integral curves may pass, for example at the point A; (ii) in the sub-case (4.432d) then $d\bar{x}/dq = \infty$ or $dq/d\bar{x} = 0$, implying that there is a double root, such as the transition from two distinct real roots to a complex conjugate pair, for example at the point B. In both cases there is a change in the number of roots, and hence a change in the qualitative behavior and both points correspond to a bifurcation, namely

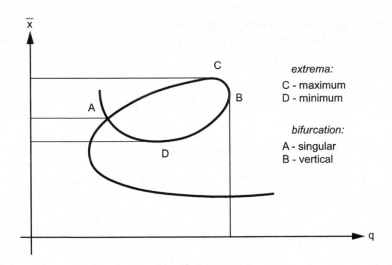

FIGURE 4.22
The path in the parameter space of the equilibrium positions $\overline{x}$ of a non-linear dynamical system versus a bifurcation parameter q may include two types of bifurcation points: (i) singular A where several curves pass, and thus lie on both sides; and (ii) vertical where a single curve passes with vertical tangent and thus lies to one side.

a **singular (vertical) bifurcation** of type I(II) at the point $A(B)$. Thus, *a stationary point of the force (4.432a) that is an inflection of the potential (4.432b) of a second-order system (4.429a–c) corresponds to a bifurcation, of singular (4.432c) [vertical (4.432d)] type if there are two roots on both sides (only on one side) as at the point $A(B)$ in Figure 4.22. Outside a bifurcation point (4.433a) then (4.431b) can be solved for the slope of position of the equilibrium point with regard to the parameter (4.433b):*

$$\frac{\partial f(\overline{x}, q)}{\partial \overline{x}} \neq 0: \qquad\qquad \frac{d\overline{x}}{dq} = -\frac{\partial f / \partial q}{\partial f / \partial \overline{x}}; \qquad\qquad (4.433a, b)$$

the extreme positions of the equilibrium points (4.434a) correspond to maxima (minima) for (4.434b) [(4.434c)]:

$$\frac{\partial f}{\partial \overline{x}} \neq \infty: \qquad 0 = \frac{d\overline{x}}{dq} \Rightarrow \frac{\partial f}{\partial q} = 0: \quad \begin{cases} \partial^2 f / \partial q^2 > 0: & minimum\ \overline{x} \\ \partial^2 f / \partial q^2 < 0: & maximum\ \overline{x}, \end{cases} \qquad (4.434a\text{–}c)$$

such as the points C (D) in the Figure 4.22.

The preceding results (Figure 4.22) are applied next to the stability boundary (4.430a) in the parametric space, relating the positions of equilibrium to the parameter; it is assumed that the stability curves $f(\overline{x}, q) = const$ correspond (Figure 4.23) to positive (negative) force $f > 0 (f < 0)$ inside

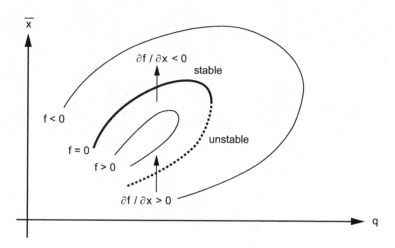

FIGURE 4.23
The theorem on stability boundaries applies in the parameter space (Figure 4.22) for a force that is zero on an equilibrium boundary, and positive (negative) inside (outside). The part of the equilibrium boundary facing upward (downward), that is the solid (dotted) line, is stable (unstable), because crossing it leads to a negative (positive) force gradient, hence return to (further deviation from) equilibrium.

(outside) the stability boundary. Then, for a point just: (i) above the stability boundary, f decreases as $\bar{x}$ increases, implying $\partial f / \partial \bar{x} < 0$ stability by (4.430c); (ii) below the stability boundary, f increases as $\bar{x}$ increases, implying $\partial f / \partial \bar{x} > 0$ instability by (4.430d). This proves the **theorem on stability boundaries (Poincaré 1881)**: *if the stability boundary (4.430a, b) of the second-order dynamical system (4.429a, b) is such (problem 137) that the force is positive (negative) inside (outside) then there is stability (instability) in the upward (downward) facing branch of the stability boundary, shown as a solid (dashed) curve in the Figure 4.23. Likewise, in Figures 4.24a–d where Figure 4.24c(b) reassembles* Figure 4.10d (4.15) concerning the non-linear resonance of a damped anharmonic oscillator forced at the natural frequency (second harmonic). Further examples concern: (i) the hard spring has only an equilibrium point d in the Figures 4.8b, c; (ii) the soft spring has one equilibrium (two bifurcation) point(s) at f (d and h) in the Figures 4.8d, e. The suspended (inverted) pendulum [Figure 4.19b(c)] is stable (unstable) because the gravity potential is minimum (maximum). The preceding examples also suggest a classification of dynamical systems (subsection 4.8.3).

4.8.3 Gradient and Conservative Dynamical Systems

The dependence of a dynamical system (4.426a, b) on the parameter q is generally continuous; that is, small variations in the parameter q lead to paths which remain close for all time and have qualitatively similar properties

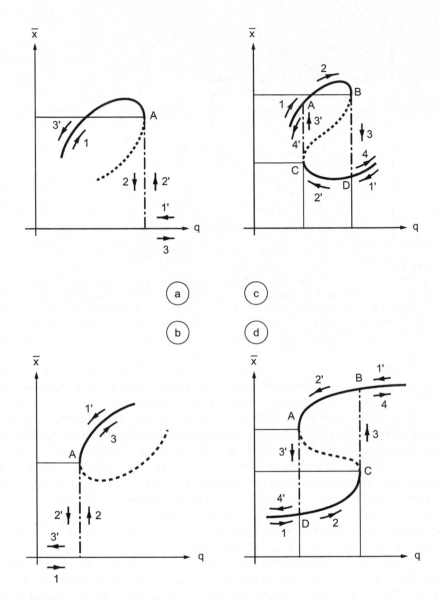

FIGURE 4.24
The examples of the application of the theorem on stability boundaries (Figure 4.23) in the parameter space (Figure 4.22) include: (a, b) amplitude jumps and suppression of oscillations, as in the Figures 4.14 and 4.15; (c, d) amplitude jumps and hysteresis loop as in the Figure 4.10d.

in the sense that there exists a differentiable map between the two paths. The values of the parameter q for which there is qualitative change in the paths are called **bifurcations**. For a bifurcation value of the parameter q_0 the paths are qualitatively different on either side $q > q_0 (q < q_0)$, and do not remain close or a differentiable map between them does not exist. The linear dynamical systems exhibit bifurcations. The simplest dynamical system is of the first-order system (4.435a) with damping $q \equiv \lambda$:

$$\dot{x} = \lambda x: \qquad x(t) = x(0)e^{\lambda t}: \quad \begin{cases} \lambda < 0 & stable, \\ \lambda = 0 & indifferent, \\ \lambda > 0 & unstable; \end{cases} \qquad (4.435a\text{–}d)$$

it has a bifurcation for the value $\lambda = 0$ of the eigenvalue λ, since it is stable (unstable) for (4.435b) [(4.435d)] and in between there is indifferent equilibrium (4.435c). A simple second-order dynamical system (section 2.2) is the harmonic oscillator (4.436c), for example with initial conditions (4.436a, b):

$$x_0 \equiv x(0) \neq 0 = \dot{x}(0) \equiv \dot{x}_0: \qquad \ddot{x} + \omega_0^2 x = 0, \qquad (4.436a\text{–}c)$$

that has a bifurcation for $q = 0$ because:

$$x(t) = \begin{cases} x_0 \cos(|\omega_0|t) & \text{if} \quad q \equiv \omega_0^2 > 0: \quad oscillation, & (4.436d) \\ x_0 & \text{if} \quad q \equiv \omega_0^2 = 0: \quad static, & (4.436e) \\ x_0 \cosh(|\omega_0|t) & \text{if} \quad q \equiv \omega_0^2 < 0: \quad monotonic, & (4.436f) \end{cases}$$

because it: (i) is oscillatory for $q \equiv \omega_0^2 > 0$ with (4.436d) natural frequency $|\omega_0|$; (ii) growing monotonically (4.436f) if $q \equiv \omega_0^2 < 0$ with growth rate $|\omega_0|$; or (iii) static in the intermediate case $\omega_0 = 0$. The bifurcation does not depend on the initial conditions used, for example (4.436a, b). The general second-order system (section 2.3) with constant coefficients (4.437b) in the case of damping (4.437a):

$$\lambda > 0: \qquad \ddot{x} + 2\lambda\dot{x} + \omega_0^2 x = 0: \quad \begin{cases} monotonic & \text{if} \quad \lambda > \omega_0, \\ transition & \text{if} \quad \lambda = \omega_0, \\ oscillatory & \text{if} \quad \lambda < \omega_0, \end{cases} \qquad (4.437a\text{–}f)$$

has a bifurcation for $q \equiv \lambda = \omega_0$ because: (i) it is oscillatory (4.437f) for $\lambda < \omega_0$; (ii) it has monotonic growth (4.437d) for $\lambda > \omega_0$; and (iii) in the intermediate case $\lambda = \omega_0$ it grows with at most a local minimum (4.437e).

The preceding examples (4.435a–d; 4.436a–f; 4.437a–f) of bifurcation of linear dynamical systems suggest a classification of dynamical systems that also applies to non-linear cases (subsections 4.8.4–4.8.9). *A dynamical system is a* **gradient system** *if it derives from a potential* (4.438a, b):

$$\frac{dx}{dt} = u = \frac{\partial\Phi}{\partial x}, \quad \frac{dy}{dt} = v = \frac{\partial\Phi}{\partial y} \qquad \Leftrightarrow \qquad \frac{\partial u}{\partial y} = \frac{\partial v}{\partial x}, \qquad (4.438a\text{–}c)$$

and a necessary and sufficient condition is that it has zero curl (4.438c) as follows from: (i) the two-dimensional form (subsection III.5.8.1) of the theorem on irrotational fields (III.5.240a–d) ≡ (4.438a–c); and (ii) the condition (subsection 3.8.1) that dΦ is an exact differential (3.153a–d) ≡ (4.438a–c). A dynamical system is **conservative** *if there is a conserved quantity (4.439a):*

$$E(x, \dot{x}) = const; \qquad \ddot{x} = -\frac{\partial \Phi}{\partial x}, \qquad E = \frac{\dot{x}^2}{2} + \Phi(x), \qquad (4.439\text{a–c})$$

for example, a force depending only on position (4.439b) leads to the conservation of energy (4.439c) as follows from:

$$0 = \dot{x}\left(\ddot{x} + \frac{\partial \Phi}{\partial x}\right) = \dot{x}\ddot{x} + \frac{\partial \Phi}{\partial x}\frac{dx}{dt} = \frac{d}{dt}\left(\frac{\dot{x}^2}{2} + \Phi\right) = \frac{dE}{dt}. \qquad (4.439\text{d})$$

Taking as an example the linear second-order system with constant coefficients (4.437b) ≡ (4.440a), it follows (4.440b):

$$\ddot{x} = -2\lambda\dot{x} - \omega_0^2 x: \qquad -4\lambda\dot{x}^2 = 2\dot{x}\left(\ddot{x} + \omega_0^2 x\right) = \frac{d}{dt}\left(\dot{x}^2 + \omega_0^2 x\right), \qquad (4.440\text{a, b})$$

that the system is conservative (4.441b) only if there is no damping or amplification (4.441a):

$$\lambda = 0: \qquad const = \dot{x}^2 + \omega_0^2 x = 2E. \qquad (4.441\text{a, b})$$

The main motivation for the consideration of bifurcations of dynamical systems (subsections 4.8.1–4.8.3) is the non-linear cases addressed next (subsections 4.8.4–4.8.9).

4.8.4 Non-Linear Perturbation of a Spiral Flow

A constant angular velocity (4.442a) and linear radial velocity (4.442b) form a linear dynamical system (4.442c) corresponding (section I.12.6) to a spiral flow (4.442d):

$$\frac{d\theta}{dt} = \Omega = const, \quad \frac{dr}{dt} = qr: \quad \frac{dr}{d\theta} = \frac{q}{\Omega}r, \quad \Omega\log r - q\theta = const. \qquad (4.442\text{a–d})$$

The general non-linear perturbation of (4.443a, b) corresponding to a gradient dynamical system is:

$$\frac{d\theta}{dt} = \Omega + \frac{\partial \Phi}{\partial \theta}, \qquad \frac{dr}{dt} = qr + \frac{\partial \Phi}{\partial r}. \qquad (4.443\text{a, b})$$

A **rotationally invariant** perturbation (4.444a) does not depend on the angle and is symmetric relative to the origin:

$$\Phi(r,\theta) = \Psi(r) = \Psi(-r) \in D^4(|R): \quad \Psi(r) = \frac{\mu}{4}r^4, \quad \frac{dr}{dt} = qr + \mu r^3, \quad (4.444a\text{--}c)$$

if the function (4.444a) has a fourth-order derivative, the leading non-linear term is a quartic in the potential (4.444b), leading to a cubic in the dynamic system (4.444c) and the azimuthal perturbation (4.443a) is unchanged. Thus only the radial perturbation (4.444c) need be considered leading to the super(sub)critical Hopf bifurcation discussed next (subsection 4.8.5).

4.8.5 Subcritical and Supercritical Bifurcation (Hopf 1943)

For small radius the second parameter in (4.444c) can be neglected, and the first parameter (4.445a) leads (4.445b) to: (i/ii) stability (instability) for $q < 0 (q > 0)$ in (4.445c) [(4.445e)]; or (iii) indifferent equilibrium (4.445d) in the intermediate case:

$$\mu = 0: \quad \frac{dr}{dt} = qr, \quad r(t) = r_0 \times \begin{cases} \exp(-|q|t) & \text{if} \quad q < 0; \text{ stable,} \\ 1 & \text{if} \quad q = 0; \text{ indifferent,} \\ \exp(qt) & \text{if} \quad q > 0; \text{ unstable.} \end{cases} \quad (4.445a\text{--}e)$$

For large radius the first parameter in (4.444c) can be neglected (4.446a) and the second parameter (4.446b) leads to (4.446c):

$$q = 0: \quad \frac{dr}{dt} = \mu r^3 \quad \Leftrightarrow \quad 2\mu t = 2\int_{r_0}^{r} \frac{dr}{r^3} = \frac{1}{r_0^2} - \frac{1}{r^2}, \quad (4.446a\text{--}c)$$

that implies (4.446c) $\equiv$ (4.446d) stability (instability) for $\mu < 0 (\mu > 0)$ in (4.446e) [(4.446f)]:

$$[r(t)]^2 = \frac{r_0^2}{1 - 2\mu r_0^2 t}: \quad \begin{cases} \text{stable} & \text{for} \quad \mu < 0, \\ \text{unstable} & \text{for} \quad \mu > 0, \end{cases} \quad (4.446d\text{--}f)$$

that is finite for all positive time [(infinite for $t = 1/(2\mu r_0^2)$]. For intermediate radius both parameters should be considered (4.444c) $\equiv$ (4.447a) and affect the existence of a **limit cycle** (4.447b) with constant radius:

$$\frac{dr}{dt} = r(q + \mu r^2) = 0: \quad r = \begin{cases} 0 & \text{if} \quad q\mu > 0: \ \textit{no limit cycle} \\ \left|\dfrac{q}{\mu}\right|^{1/2} & \text{if} \quad q\mu < 0: \ \textit{limit cycle,} \end{cases} \quad (4.447a\text{--}c)$$

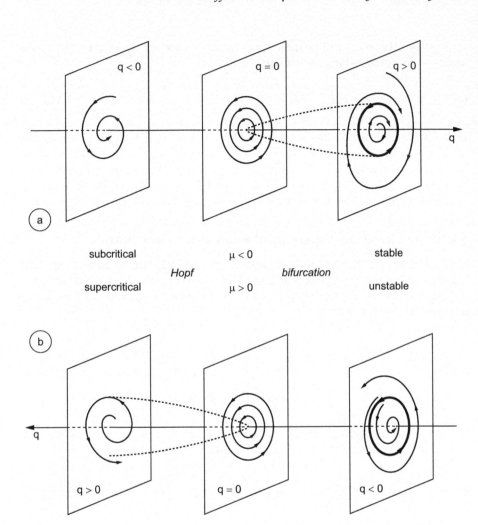

FIGURE 4.25
The Hopf (a) subcritical [supercritical (b)] bifurcation starts with a stable (unstable) spiral (Figure 4.1f) for negative (positive) parameter, goes through a center (Figure 4.1g) for the zero or bifurcation value of the parameter and ends in an stable (unstable) limit cycle for positive (negative) parameter.

namely, if $q\mu$ have the same (4.447b) [opposite (4.447c)] sign(s), there is no (there is one) limit cycle. The three conclusions (4.445a–e; 4.446a–f; 4.447a–c) are combined next.

It has been shown that *a radial perturbation (4.444c) of a spiral flow (4.442b, c) is: (i) stable for $q < 0$ or $\mu < 0$; (ii) unstable for $q > 0$ or $\mu > 0$; or (iii) there is a limit cycle if $q\mu > 0$ and not if $q\mu > 0$. This leads to two cases. The case I of (problem 138) a **Hopf subcritical bifurcation** (Figure 4.25a) corresponds $\mu < 0$ to: (I.i) for $q < 0$ to an inward spiral towards the attractor at the origin; (I.ii) for $q > 0$ since $q\mu < 0$ to a limit cycle; (I-iii/iv) inside (outside) the limit cycle there is an outward (inward) spiral* *away from the repeller at the origin (infinity); (I.v) thus the spirals both inside and outside tend to the limit cycle that is stable; (I.vi) the bifurcation $q = 0$ corresponds to a stable vortex. The case II (problem 139) of a **Hopf supercritical bifurcation** (Figure 4.25b) corresponds $\mu > 0$ to: (II.i) for $q > 0$ to an outward spiral away from the repeller at the origin; (II.i) for $q < 0$ since $q\mu < 0$ to a limit cycle; (II.iii-iv) inside (outside) the limit cycle there is an inward (outward) spiral towards the attractor at the origin (infinity); (II.v) thus the spirals both inside and outside move away from the limit cycle that is unstable; (II.vi) the bifurcation $q = 0$ corresponds to an unstable vortex.* The example of the Hopf bifurcations include the non-linear resonance of a damped anharmonic oscillator (subsections 4.6.7–4.6.8) at the second harmonic of the natural frequency (Figure 4.15) and the van der Pol oscillator considered next (subsection 4.8.5).

4.8.6 Linear Damping with Non-Linear Dependence on the Displacement (van der Pol 1922)

An example of non-linear oscillator with non-separable displacement and velocity terms is (4.448c) the **van der Pol oscillator (1922)**:

$$\gamma > 0, \quad h(x) = h(-x) \in D: \qquad \ddot{x} + \gamma h(x)\dot{x} + \omega_0^2 x = 0, \qquad \text{(4.448a–c)}$$

for which: (i) the restoring force is linear in the displacement; (ii) the friction force is linear in the velocity but the friction coefficient is a symmetric (4.448b) differentiable function of position, with magnitude of the order (4.448a). The van der Pol or **mixed non-linear oscillator** is equivalent to the dynamical system (4.449a, b):

$$\dot{x} = v, \qquad \dot{v} = -\gamma h(x)v - \omega_0^2 x; \qquad \text{(4.449a, b)}$$

using (4.449c) leads to the paths in the phase plane (4.449d):

$$\dot{v} \equiv \frac{dv}{dt} = \frac{dv}{dx}\frac{dx}{dt} = v\frac{dv}{dx}: \qquad v\,dv + \omega_0^2 x\,dx = -\gamma h(x)v\,dx. \qquad \text{(4.449c, d)}$$

Since the l.h.s. of (4.449d) is an exact differential *the mixed non-linear or van der Pol oscillator (4.448a–c) has a limit cycle if the condition (4.450a) is met:*

$$0 = \oint h(x)v\,dx = \oint h(x)\dot{x}^2\,dt: \quad \tau = \oint dt = \oint \frac{dx}{v} = -\oint \frac{dv}{\gamma h(x)v + \omega_0^2 x}, \qquad \text{(4.450a, b)}$$

in this case the period is given by (4.450b), that follows from (4.449a, d).

In the case of weak damping (4.451a), the simplest symmetric damping coefficient is (4.451b):

$$\gamma \ll 1: \quad h(x) = a - bx^2; \quad \{x(t), v(t)\} = R\{\sin(\omega_0 t), \omega_0 \cos(\omega_0 t)\}, \qquad (4.451a\text{--}d)$$

if the trajectory is approximately circular (4.451c, d) the condition (4.450a) of existence of a limit cycle:

$$\theta \equiv \omega_0 t: \quad 0 = \int_0^{2\pi} h(R \sin\theta)\, \dot{x}^2\, d\theta = \int_0^{2\pi} \left(a - b\, R^2 \sin^2\theta\right)\left(\omega_0 R \cos\theta\right)^2 d\theta$$

$$= R^2 \omega_0^2 \int_0^{2\pi} \left[a \cos^2\theta - \frac{b}{4} R^2 \sin^2(2\theta)\right] d\theta = R^2 \omega_0^2\, \pi \left(a - \frac{b\, R^2}{4}\right),$$

$$(4.452a, b)$$

specifies (4.452a, b) the radius (4.453a):

$$R = 2\sqrt{\frac{a}{b}}; \quad \gamma \ll 1: \quad \tau = -\oint \frac{dv}{\omega_0^2 x} = -\int_0^{2\pi} \frac{d(\omega_0 R \cos\theta)}{\omega_0^2 R \sin\theta} = \frac{2\pi}{\omega_0}, \qquad (4.453a\text{--}c)$$

substituting (4.451c, d) in (4.450b) specifies the period (4.453c) in the absence of damping (4.453b). Thus *the mixed non-linear oscillator (4.448c), with (problem 140) weak (4.451a) damping coefficient (4.451b) with a > 0 has a stable focus at the origin and a limit cycle of radius (4.453a) and undamped (4.453b) period (4.453c) corresponding to a sub-critical Hopf bifurcation (Figure 4.25a).*

4.8.7 Relaxation Oscillations with Mixed Damping

In the case of strong damping (4.454a) the re-scaling (4.454b) leads from a dynamical system (4.449a, b) with large coefficients:

$$\gamma \gg 1, \quad z \equiv \frac{x}{\gamma}: \qquad \dot{z} = \frac{v}{\gamma}, \qquad \dot{v} = -\gamma\left[h(\gamma z)v + \omega_0^2 z\right], \qquad (4.454a\text{--}d)$$

to a dynamical system with one small (4.454c) and one large (4.454d) coefficient. Eliminating time in the trajectory (4.454c, d) leads to the path in the phase plane (4.455a):

$$-\left[h(\gamma z)v + \omega_0^2 z\right]\frac{dz}{dv} = \frac{dz}{\gamma dt} = \frac{\dot{z}}{\gamma} = \frac{v}{\gamma^2}: \qquad \frac{dz}{dv} = 0 \quad or \quad \omega_0^2 \frac{x}{\gamma} = -h(x)\,\dot{x},$$

$$(4.455a\text{--}c)$$

that for large (4.454a) implies either (4.455b) or (4.455c). The former (4.455c) is of no interest and the latter (4.455c) leads to (4.456a) in the case (4.451b):

$$0 \simeq \frac{\omega_0^2}{\gamma} x = -\left(a - bx^2\right)\dot{x} = -\frac{d}{dt}\left(ax - \frac{bx^3}{3}\right): \qquad 3ax - bx^3 = const; \qquad \text{(4.456a, b)}$$

from (4.456a) follows a cubic equation (4.456b) like in the case (Figure 4.10d) of the non-linear resonance of the damped anharmonic oscillator forced at the natural frequency (subsections 4.6.1–4.6.3) corresponding to a hysteresis loop.

The triple roots of (4.456b) occur for (4.457a), and only two are stable, corresponding (Figure 4.26) to two fast jumps (*BC, DA*) on a time scale $1/\gamma$ and slow motion (*AB, CD*) on a time scale γ. The point *B*:

$$a > 0: \qquad\qquad \dot{x} = \infty \qquad \Rightarrow \qquad x_- = \sqrt{\frac{a}{b}} = \frac{x_+}{2}, \qquad\qquad \text{(4.457a–d)}$$

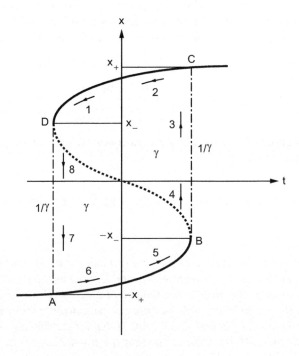

FIGURE 4.26
The one-dimensional oscillator with linear restoring force and linear damping with non-linear dependence on the displacement of van der Pol type has an hysteresis loop with two jumps, as in Figure 4.10d, with period determined by the slow motion along *BA* and *DC*, and a small effect of the phase jumps *CB* and *AB*.

corresponds (4.457c) to vertical slope (4.456b) in (4.456a), and C is at approximately double height (4.457d). The period follows from (4.456a) for the segments of slow motion:

$$\tau = 2\int_{CD} dt = 2\int_{x_-}^{x_+} \frac{dx}{\dot{x}} = -\frac{2\gamma}{\omega_0^2}\int_{x_-}^{x_+} \frac{a-b\,x^2}{x}\,dx = -\frac{2\gamma}{\omega_0^2}\left[a\log\left(\frac{x_+}{x_-}\right) - \frac{b}{2}\left(x_+^2 - x_-^2\right)\right]$$

$$= \frac{2\gamma}{\omega_0^2}\left(\frac{3b}{2}x_-^2 - a\log 2\right) = \frac{\gamma a}{\omega_0^2}\left(3 - 2\log 2\right) = 1.6137\frac{\gamma a}{\omega_0^2}.$$

$$(4.458)$$

Thus *the mixed non-linear oscillator (4.448c) with (problem 141) strong (4.454a) damping with coefficient (4.451b) has a hysteresis loop (Figure 4.26) with period (4.458)*, corresponding to the resonance of the anharmonic damped oscillator forced at the natural frequency (subsections 4.6.1–4.6.3).

4.8.8 One-Parameter Cubic Potential and Fold Catastrophe

The hysteresis cycle appears when considering the second of the sequence of the two most common bifurcations or "catastrophes": the fold and the cusp. *The **fold catastrophe** corresponds (problem 142) to (Figure 4.27a) to cubic potential (4.459a):*

$$\Phi(x;a) = \frac{x^3}{3} - ax, \qquad a - x^2 = -\frac{\partial\Phi(x;a)}{\partial x} = f(x;a) = \ddot{x}, \quad (4.459a, b)$$

and dynamical system (4.459b). The force vanishes (4.460a) at (Figure 4.27b) the equilibrium points (4.460b):

$$f(\bar{x};a) = 0: \quad \bar{x}_\pm = \pm\sqrt{a}; \quad \Phi_\pm'' \equiv \frac{\partial^2\Phi}{\partial\bar{x}_\pm^2} = 2\bar{x}_\pm = \pm 2\sqrt{a}, \quad \bar{\Phi}_\pm \equiv \Phi(\bar{x}_\pm;\lambda) = \mp\frac{2}{3}a^{3/2},$$

$$(4.460a–d)$$

the real equilibrium points exist for $a > 0$ and coincide for $a = 0$, that is the bifurcation value. The equilibrium is (Figure 4.27c) stable (unstable) for the upper (lower) sign in (4.460c), and at the equilibrium the potential takes (Figure 4.27d) the value (4.460d). Since the potential depends on only one parameter (Figure 4.27a), a plane representation with solid (dotted) line for the stable (unstable) case is sufficient to represent: (i) the (4.460b) equilibrium position (Figure 4.27b); (ii) the (4.460c) stability eigenvalues (Figure 4.27c); and (iii) the corresponding value (4.460d) of the potential (Figure 4.27d). The designation fold catastrophe is due to Figure 4.27d appearing as a folded and open sheet. Since the fold catastrophe involves only

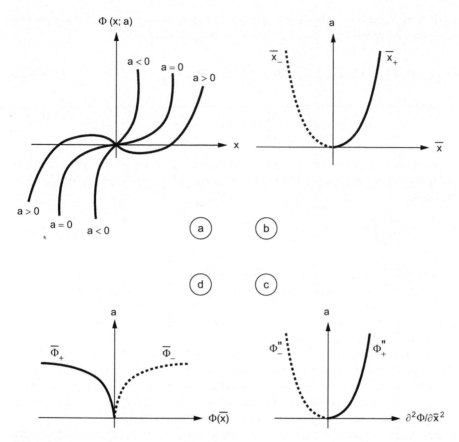

FIGURE 4.27
The anharmonic oscillator with cubic term in the potential (a) has stable (unstable) equilibrium positions (b) corresponding to positive (negative) curvature of the potential (c) and symmetric cusp of the potential at the equilibrium positions (d). The histeresis loop (Figure 4.26) and cubic potential (Figure 4.27) relate to the fold or cusp catastrophe (Figure 4.28).

one parameter a plane representation is adequate. In the case of the cusp catastrophe (subsection 4.8.9) that involves two parameters a representation in a three-dimensional space is more illuminating than its plane projections.

4.8.9 Two-Parameter Quartic Potential and Cusp Catastrophe

The **cusp catastrophe** corresponds (problem 143) to a quartic potential (4.461a):

$$\Phi(x;a,b) = -\frac{x^4}{4} + \frac{ax^2}{2} - bx: \qquad x^3 - ax + b = -\frac{d\Phi}{dx} = f(x;a,b) = \ddot{x}, \qquad \text{(4.461a, b)}$$

with two parameters that both appear in the force (4.461b). The two parameters can be reduced to one by the re-scaling (4.462a):

$$\{x,a,b,\Phi,f\}=\{\lambda\tilde{x},\lambda^2\tilde{a},\lambda^3\tilde{b},\lambda^4\tilde{\Phi},\lambda^3\tilde{f}\}: \qquad \tilde{\Phi}=-\frac{\tilde{x}^4}{4}+\frac{\tilde{a}\tilde{x}^2}{2}-\tilde{b}\tilde{x}, \qquad \text{(4.462a, b)}$$

$$-\tilde{x}^3+\tilde{a}\tilde{x}+\tilde{b}=-\frac{d\tilde{\Phi}}{d\tilde{x}}=\tilde{f}=\ddot{\tilde{x}}, \qquad 0=-\tilde{x}^3+\tilde{a}\tilde{x}+\tilde{b}, \qquad \text{(4.462c, d)}$$

applied to the potential (4.462b), force (4.462c), and equilibrium positions (4.462d). Differentiating (4.462d) with regard to b (tildes suppressed) leads to (4.463a):

$$1=\left(3\bar{x}^2-a\right)\frac{d\bar{x}}{db}: \qquad \lim_{b\to 0}\bar{x}=\pm\sqrt{a}, \qquad \lim_{b\to 0}\frac{d\bar{x}}{db}=\frac{1}{2a}, \qquad \text{(4.463a–c)}$$

also (4.462d) implies (4.463b) leading to (4.463c). Thus (Figure 4.28b) $\bar{x}$ as a function of b has at the origin vertical slope for $a=0$, positive slope for $a>0$ and negative slope for $a<0$. It follows that for a given b: (i) there is only one equilibrium position $\bar{x}$ if $a<0$; and (ii) if $a>0$ there is a range $b_-<b<b_+$ of values of b with a triple root for $\bar{x}$.

*The positions (problem 143) of the equilibrium points $\tilde{\bar{x}}$ as a function of $\tilde{b}$ are shown in Figure 4.28a for $\tilde{a}$ zero, positive, and negative; these correspond to cuts $\tilde{a}=\text{const}$ of the **cusp-catastrophe manifold** in Figure 4.28b, using coordinates (x, a, b); this involves stretching back (4.462a) to (x, a, b) from $\left(\tilde{x},\tilde{a},\tilde{b}\right)$, which has different scales along the axis, but does not modify the topology. The equilibrium points, where the force (4.461b) corresponding to the quartic potential (4.461a) vanishes, lie on the fold lines of the cusp catastrophe manifold (Figure 4.28b) and are projected on the parameter plane (Figure 4.28c) as a cusp, where coincide the stable and unstable roots; each line issuing from the cusp is a **directrix**. Hence the name cusp catastrophe; the cross-sections $a=\text{const}$ show an hysteresis loop (Figures 4.26 and 4.10d) emerging a double fold from an unfolded surface as $a<0$ crosses the bifurcation point $a=0$ into $a>0$. The discontinuous lines on the parameter plane (Figure 4.28c) are a two-dimensional projection of a smooth surface (Figure 4.28b) in three dimensions. In this case of a bifurcation with two parameters, the three-dimensional representation (Figure 4.28b) is more illuminating that its plane projections (Figures 4.28a, c). Another method of investigation of paths is the use of Poincaré maps (subsection 4.8.10).*

4.8.10 Iterated Maps and Classification of Singularities (Poincaré 1881–1886)

*Consider a dynamical system with **isolated singularities**; that is, there is a neighborhood of each singularity that does not contain other singularities. Let a line pass through a singularity, a straight line called the **directrix**, that is not tangent to any*

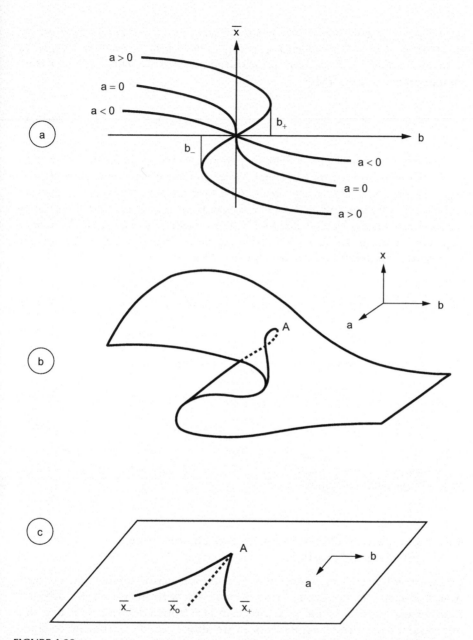

FIGURE 4.28
The potential of an anharmonic oscillator with *a* cubic term (Figure 4.28a) leads to the equilibrium positions $\bar{x}$ as function of the coefficient *b* of the linear terms that (a) have three values when the coefficient *a* of the cubic term increases. Choosing *a* as *a* third coordinate orthogonal to the $(\bar{x}, b)$ plane leads to *a* fold in the surface (b), like the hysteresis loops in Figures 4.10d, 4.24c, d, and 4.26. The projection on the plane (a, b) of the coefficients of the potential (c) leads to *a* cusp at the function of the stable and unstable branches of equilibrium positions as in Figure 4.27d.

integral curve (Figure 4.29a). The point(s) where the integral curves intersect the directrix (problem 144) are ordered according to their distance from the singularity (4.464a), and their sequence forms an **iterated or Poincare (1881–1886)** *or* **iterated or return map (4.464b):**

$$x_1 < x_2 < ... < x_{n-1} < x_n < x_{n+1} < ...: \qquad x_{n+1} = f(x_n). \qquad \text{(4.464a, b)}$$

Five cases can arise (Figure 4.29b): (i) if every directrix is cut always at the same fixed point (4.465a) the path is a **limit cycle**; *(ii) if for every directrix the iterated map (4.464b) tends to a fixed point (4.465b) then the path tends to a limit cycle; (iii/iv) if on every directrix (at least one directrix) the map tends to the singularity (infinity) the singularity is* **stable** *(4.465c)* **[unstable** *(4.465d)]; (v) if the iterated map covers densely the interval (4.465e) and never or seldom comes through the same point the* **motion is chaotic** *(Figure 4.29c):*

$$\text{Poincaré map:} \begin{cases} n = 1,2,... \infty \Rightarrow x_n = x_0: & \text{limit cycle,} & (4.465a) \\[2mm] \lim_{n \to \infty} x_n = x_0: & \text{tends to limit cycle,} & (4.465b) \\[2mm] \lim_{n \to \infty} x_n = 0: & \text{stable,} & (4.465c) \\[2mm] \lim_{n \to \infty} x_n = \infty: & \text{unstable,} & (4.465d) \\[2mm] x_- < x_n < x_+ & \text{chaotic.} & (4.465e) \end{cases}$$

For a chaotic system a small change in the initial conditions can lead to a large change in the path over a finite time.

4.8.11 Comparison of Deterministic and Chaotic Systems (Ruelle & Takens 1971)

The apparently random motion of a fluid known as **turbulence** may be interpreted as a chaotic motion. The transition of a flow from laminar to turbulent then corresponds to the passage (problem 145) from a deterministic to a chaotic system. An infinite sequence of bifurcations could clearly lead to chaos (Landau 1953). A more economical scenario (Reulle-Takens 1971) involves three bifurcations: (i) from a periodic orbit with one frequency ω_1 to a quasi-periodic orbit with frequencies (ω_1, ω_2); (ii) another quasi-periodic orbit $(\omega_1, \omega_2, \omega_3)$, which can be perturbed into (iii) a chaotic motion. In the case of three Hopf bifurcations they can be (i) all visible (Figure 4.30a), (ii) two visible (Figure 4.30b), or (iii) only one visible (Figure 4.30c). Besides a sequence of bifurcations, other properties of non-linear systems can contribute to

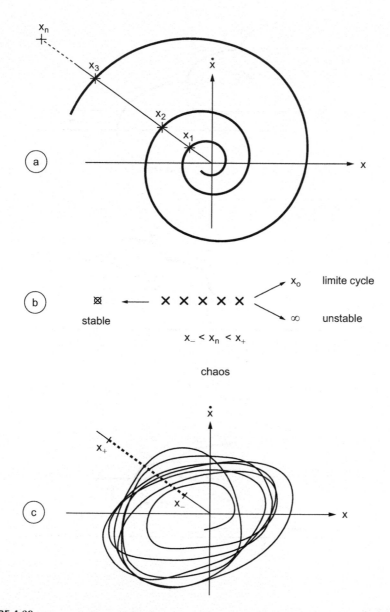

FIGURE 4.29
The **Poincaré map** is the ordered set (b) of the points of intersection with the integral curves of
a directrix, that is a straight line passing through an isolated singularity, placed at the origin of
the phase plane (a). Four cases can arise: (i) if the map stabilizes at a fixed point $x_n = x_0$ or tends
$n \to \infty$ to a fixed point $x_n \to x_0$ there is a limit cycle around a center (Figure 4.1g); (ii/iii) if the
map tends to the origin $x_n \to 0$ (infinity $x_n \to \infty$) the system is stable (unstable), for example an
inward (Figure 4.1f) [outward] spiral; or (iv) if the map does not converge and covers an interval $x_- < x_n < x_+$ in the sense it comes arbitrarily close to any point in a finite time, the system
is chaotic (c). The case (iv) implies that a small change in initial conditions can lead to a large
change in position in a finite time.

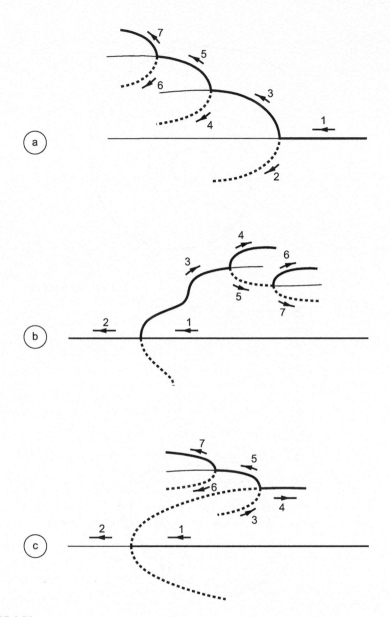

FIGURE 4.30
An example of chaotic motion is the transition from laminar to turbulent flow with three bifurcations that are: (i) all three visible (a); (ii) only two visible (b); (iii) only one visible (c). It is possible to envisage simpler scenarios of transition to turbulence with: (i) just one bifurcation to generate a second frequency; (ii) non-linear and parametric resonance to generate all combinations of harmonics (subharmonics) with high (low) frequencies corresponding to small (large) eddies, or random (coherent) flow structures; (iii) this comes arbitrarily close to a dense set of frequencies; (iv) the parametric resonance feeds energy from subharmonics to harmonics, that is large to small scales; or (v) the dissipation causes decay of the small scales.

complete the scenario of transition to turbulence. A possible definition of turbulence is a system with a dense or almost dense set of natural frequencies, which can be excited anywhere within a frequency range. Based on this definition, a third scenario, with only one or two bifurcations, might be possible; or the additional effects to be mentioned in the sequel could add to the two previous scenarios. The difference between a **deterministic (chaotic) system** *is that a small perturbation of the initial conditions always has a small (can have a large) effect at later times, so that the system can (cannot) be traced back to its origins.* The transition from laminar to turbulent flow is an example of the passage from a deterministic to a chaotic system, for which possibly the simplest scenario is outlined next (subsection 4.8.12).

4.8.12 Transition from Laminar to Turbulent Flow

The simplest scenario of **transition to turbulence** *(problem 146) can involve only one bifurcation and non-linear and parametric resonance in steps: (i) a bifurcation may cause an oscillatory system with natural frequency ω_1 to gain another natural frequency ω_2; (ii) if the starting system was non-oscillatory, a preliminary bifurcation could generate the first natural frequency, for a total of two bifurcations; (iii) it is assumed that the two frequencies are not multiples of the same fundamental frequency, that is arise from distinct physical processes; (iv) if the system is non-linear it generates harmonics that are sums (4.466b) of multiples (4.446a) of the two frequencies:*

$$m, n, p \in | Z: \qquad \omega_{m,n} = m\omega_1 + n\omega_2 ; \qquad \omega_{m,n,p} = \frac{m\omega_1 + n\omega_2}{p} , \qquad (4.466a\text{--}c)$$

(v) if the system is time dependent, parametric resonance can lead to sub-harmonics (4.466c) of (4.446b), leading to a three-parameter (4.466a) family of frequencies; (vi) given two frequencies ω_1 and ω_2 that are not multiples any third frequency, ω_3 can be approximated by (4.446c) through a suitable choice of three integers; (vii) thus is obtained a dense set of frequencies corresponding to a chaotic system. Thus the dynamics of non-linear and time dependent systems provide mechanisms for the transition to an almost dense set of oscillation frequencies like that found in turbulence. The non-linear generation of harmonics corresponds to the decay of turbulence from larger to smaller eddies that are dissipated more effectively; the excitation of parametric resonance can sustain lower frequencies corresponding to large scale flow structures found underlying turbulent flows. In very high speed or hypersonic flight of rockets the transition from laminar to turbulent flow enhance heat conduction and may increase temperatures beyond the thermal limit of structures. In low or high subsonic flight of aircraft the transition from laminar to turbulent flow can cause the aircraft to stall and to depart from steady flight into a spin (subsection 4.8.13).

4.8.13 Stall and Departure from Stable Flight

The straight, level, and steady flight of an aircraft (Figure 4.31a) requires that: (i) the **weight** W be balanced by the **lift** L generated mostly by the wing; (ii) associated with the lift there is **drag** D that must be balanced by the **thrust** T of the engine. This stable flight condition is maintained for **angle-of-attack** smaller than for stall, when the flow around the aircraft is mostly laminar. If the angle of attack exceeds the **stall** value (Figure 4.31b): (i) there is flow separation in the upper surface of the wing, increasing the pressure so that there is a loss of lift; (ii) the transition from laminar to turbulent flow also increases the drag, that may exceed the available thrust, causing the aircraft to slow down. The reduction in airspeed (ii) aggravates the (i) loss of lift, that is insufficient to balance the weight, so the aircraft falls down. The **departure** from stable flight is quite sensitive to initial conditions, for example a slight asymmetry may cause the port or left wing to drop first or the reverse. The velocity

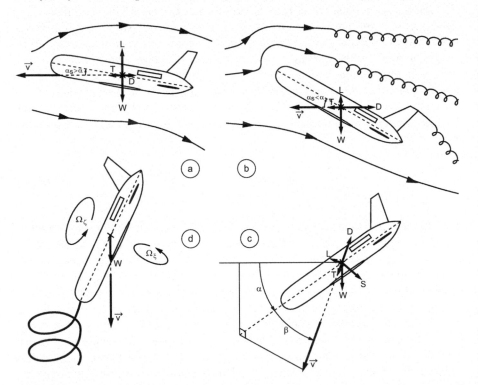

FIGURE 4.31
In the controlled flight of an aircraft at angle-of-attack below stall (a) the flow is mostly laminar. If the stall angle-of-attack is exceeded, flow separation occurs (b), and the transition from laminar to turbulent flow reduces lift L and increases drag D. The lift no longer compensates weight W and the thrust T of the engine does not balance drag D, leading to a departure from controlled flight that may (c) include side-slip and a side force S. This combination of forces and moments causes a spiraling falling motion or spin with rapid loss of altitude that may be a chaotic motion, and requires recovery action to escape this hazardous situation.

vector can diverge significantly from the aircraft axis (Figure 4.31c) both in the vertical plane (angle-of-attack) and in the horizontal plane (**angle-of side-slip**), leading to the appearance of a **side forces**. The combination of (i) side force, (ii) imbalance between weight and lift, and (iii) mismatch of drag and thrust creates unbalanced force and moments that cause the aircraft to rotate around one or more axis (Figure 4.31d); the resulting tumbling chaotic motion (problem 147) is called **spin,** and is one of the most dangerous flight conditions, with rapid altitude loss that can be fatal near the ground.

The safest approach is to not allow a spin to develop, by preventing that the stall angle of attack be exceeded. Many aircraft have a natural stall warning, in the form of vibration or buffet as the stall angle-of-attack is approached. For aircraft without natural **stall warning**, the certification authorities may require a **stick shaker**, and sometimes also a horn, to warn the pilot of the approach to stall. If the pilot persists in increasing the angle of attack toward the stall a **stick pusher** may apply opposing force and prevent the stall. The **departure protection** system may be more sophisticated than just **stall prevention** and include other features to keep the airplane within its safe flight envelope. Since airliners are flown gently for passenger comfort and reassurance, there is no need for large angles-of-attack or sideslip that could lead to a stall followed by spin. However, there may be emergency situations, like (i) violent atmospheric winds or storms, or (ii) sudden loss of thrust from engines, that can cause a stall and spin entry if not countered rapidly. In the case of fighters or aerobatic aircraft that perform aggressive or violent maneuvers, the risk of stall and spin is higher (subsection 4.8.14).

4.8.14 Spin Prediction, Testing, and Recovery

A typical air combat situation that can lead to stall and spin is described next: (i) a fighter aircraft performs a **high-speed turn** with small radius to escape a threat or to gain a firing position; (ii) the large centrifugal force must be countered by the horizontal component of the lift with a large **bank angle**; (iii) the high lift will require a large angle-of-attack close to the stall to be achieved in an **instantaneous turn**; (v) the high lift implies high drag, and to perform a **sustained turn** a high matching thrust is needed; (vi) the engine may flameout in a turn due to **inlet flow distortion** or a rapid throttle movement; (vii) thus the conditions for a spin are created. The stall and spin protection of a military fighter aircraft must allow for high maneuverability and survivability, and the risk of stall followed by spin cannot be excluded. The flight manual gives instructions on: (i) how to avoid stall and spin; (ii) the types of spin the particular aircraft can have; (iii) the control actions that can lead to recovery from each type of spin; (iv) the conditions in which no further attempt should be made to counter the spin, and ejection is advisable. In order to prepare the flight manual it is necessary to predict, test, and demonstrate spin recovery.

Spin simulations are performed in a **vertical wind tunnel** with the airflow blowing upward: (i) **aircraft model** is released with various initial conditions to try to find all possible **spin modes**; (ii) for each spin mode identified the most effective **control action** for spin recovery is sought. The vertical wind tunnel tests may be supplemented by **free flight** tests, with the aircraft model dropped from some altitude, for example from an helicopter; during the fall of the model the controls are actuated remotely to demonstrate **spin recovery** techniques. The spin recovery must be validated by **flight tests** in the real aircraft, usually fitted with two safety systems; (i) a large **spin parachute** with precursors to extract it, that virtually stops the aircraft in the air; (ii) the spin parachute is ejected, and if the engine has flamed-out a powerful **in-flight starter** relights the engines to fly away. A spin test starts at high altitude because of the rapid altitude loss, and is terminated by spin parachute deployment at a **safety altitude** sufficiently above the ground. Spin testing is the most dangerous flight testing and can subject the pilots to high accelerations, with ejection recommended before these become too high.

The control actions for spin recovery are not obvious: (i) a control surface deflection may be ineffective if the flow over its surface is already separated; (ii) the large amplitude motion couples the three axis of the aircraft, so that acting on one axis can affect the others. Thus it is not straightforward to: (i) find all spin modes; (ii) identify corrective actions for each of them. Besides the risk of **loss of consciousness** due to high accelerations there is also the risk of spatial disorientation due to prolonged rotations. The human **sense of orientation** comes from the otoliths in the inner ear that are set in motion by rotation of the body; after the human body stops rotating the otoliths still rotate three times. Thus a pilot may have succeeded in stopping the spin but still have the sensation that the rotation goes on, and the **spatial disorientation** may lead to continued use of the controls that could re-start the spin. The solution is to fix a point on the ground, and when its position is fixed the spin has ended. The most common spin is the **vertical spin** where the aircraft spirals nose down. The **flat spin**, where the aircraft rotates horizontally as it falls, can be the most dangerous because it generates eyeballs out accelerations that can affect vision. All efforts to detect spin modes and find recovery techniques for each of them do not exclude the possibility that a new, previously unknown spin mode appears later in regular service. This brief account of stall and spin shows that chaotic motions exist and can be quite difficult to master.

4.9 Competing Populations (Volterra 1931) and Limit Cycles

The theory of dynamical systems (chapter 4) applies not only to mechanical oscillators and electrical circuits (chapter 2), but also to a variety of other phenomena in the physical, biological, and social sciences and economics. An example is the problem of competing populations of predators and prey (subsection 4.9.1); it leads to a dynamical system that is not of gradient type but has a limit cycle (subsection 4.9.2). The existence of a limit cycle (subsection 4.9.3) is one aspect of the stability of orbits of non-linear dynamical systems (subsection 4.9.4) that can be deduced from the stability of the linear part using the Lyapunov theorem (section 9.3). The existence of a limit cycle (subsection 4.9.3) can be also be addressed as one aspect of the location of singularities of dynamical systems (subsection 4.9.10), such as multipoles (subsection 4.9.9) or other special points. One method of location uses the index or winding number of a singularity (subsection 4.9.5) that can be extended to the point at infinity (subsection 4.9.7). The sum of the indices of singularities in the finite plane (also at infinity) provides the original (extended) form of the winding theorem [subsection 4.9.6 (4.9.8)].

4.9.1 Growth, Decay, and Mutual Interactions

The problem of competing populations can be formulated in several ways, one of which is (problem 148) the interaction of prey x (predators y): (i) assuming an unlimited supply of food, the population of prey x increases at a birth rate a if unaffected by the predators (4.467b); (ii) the population of predators y declines at a death rate c, unless they find food (4.467c):

$$a,b,c,d > 0: \qquad \dot{x} = ax - bxy, \qquad \dot{y} = -cy + dxy. \qquad (4.467a\text{–}c)$$

and (iii) the encounters of prey and predator animals are proportional to the product of the two populations, and decrease (increase) the number of the former (latter) at a rate b (d) in (4.467b) [(4.468c)]. All the coefficients in the non-linear second-order dynamical system (4.467b, c) are positive (4.467a). The coupled system of non-linear differential equations (4.467b, c) may be solved: (i) cancelling the non-linear terms, multiplying (4.467b) [(4.467c)] by d (b) and adding (4.468a):

$$\dot{x}d + b\dot{y} = xad - ycb; \qquad c\frac{\dot{x}}{x} + a\frac{\dot{y}}{y} = -bcy + xad, \qquad (4.468a, b)$$

(ii) the linear terms on the r.h.s. are cancelled, multiplying (4.467b) [(4.467c)] by $c/x(a/y)$ and adding (4.468b); (iii) the difference of (4.468a, b) separates the variables (4.469a):

$$\dot{x}\left(d - \frac{c}{x}\right) + \dot{y}\left(b - \frac{a}{y}\right) = 0; \qquad \frac{\dot{y}}{\dot{x}} = \frac{-cy + dxy}{ax - bxy} = \frac{d - c/x}{-b + a/y}; \qquad (4.469a, b)$$

(iv) the ratio of (4.467b, c) leads to the same result (4.469b) $\equiv$ (4.469a); and (v) the integration of (4.469a) $\equiv$ (4.469b) is immediate (4.470a):

$$xd + by - c\log x - a\log y = const \equiv \log E: \qquad e^{xd}x^{-c} = Ee^{-by}y^{a}, \qquad (4.470a, b)$$

and leads to (4.470b). Thus *the autonomous non-linear differential system (4.467a–c) representing competing populations has (problem 148) solution (4.470b) where E is constant.* It is shown next (subsection 4.9.2) that the evolution of the two populations is periodic with a phase delay.

4.9.2 Coupled Oscillations with Phase Delay

The solution (4.470b) can be written (4.471a) in terms of the variables (4.471b) [(4.471c)]:

$$X = EY: \qquad\qquad X \equiv e^{xd}x^{-c}, \qquad Y = e^{-by}y^{a}, \qquad\qquad \text{(4.471a–c)}$$

whose: (i) first two derivatives are (4.472a, b) [(4.473a, b)]:

$$\frac{dX}{dx} = X\left(d - \frac{c}{x}\right), \qquad\qquad \frac{d^2X}{dx^2} = \left(d - \frac{c}{x}\right)^2 X + \frac{c}{x^2}X; \qquad \text{(4.472a, b)}$$

$$\frac{dY}{dy} = Y\left(\frac{a}{y} - b\right), \qquad\qquad \frac{d^2Y}{dy^2} = \left(\frac{a}{y} - b\right)^2 Y - \frac{a}{y^2}Y; \qquad \text{(4.473a, b)}$$

(ii) extrema are a minimum (4.474a–d) [maximum (4.475a–d)]:

$$x = \frac{c}{d}: \qquad X' = 0, \qquad X'' = \frac{d^2}{c}X > 0, \qquad X_{min} = X\left(\frac{c}{d}\right) = e^c\left(\frac{c}{d}\right)^{-c} = \left(\frac{ed}{c}\right)^c,$$

$$\text{(4.474a–d)}$$

$$y = \frac{a}{b}: \qquad Y' = 0, \qquad Y'' = -\frac{b^2}{a}Y < 0, \qquad Y_{max} = Y\left(\frac{a}{b}\right) = e^{-a}\left(\frac{a}{b}\right)^a = \left(\frac{a}{be}\right)^a;$$

$$\text{(4.475a–d)}$$

and (iii) also (4.471b) diverges [(4.471c) vanishes] both at the origin and infinity (4.476a, b) [(4.476c, d)]:

$$\lim_{x\to 0} X = \infty = \lim_{x\to\infty} X, \qquad\qquad \lim_{y\to 0} Y = 0 = \lim_{y\to\infty} Y. \qquad \text{(4.476a–d)}$$

This suggests the following method (Figure 4.32a) to construct the integral curves: (i) the (X, Y) axis are drawn in the third quadrant opposite to $y(x)$, together with the characteristic curve that is the straight line (4.471a) through the origin; (ii) the function $X(x)$ $[Y(y)]$ is drawn in the fourth (second) quadrant noting (4.471b) [(4.471c)] that it has a minimum (4.474a–d) [maximum (4.475a–d)] and tends to infinity (4.476a, b) [zero (4.476c, d)]

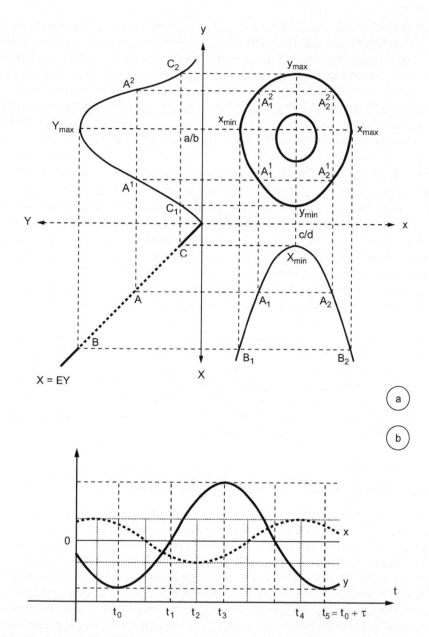

FIGURE 4.32
An example of limit cycle oscillation is the case of competing populations, for example:
(i) herbivorous animals x with an unlimited supply of food; (ii) carnivorous animals y that
can only feed on the herbivorous. The limit cycle occurs (a) because: (i) when there are too
many herbivorous they are easy to find and the carnivorous population increases; (b) a large
carnivorous population cannot be sustained by a declining herbivorous population. Thus
the carnivorous and herbivorous population decrease and increase in opposition with some
time lag between the extrema (b).

both at the origin and at infinity; (iii) a horizontal line is drawn through X_{min} that intersects the characteristic curve at the point C, and the vertical line passing through C intersects Y at $C_1(C_2)$ that specifies the minimum y_{min} (maximum y_{max}) of y; (iv) likewise, a vertical line is drawn through Y_{max} that intersects the characteristic curve at the point B and the horizontal line passing through B intersects X at $B_1(B_2)$ that specifies the minimum x_{min} (maximum x_{max}) of x; (v) all possible states of the system correspond to a point A on the characteristic curve between B and C, and its horizontal (vertical) projections intersects X at (A_1, A_2) [intersects Y at (A^1, A^2)]; (vi) thus the horizontal lines through (A^1, A^2) and the vertical lines through (A_1, A_2) intersect at four points $(A_1^1, A_1^2, A_2^2, A_2^1)$ that lie on the integral curve. *The competing populations (4.467a–c) have (problem 148) an integral curve (4.471a–c) with constant E corresponding to: (i) a limit cycle in Figure 4.32a; and (ii) oscillation of the two populations with the same period and a phase delay in Figure 4.32b.* The latter (Figure 4.32b) may be explained: (i) at the time t_0 the population of prey is at a minimum and the population of predators is neither at a maximum nor at a minimum; (ii) the predators cannot easily find prey, so that the population of predators decreases, allowing the population of prey to increase $t_0 < t < t_1$; (iii) at the time t_2 that the population of predators is at a minimum and the population of prey is still increasing; (iv) at the time t_3 that the population of prey is at a maximum the predators can easily find prey so the population of predators continues to increase; (v) at the time t_4 the population of predators is maximum and the population of prey continues to decrease; and (vi) the population of prey comes back to a minimum at a time $t_5 - t_0 = \tau$ separated by a period τ, and the cycle starts again.

4.9.3 Existence and Identification of Limit Cycles

The following two criteria may serve to indicate limit cycles: (i) the iterated map (subsection 4.8.10) is a fixed point (4.465a) or tends to a fixed point (4.465b); and (ii) the linearized system has a pair of complex conjugate eigenvalues (subsection 4.9.4). Thus, if the complex eigenvalues are calculated as a function of the parameters of a dynamical system with real coefficients, when they reach the imaginary axis a limit cycle can exist, provided that it is stable. The following theorem concerns the non-existence of a limit cycle: *the dynamical system (4.477a, b) with continuously differentiable forcing vector whose divergence (problem 149) does not vanish in a domain (4.477c) cannot have a limit cycle in the domain:*

$$\{\dot{x}, \dot{y}\} = u, v \in C^1(D): \qquad\qquad R \equiv \frac{\partial u}{\partial x} + \frac{\partial v}{\partial y} \neq 0 \ in \ D. \qquad (4.477a\text{–}c)$$

The proof is based on the following contradiction: (i) the divergence has a fixed sign (4.478a) in D, and by the divergence theorem the (III.5.220c) integral (4.478b) also has the same fixed sign:

$$R \gtrless 0: \qquad 0 \gtrless \iint_D \left(\frac{\partial u}{\partial x} + \frac{\partial v}{\partial y} \right) dS = \int_{\partial D} \left(v\, dx - u\, dy \right); \qquad (4.478\text{a, b})$$

and (ii) the integral (4.478b) is zero (4.479) if a limit cycle exists:

$$\int_{\partial D} \left(v\, dx - u\, dy \right) = \oint \left(vu - uv \right) dt = 0. \qquad (4.479)$$

Thus (4.479) is incompatible with (4.478a, b). For example, *a compressible flow with velocity $\vec{v}$ corresponding to expansion or contraction at all points (4.480a) cannot have a limit cycle; a limit cycle is possible only if there are expansions and contractions at different points:*

$$\dot{D} = \nabla . \vec{v} \gtrless 0; \qquad q/\varepsilon = \nabla . \vec{E} \gtrless 0, \qquad (4.480\text{a, b})$$

likewise, (4.480b) an electrostatic field $\vec{E}$ in a dielectric of permittivity ε with only positive or negative electric charge density q cannot have a limit cycle; a limit cycle is possible only in the presence of electric charges with opposite signs. The interpretation is that electric charges with the same (opposite) sign repel (attract) each other, so a limit cycle is not (is) possible. The existence of a limit cycle raises the question of its stability. The stability of a non-linear system can be inferred from the corresponding linear dynamical system as shown next (subsection 4.9.4).

4.9.4 Stability of Orbits of Dynamical Systems

The third Lyapunov theorem (subsection 9.2.16) can be stated in the particular case of *stability (problem 150)* of **non-linear dynamical systems** *(4.1a, b)* $\equiv$ *(4.481b, d) with* **forcing vector** *(4.481a, c) with continuous second-order derivatives:*

$$u \in C^2 \left(| R^2 \right): \qquad \frac{dx}{dt} = u(x,y) = ax + by + O\left(x^2, y^2, xy \right), \qquad (4.481\text{a, b})$$

$$v \in C^2 \left(| R^2 \right): \qquad \frac{dx}{dt} = v(x,y) = cx + dy + O\left(x^2, y^2, xy \right), \qquad (4.481\text{c, d})$$

is the same as that of the corresponding linear system; it is specified by the eigenvalues (4.38a, b) ≡ (4.482a–d):

$$0 = \begin{vmatrix} a-\lambda & b \\ c & d-\lambda \end{vmatrix} = (a-\lambda)(d-\lambda) - bc = \lambda^2 - (a+d)\lambda + ad - bc = (\lambda - \lambda_-)(\lambda - \lambda_+),$$

$$(4.482a\text{–}d)$$

implying that the system is stable (unstable) if (4.483a) both eigenvalues (at least one eigenvalue) have (has) negative (positive) real parts (4.483b) [part (4.483c)]:

$$2\lambda_\pm = a + d \pm \left| (a-d)^2 + bc \right|^{1/2} = \begin{cases} 0 > \operatorname{Re}(\lambda_+) \geq \operatorname{Re}(\lambda_-): & stable; \\ 0 < \operatorname{Re}(\lambda_+) \geq \operatorname{Re}(\lambda_-): & unstable. \end{cases}$$

$$(4.483a\text{–}c)$$

The theorem can be applied to a second-order system (4.484b) specified (problem 151) by a continuously differentiable function (4.484a):

$$f(x, \dot{x}, t) \in C^1(|R^3): \qquad \ddot{x} = f(x, \dot{x}, t) = -ax + b\dot{x} + O(x^2, \dot{x}^2, \dot{x}x); \qquad (4.484a, b)$$

this corresponds to the linear second-order system (2.81b) ≡ (4.484b) ≡ (4.484d):

$$\ddot{x} + 2\lambda\dot{x} + \omega_0^2 x = 0: \qquad a \equiv \omega_0^2 \begin{cases} > 0 & atractor \\ < 0 & repeller \end{cases}, \qquad b = -2\lambda \begin{cases} < 0 & damping \\ > 0 & amplification \end{cases}$$

$$(4.484d\text{–}f)$$

including four combinations of attractor or repeller (4.484e) and damping or amplification (4.484f). The system (4.484c) is equivalent to the corresponding autonomous system (4.485a, b):

$$v = \dot{x}, \dot{v} = ax + by; \qquad 0 = \begin{vmatrix} -\lambda & 1 \\ a & b-\lambda \end{vmatrix} = \lambda^2 - \lambda b + a = (\lambda - \lambda_+)(\lambda - \lambda_-);$$

$$(4.485a\text{–}c)$$

its eigenvalues (4.485c) are (4.486a):

$$2\lambda_\pm = b \pm \sqrt{b^2 - 4a}; \qquad a < 0: \qquad \lambda_\pm > 0 > \lambda_-. \qquad (4.486a\text{–}c)$$

In the case of a repeller (4.486b) one eigenvalue is positive (4.486c), leading to instability.

In the case of an attractor (4.487a), both eigenvalues have positive (4.487b) [negative (4.487c)] real parts for amplification (damping) leading to instability (stability):

$$a > 0: \qquad \text{Re}(\lambda_\pm) \begin{cases} > 0 & \text{if} & b > 0, \\ < 0 & \text{if} & b < 0. \end{cases} \qquad (4.487a\text{–}c)$$

The result (4.487b) [(4.487c)] follows from (4.448a–c) [(4.448d–f)]:

$$b > 0: \qquad \text{Re}(\lambda_\pm) = \begin{cases} \text{Re}\left\{ b \pm i \left| 4a - b^2 \right|^{1/2} \right\} = b > 0 & \text{if} & b^2 < 4a, \\ b \pm \left| b^2 - 4a \right|^{1/2} > 0 & \text{if} & b^2 > 4a; \end{cases}$$

$$(4.488a\text{–}c)$$

$$b < 0: \qquad \text{Re}(\lambda_\pm) = \begin{cases} \text{Re}\left\{ b \pm i \left| 4a - b^2 \right|^{1/2} \right\} = b < 0 & \text{if} & b^2 < 4a, \\ b \pm \left| b^2 - 4a \right|^{1/2} < 0 & \text{if} & b^2 > 4a. \end{cases}$$

$$(4.488d\text{–}f)$$

Thus, *a non-linear second-order differential equation (4.484b) specified (problem 151) by a continuously differentiable function is: (i) always unstable (4.486b, c) for a repeller; (ii) is unstable (4.487b) [stable (4.487c)] for an attractor (4.487a) in the case of amplification (damping).* Another aspect of a dynamical system is the existence of singularities in the finite plane (subsections 4.9.5–4.9.6) or at infinity (subsections 4.9.5–4.9.8).

4.9.5 Index of a Singularity or Equilibrium Point

Consider an isolated singularity or equilibrium point (4.489a) and draw a circle (4.489b, c) around it (Figure 4.33a) not containing any other singularity or equilibrium point:

$$z_0 \equiv x_0 + iy_0, \qquad z \equiv x + iy: \qquad\qquad z - z_0 = re^{i\phi}; \qquad (4.489a\text{–}c)$$

on each point of the circle (4.489c) draw the tangent vector (4.490b) to the differentiable integral curve (4.490a):

$$\{\dot{x}, \dot{y}\} = u, v(x, y) \in C^1\left(|R^2\right): \qquad\qquad u + iv = we^{i\theta}. \qquad (4.490a, b)$$

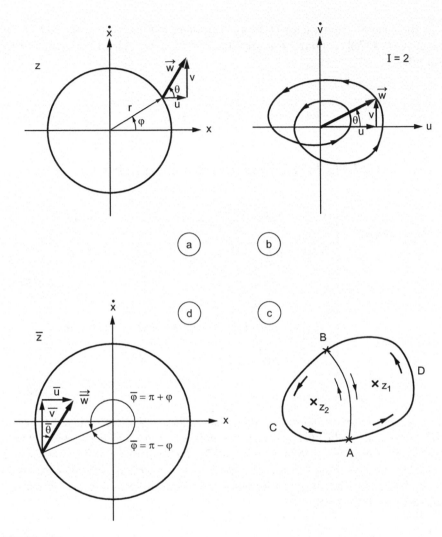

FIGURE 4.33
The index of a singularity of a dynamical system is the number of times the vector of coefficients performs a full rotation when the path of the system goes around the origin once (a); for example, the index or winding number is $I = 2$ in (b). The index of a set of singularities is the sum of the indices of each of them (c) because the path can be subdivided into sub-paths around each singularity with cancellation in all internal paths since they are taken twice in opposite directions. The inversion relative to the origin specifies the index at infinity (d).

The **index** of the singularity or equilibrium point (4.491b) is the integral of the angle of the vector tangent to the integral curves along the circle (4.491a) divided by 2π:

$$\partial D: \quad |z - z_0| = r: \quad 2\pi I \equiv \int_{\partial D} d\theta = \int_{\partial D} d\left[\arctan\left(\frac{v}{u} \right) \right] = \int_{\partial D} \frac{u\,dv - v\,du}{u^2 + v^2}, \quad (4.491a, b)$$

where (II.7.124a) was used:

$$d\left[\arctan\left(\frac{v}{u}\right)\right] = \frac{d\left(\dfrac{v}{u}\right)}{1+\dfrac{v^2}{u^2}} = \frac{\dfrac{dv}{u}-\dfrac{v\,du}{u^2}}{1+\dfrac{v^2}{u^2}} = \frac{u\,dv-v\,du}{u^2+v^2}. \tag{4.491c}$$

The index I coincides with the **winding number**; that is, the number of turns of the vector tangent to the integral curves as it describes the circle around the singularity, for example $I=2$ in Figure 4.33b.

4.9.6 Theorems of the Index or Winding Number

If there are several singularities or equilibrium points in a domain, their indices add to each other, as stated more precisely in the **theorem of addition of indices**: *the autonomous system (4.492a) has a continuously differentiable forcing vector that does not vanish (4.492c) on a closed simple curve ∂D, nor in its interior, except (4.492b) possibly at a finite set of points z_n:*

$$\{\dot{x},\dot{y}\} = \{u,v\} \in C^1\!\left(|R^2\right): \quad D - \partial D - \sum_{n=1}^{N} z_n: \quad u^2+v^2 \neq 0: \quad \sum_{n=1}^{N} I_n = I = \frac{[\theta]}{2\pi}. $$

$$\tag{4.492a–d}$$

then (problem 152) the sum of the indices of each point equals (4.492d) the index of the domain, that is $1/2\pi$ times the variation of the angle of the tangent vector to the integral curves along the boundary of the domain. It is sufficient to prove the theorem for two singular points (Figure 4.33c). The domain is divided by a curve AB joining the two boundaries and leaving one singularity on each side. The index of each singularity is (4.493a, b):

$$2\pi I_1 = \int_{ADBA} d\theta; \, 2\pi I_2 = \int_{ABCA} d\theta: \quad 2\pi(I_1+I_2) = \int_{ADB} d\theta + \int_{BA} d\theta + \int_{AB} d\theta + \int_{BCA} d\theta,$$

$$\tag{4.493a–c}$$

and their sum is (4.493c). The integrals along AB in opposite directions cancel (4.494a):

$$\int_{AB} d\theta = -\int_{BA} d\theta: \qquad I_1+I_2 = \frac{1}{2\pi}\int_{ADBCA} d\theta = \int_{\partial D} d\theta = I, \tag{4.494a, b}$$

and the remaining two integrals in (4.493c) add to the integral (4.494b) around the boundary. *In particular if the boundary is a closed integral curve, then the index is at least one,* because there is at least one singularity or equilibrium point inside. The preceding theorems can be extended to the index (subsections 4.9.8) at the singularity at infinity (subsections 4.9.7).

4.9.7 Singularity at Infinity and Inversion/Stereographic Projection

The singularity at infinity can be obtained via the stereographic projection (sections I.9.2–I.9.3) or by inversion (4.495a) of a complex number (4.489b) ≡ (4.495b) relative to the origin (4.495a), implying (4.495d–g):

$$\bar{z} \equiv \frac{1}{z}: \qquad z \equiv x + iy = re^{i\phi}, \qquad \bar{z} \equiv \bar{x} + i\bar{y} = \bar{r}e^{i\bar{\phi}}: \qquad (4.495a\text{–}c)$$

$$\bar{x} + i\bar{y} = \frac{1}{x + iy} = \frac{x - iy}{x^2 + y^2} = \frac{x - iy}{r^2}, \qquad x + iy = \frac{\bar{x} - i\bar{y}}{\bar{r}^2}, \qquad (4.495d, e)$$

$$r = |z| = |\bar{z}|^{-1} = \bar{r}^{-1}, \qquad\qquad \bar{\phi} = \arg(\bar{z}) = -\arg(z) = -\phi. \qquad (4.495f, g)$$

This leads to the direct (4.496a) inverse (4.496b)] coordinate transformations (4.496c, d):

$$\{\bar{x}, \bar{y}\} = r^{-2}\{x, -y\}, \qquad \{x, y\} = \bar{r}^{-2}\{\bar{x}, -\bar{y}\}, \qquad r\bar{r} = 1, \qquad \phi + \bar{\phi} = 0, \qquad (4.496a\text{–}d)$$

and the corresponding direct (4.497a) [inverse (4.497b)] transformation matrices:

$$\frac{\partial(\bar{x}, \bar{y})}{\partial(x, y)} = \begin{bmatrix} \dfrac{\partial \bar{x}}{\partial x} & \dfrac{\partial \bar{x}}{\partial y} \\[2mm] \dfrac{\partial \bar{y}}{\partial x} & \dfrac{\partial \bar{y}}{\partial y} \end{bmatrix} = r^{-4} \begin{bmatrix} y^2 - x^2 & -2xy \\[2mm] 2xy & y^2 - x^2 \end{bmatrix}, \qquad (4.497a)$$

$$\frac{\partial(x, y)}{\partial(\bar{x}, \bar{y})} = \begin{bmatrix} \dfrac{\partial x}{\partial \bar{x}} & \dfrac{\partial x}{\partial \bar{y}} \\[2mm] \dfrac{\partial y}{\partial \bar{x}} & \dfrac{\partial y}{\partial \bar{y}} \end{bmatrix} = \bar{r}^{-4} \begin{bmatrix} \bar{y}^2 - \bar{x}^2 & -2\bar{x}\bar{y} \\[2mm] 2\bar{x}\bar{y} & \bar{y}^2 - \bar{x}^2 \end{bmatrix}, \qquad (4.497b)$$

for example:

$$\frac{\partial r}{\partial x} = \frac{\partial}{\partial x}\left(|x^2 + y^2|^{1/2}\right) = x|x^2 + y^2|^{-1/2} = \frac{x}{r}, \qquad (4.497c)$$

$$\frac{\partial \bar{x}}{\partial x} = \frac{1}{r^2} - \frac{2x}{r^3}\frac{\partial r}{\partial x} = \frac{r^2 - 2x^2}{r^4} = \frac{y^2 - x^2}{r^4}, \qquad (4.497d)$$

$$\frac{\partial \bar{x}}{\partial y} = -\frac{2x}{r^3}\frac{\partial r}{\partial y} = -\frac{2xy}{r^4}. \qquad (4.497e)$$

The coordinate transformations (4.496a, b) and partial derivatives (4.497a, b) transform the autonomous dynamical system (4.1a, b):

$$\dot{x} \equiv \frac{dx}{dt} = \frac{\partial \bar{x}}{\partial x}\frac{dx}{dt} + \frac{\partial \bar{x}}{\partial y}\frac{dy}{dt} = \frac{\left(y^2 - x^2\right)\dot{x} - 2xy\dot{y}}{r^4} = \left(\bar{y}^2 - \bar{x}^2\right)u + 2\bar{x}\bar{y}v, \qquad (4.498a)$$

$$\dot{y} \equiv \frac{dy}{dt} = \frac{\partial \bar{y}}{\partial x}\frac{dx}{dt} + \frac{\partial \bar{y}}{\partial y}\frac{dy}{dt} = \frac{2xy\dot{x} + \left(y^2 - x^2\right)\dot{y}}{r^4} = -2\bar{x}\bar{y}u + \left(\bar{y}^2 - \bar{x}^2\right)v, \qquad (4.498b)$$

to (4.498a, b) ≡ (4.498c, d):

$$\frac{dx}{dt} = \left(\bar{y}^2 - \bar{x}^2\right)u + 2\bar{x}\bar{y}v \equiv \bar{u}, \qquad \frac{d\bar{y}}{dt} = -2\bar{x}\bar{y}u + \left(\bar{y}^2 - \bar{x}^2\right)v \equiv \bar{v}. \qquad (4.498c, d)$$

Thus *(problem 153) the inversion at the origin (4.495a–c) transforms (4.496a–d; 4.497a, b) the autonomous dynamical system (4.1a, b) to (4.498c, d) at infinity.*

4.9.8 Index and Angle at the Singularity at Infinity

From (4.498c, d) it follows that the forcing vector transforms under inversion (4.495a) ≡ (4.499a) like (4.499b):

$$\bar{z} = \frac{1}{z}: \qquad \bar{w} = \bar{u} + i\bar{v} = \left(\bar{y}^2 - \bar{x}^2\right)(u + iv) + 2\bar{x}\bar{y}(v - iu)$$

$$= \left[\left(\bar{y}^2 - \bar{x}^2\right) - 2i\bar{x}\bar{y}\right](u + iv) = -\left(\bar{x} + i\bar{y}\right)^2 w = -\bar{z}^2 w; \qquad (4.499a, b)$$

using (4.500a, b) leads to (4.500c):

$$w \equiv Re^{i\theta}, \bar{w} = \bar{R}e^{i\bar{\theta}}: \qquad \bar{R}e^{i\bar{\theta}} = \bar{w} = -\bar{z}^2 w = -\bar{r}^2 e^{-2i\bar{\phi}}Re^{i\theta} = \bar{r}^2 Re^{i\pi + i\theta - i2\bar{\phi}}, \qquad (4.500a-c)$$

that specifies to within a multiple (4.501a) of 2π the relation between the angles (4.501b, c):

$$\mathrm{mod}(2\pi): \qquad \bar{\theta} - \theta = \pi - 2\bar{\phi} = 2\bar{\phi} - \pi. \qquad (4.501a-c)$$

Both relations (4.499b) ≡ (4.502a) [(4.501b) ≡ (4.502b)] could be obtained directly from the complex form:

$$\bar{w} \equiv \frac{d\bar{z}}{dt} = \frac{d}{dt}\left(\frac{1}{z}\right) = -\frac{1}{z^2}\frac{dz}{dt} = -\bar{z}^2 w, \qquad (4.502a)$$

$$\bar{\theta} = \arg(\bar{w}) = \pi - 2\arg(\bar{z}) + \arg(w) = \pi + \theta - 2\bar{\phi}. \qquad (4.502b)$$

When going round the loop, ϕ varies by 2π in (4.503a) and $\bar{\phi}$ by -2π because (4.503b) the loop at infinity is described in the opposite direction (4.496d). Since θ increases by Θ in (4.503c) then $\bar{\theta}$ decreases by (4.503d):

$$[\phi] = 2\pi = -[\bar{\phi}], \qquad [\theta] \equiv \Theta, \qquad [\bar{\theta}] = -2\pi\bar{I}; \qquad -2\pi\bar{I} - \Theta = -4\pi, \qquad (4.503\text{a--e})$$

substitution of (4.503b–d) in (4.501b) ≡ (4.502b) yields (4.503e). Thus, *the index of the point at infinity is (4.504a), where* $[\bar{\theta}]$ *is the variation of the angle* θ *along a loop around the inverse point of infinity in the finite part of the plane:*

$$\Theta \equiv [\bar{\theta}] \equiv -2\pi\bar{I} = \Theta - 4\pi = 2\pi I - 4\pi; \qquad 2 = I + \bar{I} = \bar{I} + \sum_{n=1}^{N} I_n. \qquad (4.504\text{a, b})$$

Substituting (4.492d) in (4.504a) yields (4.504b) the **theorem of winding numbers**: *the sum (problem 154) of the indices at all singular points including the point at infinity is two (4.504b).*

4.9.9 Index at Multipoles and Other Singularities

Changing the sign of time (4.505a) also changes the sign of the forcing vector (4.505b) and hence adds π its angle (4.505c): the addition of a constant (4.505d) does not change the variation of the angle (4.505e) nor the index (4.505f):

$$t \rightarrow -t, \quad \{u, v\} \rightarrow \{-u, -v\}, \quad \theta \rightarrow \theta + \pi; \quad t \rightarrow t + t_0: \quad [\theta] \rightarrow [\theta], \quad I \rightarrow I.$$
$$(4.505\text{a--f})$$

Thus *the index cannot distinguish a source from a sink, nor a point from a stable or unstable equilibrium. Some typical values (problem 155) of the index: (i–iv) a 2^N – multipole has index $I = N + 1$, for example $I = 3$ for a quadrupole $N = 2$ (Figure 4.34a), $I = 2$ for a dipole $N = 1$ (Figure 4.34b) and $I = 1$ for a monopole $N = 0$, such as $I = 1$ for a source/sink (Figure 4.34c) and vortex (Figure 4.34d); (v–vi) the latter corresponds to a center (Figure 4.34d) that has index $I = 1$ as the tangent (Figure 4.34e) and inflection (Figure 4.34f); (vii) the spiral (Figure 4.34g) also has index unity $I = 1$; (viii) the index zero $I = 0$ is an simple point (Figure 4.34h); and (ix) a saddle point (Figure 4.34i) has negative index $I = -1$.* The preceding statements can be justified as follows: (i) an ordinary point (Figure 4.34h) involves no change of direction hence index zero $I = 0$; (ii) going around the saddle point (Figure 4.34i) the forcing vector comes in opposite direction hence the minus unity $I = -1$; (iii) the index plus unity $I = 1$ corresponds to one complete rotation for the spiral (Figure 4.34g), center or vortex (Figure 4.34d), inflection (Figure 4.34f) or tangent (Figure 4.34e), and source (Figure 4.34c) or sink; (iv) the dipole (Figure 4.34b) involves two rotations hence the index $I = 2$; (v) the quadrupole (Figure 4.34a) has

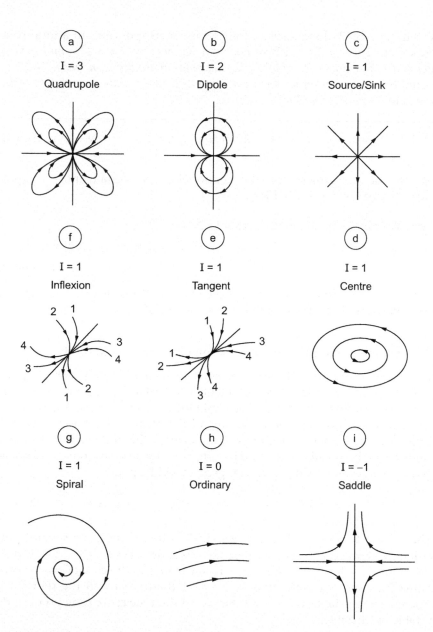

FIGURE 4.34

The sum of all indices including the point at infinity is two. This result can be used to obtain compatible sets of singularities or to find singularities missing in an incomplete set. The singularities include: (i–iii) a multipole of order n, or $2n$ – pole, that has index $I = n + 1$, for example $I = 1$ for $n = 0$ monopole (c) or source/sink (Figure 4.1c), $I = 2$ for $n = 1$ a dipole (b), and $I = 3$ for $n = 2$ a quadrupole (a); (iv–vi) the inflexion (f) [tangent (e)] point [Figure 4.1a(b)] has index $I = 1$, as the center (d) that corresponds (Figure 4.1g) to a vortex and hence a monopole (iii); (vii) the spiral point (f) has (Figure 4.1f) index $I = 1$ as the center; (viii) the simple point (h) has (Figure 4.1d) index $I = 0$; (ix) the negative index $I = -1$ corresponds (i) to the saddle point (Figure 4.1e).

a rotation of $3\pi/2$ along each of four lobes, corresponding to a total rotation $4 \times 3\pi/2 = 6\pi = 2\pi I$ with index $I = 3$. Generally, *for a 2^N – multipole of order N: (i) there are (4.506a) lobes; (ii) the rotation around each lobe is (4.506b); and (iii) the total rotation (4.506c) corresponds to the winding number (4.506d) equal to unity plus the multipole order:*

$$M = 2N, \quad \theta_N = \pi + \frac{\pi}{N}: \qquad \theta = \theta_N \quad M = (N+1)2\pi = 2\pi I, \quad I = N+1; \quad (4.506\text{a–d})$$

this can be confirmed for the monopole/dipole/quadrupole/octupole (4.506e, f) that have index (4.506g):

monopole/dipole/quadrupole/octupole: $N = 0,1,2,3$;

$$2^N = 1,2,4,8, N+1 = 1,2,3,4.$$

$$(4.506\text{e–g})$$

The theorem on winding numbers (4.504b) limits the possible combinations of singularities in the phase plane (subsection 4.9.10).

4.9.10 Location and Identification of Singularities and Stability

The theorem of addition of indices without (4.492a–d) or with (4.504b) the point at infinity can be used to locate and identify singularities and points of equilibria by determining the index of several loops. As an example, suppose that there is a saddle point in the finite part of the plane (4.507a) and a focus at infinity (4.507b). If there is one more singularity in the finite part of the plane its index must be (4.507c), so it could be a dipole (Figure 4.34b), but could not be a quadrupole, center, focus, spiral, tangent, or saddle point:

$$I_1 = -1, \overline{I} = 1, \quad I_2 = 2 - \overline{I} - I_1 = 2; \qquad I_3 + I_2 = 2, \qquad (4.507\text{a–d})$$

if there were two more singularities (4.507d) there would be several possibilities: (i) two centers, spiral, foci, tangents, or inflexions $I_2 = 1 = I_3$; (ii) a quadrupole $I_2 = 3$ and a saddle $I_3 = -1$. Returning to the case of two known singularities and one unknown, if $I_1 + \overline{I} \neq 1$ then $I_2 \neq 1$ and the remaining point must be unstable; if $I_1 + \overline{I} = 1$ then $I_2 = 1$ then the remaining point could be stable or unstable.

NOTE 4.1: Multi-Dimensional Dynamical Systems

The **multi-dimensional dynamical system** is the extension of (4.1a, b) to N dimensions (4.508a):

$$n, m = 1,...,N: \qquad \dot{x}_n = f(x_m, t) = g_n(x_m) = \frac{\partial \Phi}{\partial x_n}, \qquad (4.508\text{a–d})$$

and may be **unsteady** (4.508b), **time independent** or steady (4.580c), or a **gradient** (4.508d) system. The latter has a zero curl (4.509a) ≡ (3.269c):

$$\frac{\partial g_n}{\partial x_m} = \frac{\partial g_m}{\partial x_m}, \qquad E = \Phi + \sum_{n=1}^{N} \dot{x}_n^2, \qquad (4.509a, b)$$

and a conserved first integral (4.509b). The iterated map (Figure 4.29a) or chaos (Figure 4.29c) extend to any dimension cutting the integral curves by a directrix hypersurface not tangent to them. The dependence on P parameters (4.510a) can be considered (4.510b):

$$p = 1,....,P: \qquad \dot{x}_n = f\left(x_m, t; q_p\right), \qquad (4.510a, b)$$

using a space of $P + 2N$ dimensions consisting of the product of: (i) the phase space $(x_n, \dot{x}_n)$ with $2N$ dimensions (4.508a–d); and (ii) the parameter space with P dimensions (4.510a, b). The exposition on dynamical systems has emphasized mechanical devices, electrical circuits, and other applications; of the many theorems in dynamical systems theory only a few that are simpler and more intuitive have been convered. The material covered includes: (i) decay, amplification, and oscillations in linear and non-linear, damped and undamped, free and forced systems; (ii) multiple examples of singularities of linear and non-linear differential equations and autonomous systems of the second-order; and (iii) a few examples of the most frequent bifurcations, with no attempt at a systematic classification. The stagnation point of first degree of a flow (sections 4.1–4.2 and Figures 4.1a–f) serve as examples of singularities (section 4.9 and Figures 4.34a–i). The non-linear resonance (sections 4.4–4.5) and damping or amplification (sections 4.6–4.7) give examples of bifurcations (section 4.8). The simplest scenario of transition to turbulence (subsection 4.8.9) also involves parametric resonance (section 4.3). The bifurcations of dynamical systems (sections 4.8–4.9) are an example of the qualitative theory of differential equations (NOTE 4.2), that is complementary (NOTE 4.6) with regard to (Table 4.6): (i, ii) the analytical exact (approximate) solution of differential equations [NOTES 4.3(4.4)]; (iiii) the numerical methods (NOTE 4.5).

NOTE 4.2: Qualitative Theory of Differential Equations

The aim of the qualitative theory of differential equations is to obtain some properties of the solution without explicitly solving the differential equation. The main advantage of this approach is that the qualitative theory applies to much wider classes of differential equations than those that can be solved. The examples of the qualitative theory of differential equations (section 9.1) include: (i) existence theorems stating that a certain class of functions provides at least one solution; (ii) unicity theorems stating initial or

TABLE 4.6

Methods of Solution of Differential Equations

Method Name	I Qualitative	II Exact Analytical	III Approximate Analytical	IV Numerical
error of computation or accuracy	not quantifiable	truncation	truncation + systematic	truncation + systematic + discretization
copes with singularities	generally	always	sometimes	seldom
gives properties of solution	most	all	some	depends on interpretation
domain of application	very large	small	intermediate	large

Note: The solution of differential equations is based on qualitative and quantitative methods; the quantitative methods may be numerical or analytic; the analytic methods may be exact or approximate.

boundary conditions that ensure that the solution is unique; and (iii) regularity theorems indicating properties of the solution, such as differentiability or continuity with regard to parameters. The bifurcations are an example of non-regularity and/or non-unicity, since beyond the bifurcation point several solutions may exist and be stable or unstable. The stability or instability of solutions can be established (section 9.2) by the qualitative theory for fairly general classes of differential equations. The singularities of the differential equation or of its solutions may limit the methods of solution that can be applied. Thus the qualitative methods are very useful for preliminary investigation of the properties of fairly wide classes of differential equations; the inherent limitation is the lack of quantitative results that are essential in many scientific and technical applications.

NOTE 4.3: Exact Analytical Solutions of Differential Equations

In principle, the exact analytical solution of a differential equation should contain all properties with arbitrary accuracy. It must be ensured that what is obtained is the general integral that contains all particular integrals; if there are special integrals not included in the general integral they need to be investigated. An exact analytical solution has to be computed, and thus there are **truncation errors** that can be reduced by using more digits. The exact analytical solutions of some differential equations may be obtained in finite terms using only elementary functions. Wider classes of differential equations may still be amenable to exact analytical solution but only in terms of series, parametric integrals, continued fractions, infinite products, or other representations raising convergence issues (section 9.4–9.9). These solutions enlarge the class of ordinary functions to the special functions, e.g. hypergeometric, Legendre, Hermite, Chebychev, Bessel, etc., each having an associated theory that may be quite extensive (notes 9.1–9.47). The exact analytical solutions provide the most detailed understanding of the phenomena and processes described by a differential equation; however, the classes of differential equations amenable to exact analytical solution are limited and other methods are needed outside those classes.

NOTE 4.4: Analytical Approximations to the Solution of Differential Equations

If a differential equation in its general form has no known exact analytical solution, it may be possible to simplify some terms so that it falls into one of the classes to which analytical methods apply. This introduces a **systematic error** associated with the approximation(s) made or term(s) omitted, in addition to the truncation error of computation. An example is the method of parametric expansions (section 4.5) that transforms a nonlinear differential equation into a sequence of linear differential equations solvable by recurrence. The analytical approximation may also have limitations, for example the parametric solutions often can be obtained without excessive labor only for the two lowest orders. They may also raise questions of validity, such as convergence. It may happen that the approximate analytical solution of a differential equation can be obtained in finite terms using elementary functions whereas the exact analytical solution would require infinite expansions or special functions. The interpretation of the approximate analytical solution of a differential equation should take into account all the consequences of the approximation such as: (i) properties that are not generally valid but rather due to the approximation; and (ii) the possibility that other solutions are excluded by the assumptions made. The approximate analytical solution of a differential equation is an intermediate compromise between: (a) the exact analytics solution (note 4.3) that is more accurate but may be more complex or unavailable for the differential equation in question; and (b) the numerical solution of the differential equation (note 4.5) that will apply to wider classes of differential equations, and in this sense being the only alternative, while raising still more issues.

NOTE 4.5: Numerical Solution of Differential Equations

The chief attraction of numerical methods is that they provide quantitative solutions of the widest classes of differential equations, and in many cases are the only option available. Numerical methods can be: (i) deceptively simple to apply; (ii) often much harder to justify fully than an analytic method; and (iii) almost always lead to some numbers at the end (except for overflows or underflows) whose accuracy and interpretation may be the biggest issue. The numerical methods are usually based on a discretization of a continuum into a sequence of discrete steps; this introduces a **discretization error**, that adds to the systematic (truncation) errors associated [note 4.4 (4.3)] with approximations (computations). When applying a numerical method, ensure that: (i) the problem has a unique solution; (ii) the method converges to that solution; and (iii) the sum of discretization, systematic, and truncation errors does not call into question the accuracy of the results. It may be that the steps (i) to (iii) are harder than the application of the numerical method "ad hoc"; the latter approach may lead to overflows or underflows, numerical "instabilities", dispersion of results, non-convergence, or simply to an apparently neat convergence that is no guarantee of correctness of the final result. If the final result is correct there may still be the question of interpretation of its meaning, that may be helped by some of the earlier approaches.

NOTE 4.6: Complementarity of the Four Methods of Solution

The preceding account on qualitative (note 4.2) and quantitative analytical exact (note 4.3) [approximate (note 4.4)] and numerical (note 4.5) methods of solution of differential equations shows that each approach has strengths and weaknesses, and the best approach is to combine all four methods to use the strengths of some to overcome the weaknesses of the others. Starting with a differential equation describing a phenomenon or process, the qualitative theory (A) can establish: (A.1) the existence of solution; (A.2) the boundary and/or initial conditions for the unicity of solution; (A.3) the stability conditions; and (A.4) the singularities or bifurcations and associated properties. Next the differential equation is restricted to a form amenable to exact analytical solution (B): (B.1) the qualitative theory (A) indicates the general form of the solution; (B.2) the method chosen should specify the solution completely; (B.3) any issues of validity, like convergence of an infinite representation, need to be settled; and (B.4) the interpretation of the solution should lead to a good understanding of the phenomena and processes it describes.

The exact analytical solution (B) may be too restricted in its domain of application, so a more general form of the differential equation is considered, that is still amenable to approximate analytical solution (C): (C.1) the approximation(s) must be carefully chosen so as not to exclude some of the essential aspects being investigated; (C.2) once the approximation(s) is(are) made the solution should be exact, without any simplifications except those that are a consequence of previous approximations; (C.3) the results should be interpreted, bearing in mind that some properties may not be general and rather are a by-product of the approximations made at the beginning; and (C.4) it should be borne in mind that other solutions and properties may have been excluded by the approximations. The "real" problem in question, due to a complex geometry, or/and inhomogeneous media, complex forcings, etc., may not be amenable to exact (B) or approximate (C) analytical solution without excessive "idealization", so a numerical solution (D) is sought: (D.1) the numerical method should be chosen so as to converge to the unique solution specified by the qualitative theory (A); (D.2) the presence of singularities or bifurcations may require separate treatment to avoid numerical instabilities or wrong matchings across a special points; (D.3) the discretization, systematic, and truncation errors should be consistent with the required accuracy; and (D.4) the results should be interpreted with as much insight as could be gained from the preceding three approaches (A, B, C).

NOTE 4.7: Comparison of Methods in a Sample Case

The thorough $4 \times 4 = 16$ step approach (note 4.6) to the solution of a differential equation is not always or often applied. In the cases of some standard differential equations, e.g., linear with constant coefficients, the general properties are already well-known; in the case of non-linear differential equations and/or variable coefficients, starting a numerical algorithm while ignoring singularities and/or bifurcations could: (i) lead to instabilities or overflows, signaling that something amiss; (ii) lead smoothly to a wrong result, without any warning; or (iii) lead to the right result if the chosen method fortuitously matched the properties of the solution. Next is considered the numerical solution of a rather simple differential equation so that the exact analytical and approximate numerical approaches can be compared in detail. As a numerical method this is the simplest of samples.

NOTE 4.8: Definition and Discretization of Derivatives of Any Order

The first-order **derivative** is defined as the slope (Figure 4.35a) or limit of the **incremental ratio** as the **step size** tends to zero:

$$f'(x) \equiv \lim_{h \to 0} \frac{f(x+h) - f(x)}{h}, \tag{4.511}$$

if it exists. It follows that the second-order derivative (Figure 4.35b) is given by:

$$f''(x) = \lim_{h \to 0} \frac{f'(x+h) - f'(x)}{h} = \lim_{h \to 0} \frac{f(x+2h) - 2f(x+h) + f(x)}{h^2}. \tag{4.512a, b}$$

Applying the definition of derivative (4.511) iteratively it follows that the derivative of order n is given by:

$$f^{(n)}(x) = \lim_{h \to 0} h^{-n} \sum_{k=0}^{n} (-)^{n-k} \binom{n}{k} f(x+kh). \tag{4.513}$$

The proof of (4.513) is made by induction: (i) it holds for $n = 1 (n = 2)$ when it reduces to (4.511) [(4.512b)]; (ii) if it holds for n it also holds for $n + 1$. The proof

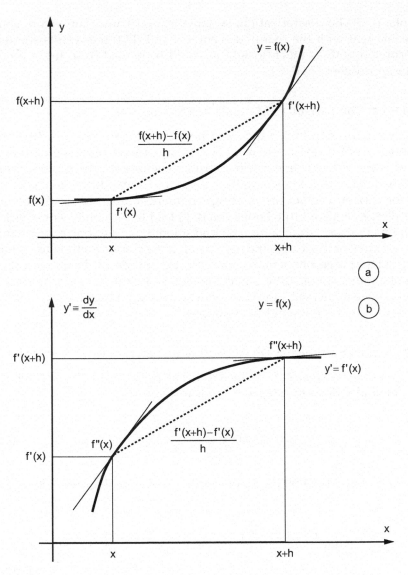

FIGURE 4.35
The numerical solution of a differential equation is based on the discretization of the derivatives leading to a finite difference equation. The first-order derivative or tangent is replaced by a secant (a), the second-order derivative or curvature is replaced by a secant of secants (b), and so on for higher orders.

of item (ii) follows starting from the definition of derivative (4.511) applied to $f^{(n)}(x)$:

$$f^{(n+1)}(x) \equiv \frac{df^{(n)}}{dx} = \lim_{h \to 0} \frac{f^{(n)}(x+h) - f^{(n)}(x)}{h}. \tag{4.514a}$$

Substitution of (4.513) that is assumed to hold leads to:

$$f^{(n+1)}(x) \equiv \lim_{h\to 0} h^{-1} \left[f^{(n)}(x+h) - f^{(n)}(x) \right]$$

$$= \lim_{h\to 0} h^{-1-n} \sum_{k=0}^{n} (-)^{n-k} \binom{n}{k} \left[f(x+kh+h) - f(x+kh) \right]$$

$$= \lim_{h\to 0} h^{-1-n} \left\{ f(x+nh+h) + \sum_{k=0}^{n-1} (-)^{n-k} \binom{n}{k} f(x+kh+h) \right.$$

$$\left. - \sum_{k=1}^{n} (-)^{n-k} \binom{n}{k} f(x+kh) - (-)^n f(x) \right\}.$$

(4.514b)

The latter expression is simplified next.

Using the change of summation variable (4.515a), the first sum on the r.h.s. of (4.514) combines with the second sum, leading to (4.515b):

$$m = k+1: \quad \sum_{k=0}^{n-1} (-)^{n-k} \binom{n}{k} f(x+kh+h) - \sum_{m=1}^{n} (-)^{n-m} \binom{n}{m} f(x+mk)$$

$$= \sum_{m=1}^{n} (-)^{n+1-m} f(x+mh) \left[\binom{n}{m-1} + \binom{n}{m} \right].$$

(4.515a, b)

The term in square brackets simplifies to:

$$\binom{n}{m-1} + \binom{n}{m} = n! \left[\frac{1}{(n-m+1)!(m-1)!} + \frac{1}{(n-m)!m!} \right]$$

$$= \frac{n!}{(n-m)!(m-1)!} \left(\frac{1}{n-m+1} + \frac{1}{m} \right) = \frac{(n+1)!}{(n-m+1)!\,m!} = \binom{n+1}{m}.$$

(4.516)

Substitution of (4.516) in (4.515b) and (4.514) gives:

$$f^{(n+1)}(x) = \lim_{h\to 0} h^{-1-n} \left[f(x+nh+h) + (-)^{n+1} f(x) + \sum_{m=1}^{n} (-)^{n+1-m} \binom{n+1}{m} f(x+mh) \right]$$

$$= \lim_{h\to 0} h^{-1-n} \sum_{m=0}^{n+1} (-)^{n+1-m} \binom{n+1}{m} f(x+mh),$$

(4.517)

that proves (5.513) for *n*+1. QED. It has been shown that *the definition of first-order derivative (4.511), leads to the* **discretization** *(problem 156) second-order derivative (4.512a, b) and (4.513)* **derivative of arbitrary positive integer order** *n. The latter (4.513), apart from alternating sign, is similar to the* **binomial expansion** *(4.518):*

$$(a+b)^n = \sum_{m=0}^{n} \binom{n}{m} a^m b^{n-m}. \tag{4.518}$$

The method (4.514a, b) through (4.517) could be used as an algebraic proof of the binomial expansion (4.518) ≡ (I.25.38) instead of the proof (section I.25.9) using the Taylor series. The derivative of arbitrary order (4.513) without the limit $h \to 0$ is used to discretize a differential equation (note 4.9) with step h.

NOTE 4.9: Relation between Ordinary Differential and Finite Difference Equations

An ordinary differential equation (4.519b) of order (4.519a):

$$n = 0, 1, ..., N: \qquad 0 = F\left(x; y(x), y'(x), y''(x), ..., y^{(n)}(x), ..., y^{(N)}(x)\right),$$
$$\tag{4.519a, b}$$

can (problem 157) be transformed (4.520a, b) to a finite difference equation (4.520c) with **step** *h of the same order (4.519a):*

$$y_n = y(x), y_{n+m} \equiv y(x+mh):$$
$$0 = G\left(x; y(x), y(x+h)\right), ... y(x+nh) \equiv G\left(x, y_n, y_{n+1}, ... y_{n+N}\right),$$
$$\tag{4.520a-c}$$

by: (i) discretizing the derivatives by (4.513); and (ii) using the notation (4.520a, b) in (4.520c). The N single-point boundary conditions (1.5a, b) ≡ (4.521b) for the differential equation (4.519b) become (4.513) ≡ (4.521b) for the finite difference equation (4.520c):

$$m = 0, ..., n-1: \qquad y^{(m)}(x_0) = a_m = \sum_{k=0}^{m} (-)^{m-k} \binom{m}{k} y_k. \tag{4.521a-c}$$

For example, the linear second-order ordinary differential equation with constant coefficients (4.522a) is discretized (4.522b) using (4.512b) with step h:

$$y'' + y = 0: \qquad 0 = y_{n+2} - 2y_{n+1} + \left(1+h^2\right) y_n. \tag{4.522a, b}$$

The general integral of the differential equation (4.522a) is (4.523a), and the initial conditions (4.523b, c) lead to (4.523d):

$$y(x) = A\cos x + B\sin x: \qquad 1 = y(0) = A, \quad 0 = y'(0) = B, \qquad y(x) = \cos x.$$
$$\text{(4.523a–d)}$$

The finite difference equation (4.522b) will be solved next (note 4.10) to compare with the exact solution (4.523d).

NOTE 4.10: General Solution of a Finite Difference Equation

The finite difference equation (4.522b) has (subsection 1.9.1) solution of the form (1.444a) ≡ (4.524a) where λ is a root of the characteristic polynomial (4.524b):

$$y_n = \lambda^n: \qquad 0 = \lambda^2 - 2\lambda + \left(1 + h^2\right)\lambda = \left(\lambda - \lambda_+\right)\left(\lambda - \lambda_-\right). \qquad \text{(4.524a–c)}$$

The roots (4.524c) of the characteristic polynomial (4.524b) are (4.525a) leading to the general solution (4.525b) ≡ (4.525c):

$$\lambda_\pm = 1 \pm ih: \qquad y_n = A\left(\lambda_+\right)^n + B\left(\lambda_-\right)^n = A\left(1 + ih\right)^n + B\left(1 - ih\right)^n, \qquad \text{(4.525a–c)}$$

where A, B are arbitrary constants determined from initial conditions. The initial conditions (4.523b, c) for the differential equation (4.523a) correspond (4.511) to (4.526a, b) for the finite difference equation, and imply (4.526c, d):

$$1 = y(0) = y_0, 0 = y'(0) = \frac{y_1 - y_0}{h}: \qquad y_1 = y_0 = 1. \qquad \text{(4.526a–d)}$$

Substituting (4.525c) in (4.526c, d) leads to (4.527a, b) implying (4.527c), which determines the two constants of integration (4.528d, e):

$$1 = y_0 = A + B, \qquad 1 = y_1 = A + B + ih(A - B): \quad ih(A - B) = 0. \quad A = B = \frac{1}{2}.$$
$$\text{(4.527a–e)}$$

Substitution of (4.527d, e) in (4.525b) specifies the solution of the finite difference equation (4.522b) with initial conditions (4.526c, d):

$$y_n = \frac{1}{2}\left[\left(1 + ih\right)^n + \left(1 - ih\right)^n\right]. \qquad \text{(4.528)}$$

Using the binomial expansion (4.518) for the two terms on the r.h.s. of (4.529):

$$(1 \pm ih)^n = \sum_{k=0}^{n} \binom{n}{k} (\pm ih)^k , \tag{4.529}$$

yields by addition:

$$y_n = \sum_{k=0}^{\leq n/2} \binom{n}{2k} (\pm ih)^{2k} = \sum_{k=0}^{\leq n/2} \frac{n!(-)^k}{(2k)!(n-2k)!} h^{2k} , \tag{4.530}$$

involving only even powers of h. It has been shown that *the solutions (problem 158) of the linear differential (4.522a) [finite difference (4.522b)] equations of second-order with constant coefficients are related (4.523a) [(4.525c)] by the discretization (4.512b) with step h; the corresponding initial conditions (4.523b, c) [(4.526c, d)] lead to the exact (4.523d) [approximate (4.528) $\equiv$ (4.530)] solutions, that are compared* next (note 4.11).

NOTE 4.11: Comparison of the Exact and Approximate Solutions

The exact solution (4.523d) of the differential equation (4.522a) with boundary conditions (4.523b, c) is given by (4.531b) at the points (4.531a) separated by the step h:

$$x_n = nh: \qquad \cos(x_n) = \cos(nh) = \sum_{k=0}^{\infty} \frac{(-)^k}{(2k)!} (nh)^{2k} , \tag{4.531a–c}$$

where the power series for the cosine (II.7.14a, b) = (4.531c) was used. *Comparison of the exact (4.531c) [approximate (4.530)] solutions shows the substitution of coefficients (4.532a) [(4.532b)]:*

$$n^{2k} \qquad \leftrightarrow \qquad \frac{n!}{(n-2k)!} = n(n-1)....(n-2k+1). \tag{4.532a, b}$$

The approximate solution (4.530) may be taken to order $h^2 \left(h^4 \right)$ in (4.533a) [(4.534a)]:

$$y_n^{(1)} = 1 - \frac{n(n-1)}{2} h^2 + \varepsilon_n^{(1)} , \qquad\qquad \varepsilon_n^{(1)} \sim O\left(h^4 \right), \tag{4.533a, b}$$

$$y_n^{(2)} = y_n^{(1)} + \frac{n(n-1)(n-2)(n-3)}{24} h^4 + \varepsilon_n^{(2)} , \qquad\qquad \varepsilon_n^{(1)} \sim O\left(h^6 \right), \tag{4.534a, b}$$

with decreasing error (4.533b) [(4.534b)]. The effects of step size h and order of approximation n on the accuracy of the approximate (4.533a, b; 4.534a, b) versus the exact (4.531a–c) solution are considered next (note 4.12).

NOTE 4.12: Effects of Step Size and Order of Approximation

The circular cosine has period 2π in (4.535a), and is skew-symmetric (symmetric) in the whole (4.535b) [half (4.535c)] interval:

$$\cos(x+2\pi)=\cos x, \quad \cos(x+\pi)=-\cos x, \quad \cos\left(\frac{\pi}{2}-x\right)=-\cos\left(\frac{\pi}{2}+x\right),$$

(4.535a–c)

and thus it is sufficient to calculate it in one-quarter of the interval (4.536c) with equal steps:

$$n=0,...,N\equiv 5,10: \quad 0\le x\le\frac{\pi}{2}, \quad x_n=nh, \quad h=\frac{\pi}{2N}=\frac{\pi}{10},\frac{\pi}{20}=0.31416,0.15708.$$

(4.536a–f)

In (problem 159) Table 4.7 (4.8) are used 5 (10) steps (4.536a) [(4.536b)] of length (4.536e) [(4.536f)] corresponding to angles $\theta=18°(\theta=9°)$. Thus Table 4.7 compares at five points the exact solution (4.523d) with the approximate solution (4.530), using the first (second) order (4.533a) [(4.534a)] and the corresponding error (4.533b) [(4.534b)]. The lower (higher) approximation (4.533a) [(4.534a)] applies only for $n\ge 2(n\ge 4)$. The error increases with distance nh from the initial point, and the approximate solutions

TABLE 4.7

Comparison of Exact and Approximate Solutions

n	θ	$x_n=n\pi/10$	$\cos(x_n)$	$y_n^{(1)}$	$\varepsilon_n^{(1)}$	$y_n^{(2)}$	$\varepsilon_n^{(2)}$
0	0°	0.000 00	1.000 00	1.000 00	0.000 00	1.000 00	0.000 00
1	18°	0.314 16	0.951 06	1.000 00	+0.048 94	1.000 00	+0.048 34
2	36°	0.628 32	0.809 02	0.901 30	+0.092 29	0.901 30	+0.092 29
3	54°	0.942 48	0.587 79	0.703 91	+0.116 13	0.703 91	+0.116 13
4	72°	1.256 64	0.309 02	0.407 82	+0.098 81	0.417 56	+0.108 55
5	90°	1.570 80	0.0000	0.013 04	+0.013 04	0.061 74	+0.061 74

Note: The numerical solution of a differential equation leads to errors that: (i) increase with distance from the initial point; (ii) for the same distance from the initial point decrease for higher orders of the step size, e.g., $O(h^2)$ are more generally but not always more accurate that $O(h)$ methods.

TABLE 4.8

Higher Accuracy with Smaller Step Size

n	θ	$x_n = n\pi/20$	$\cos(x_n)$	$y_n^{(1)}$	$\varepsilon_n^{(1)}$	$y_n^{(2)}$	$\varepsilon_n^{(2)}$
0	0°	0.000 00	1.000 00	1.000 00	1.000 00	1.000 00	0.000 00
1	9°	0.157 08	0.987 69	1.000 00	0.012 31	1.000 00	0.012 31
2	18°	0.314 16	0.951 06	0.975 33	0.024 27	0.975 33	0.024 27
3	27°	0.471 24	0.891 01	0.925 98	0.034 97	0.925 98	0.034 97
4	36°	0.628 32	0.809 02	0.851 96	0.042 94	0.852 57	0.043 55
5	45°	0.785 40	0.707 11	0.753 26	0.046 15	0.756 30	0.049 20
6	54°	0.942 48	0.587 79	0.629 89	0.042 10	0.639 02	0.051 24
7	63°	1.099 56	0.453 99	0.481 85	0.027 86	0.503 15	0.049 16
8	72°	1.256 64	0.309 02	0.309 13	0.000 11	0.351 74	0.042 73
9	81°	1.413 72	0.156 43	0.111 74	−0.044 70	0.188 45	0.032 01
10	90°	1.570 80	0.000 00	−0.110 33	−0.110 33	0.017 52	0.017 52

Note: The accuracy of a numerical method improves from a larger (Table 4.7) to a smaller (Table 4.8) step size. The smaller step size requires a larger number of steps to cover the same range and leads to a larger number of computations that requires more digits to preserve accuracy.

do not reproduce accurately the zero at the end point. The second-order approximation may or may not lead to greater accuracy than the first-order approximation. For example, in Table 4.7, with five steps of (4.536e) corresponding to angles of $\theta = 18°$: (i) the first (second)-order approximation (4.533a) [(4.533b)] applies only for $n \geq 2 (n \geq 4)$; (ii) thus for $n = 1, 2, 3$ the error is the same in the first and second order approximations; (iii) for $n = 4$ the first-order approximation is more accurate, because the second-order increases the error; (iv) for $n = 5$ the second-order approximation reduces the error of the first-order approximation, through both fail to give the exact value zero. In general, a higher-order approximation gives higher accuracy, but this may not be the case at every point from one order of approximation to the next. Similarly, the accuracy improves with smaller step size, for example halving the step size in Table 4.8; this doubles number of steps needed to cover the same interval. The second-order approximation (4.534a) adds a positive term to the first-order approximation (4.534b) and thus does not always improve accuracy. This was already shown with regard to the five steps in Table 4.7 and also applies to the ten steps in Table 4.8; that is, the second-order approximation (4.534a) relative to the first-order approximation (4.533a): (i) degrades the accuracy by a small amount when the first-order approximation is an overestimate for $n \leq 8$; (ii) improves the accuracy when the lower order approximation is an underestimate for $n \geq 9$.

NOTE 4.13: **Comparison of Truncation and Discretization Errors**

The exact analytical solution (4.531c) of the differential equation may be approximated to the lowest (4.537a) [next (4.538a)] order with error (4.537b) [(4.538b)]:

$$y^{(1)}(nh) = 1 - \frac{n^2h^2}{2} + \delta^{(1)}, \qquad \delta^{(1)} \sim O(h^4), \qquad \text{(4.537a, b)}$$

$$y^{(2)}(nh) = y^{(1)}(nh) + \frac{n^4h^4}{24} + \delta^{(2)}, \qquad \delta^{(2)} \sim O(h^6). \quad \text{(4.538a, b)}$$

The truncation error (problem 160) of the exact solution is indicated in Table 4.9 for the same steps (4.536a, c, d, e) as the direct numerical solutions of the differential equation with lowest (4.533a, b) [next order (4.534a, b)] accuracy in Table 4.7. The comparison of the approximations to the exact analytical solution (Table 4.9) [with the approximations to the direct numerical solution (Table 4.7)] shows that: (i) the lowest order approximation underestimates (overestimates) the exact value, with a smaller (larger) error; (ii) the second-order correction is positive in both cases, and thus mostly improves (degrades) the accuracy. An approximation, either analytic or numerical, becomes less accurate far from the starting point, and may be unsuitable for large x; this can be avoided by using the periodicity or symmetry properties of the solution (4.535a–c) to map x back to the range $0 \le x \le \pi/2$ where accuracy is better. If the solution is not periodic and symmetry properties do not apply, then an asymptotic solution for large x may be necessary. At a given point x in general a (α) numerical approximation of the exact analytic solution of the differential equation is more accurate to the same order than (β) the direct numerical

TABLE 4.9

Numerical Approximations of the Analytical Solution

n	θ	x_n	$\cos(x_n)$	$y^{(1)}(x_n)$	$\delta_n^{(1)}$	$y^{(2)}(x_n)$	$\delta_n^{(2)}$
0	0°	0.000 00	1.000 00	1.000 00	0.000 00	1.000 00	0.000 00
1	18°	0.314 16	0.951 06	0.950 65	−0.000 41	0.951 06	0.000 00
2	36°	0.628 32	0.809 02	0.802 61	−0.006 41	0.809 10	0.000 08
3	54°	0.942 48	0.587 79	0.555 87	−0.031 92	0.588 74	0.000 96
4	72°	1.256 64	0.309 02	0.210 43	−0.098 59	0.314 33	0.005 32
5	90°	1.570 80	0.000 00	−0.233 70	−0.233 70	0.019 97	0.019 97

Note: An exact analytical solution approximated or truncated numerically (Table 4.9) is generally more accurate that a direct numerical solution of the differential equation (Table 4.7), when compared for the same step size h and order $O(h^n)$.

TABLE 4.10

Numerical Solution versus Truncated Analytical Solution to $O(h^4)$

n	$x_n = nh$	y_n	$\cos(x_n)$
0	$x_0 = 0$	$y_0 = 1$	$\cos(0) = 1$
1	$x_1 = h$	$y_1 = 1$	$\cos(h) = 1 - \dfrac{h^2}{2} + \dfrac{h^4}{24}$
2	$x_2 = 2h$	$y_2 = 1 - h^2$	$\cos(2h) = 1 - 2h^2 + \dfrac{2h^4}{3}$
3	$x_3 = 3h$	$y_3 = 1 - 3h^2$	$\cos(3h) = 1 - \dfrac{9h^2}{2} + \dfrac{81h^4}{24}$
4	$x_4 = 4h$	$y_4 = 1 - 6h^2 + h^4$	$\cos(4h) = 1 - 8h^2 + \dfrac{32}{3}h^4$

Note: Comparison of analytical formulas for the first five steps of: (i) numerical finite difference approximation that involves terms up to $O(h^4)$; (ii) truncation at the same order of the exact power series solution.

solution of the differential equation. Thus: (i) the numerical approximation of the exact analytical solution (α) is preferable if it is available; (ii) if the former (α) is not available then a direct numerical solution (β) must be used. Table 4.10 compares for the first five points: (i) the numerical solution (4.534a) that involves only terms to $O(h^4)$; (ii) the first three terms (4.538a) of the exact analytical solution that is truncated at the same $O(h^4)$. The higher accuracy of the (ii) truncated analytical solution versus (i) the numerical approximation involves algebraic coefficients that are less simple.

Conclusion 4

The integral curves of the first-order differential equation, with a singularity of first degree, represent (i) the paths in a plane motion, in the neighborhood of a stagnation point of first degree or (ii) the trajectories resulting from the composition of two identical orthogonal generalized oscillators (Table 4.1). There are three cases (i–iii) and seven sub-cases (I–VII) as follows (Figure 4.1): (i) for critically damped oscillators, or paths with one principal line, the singularity is (I) an inflexion point (a); (ii) for oscillators with over-critical damping, or paths with two principal lines, the singularity may be a (II) nodal point (b), a (III) focus (c), a (IV) simple point (d) or (V) a saddle-point (e); (iii) for oscillators with no (sub-critical) damping or imaginary principal lines, the singularity is (VI) a center (f) [(VII) an asymptotic point (g)]. The composition of two harmonic oscillators leads to a path (Figure 4.2) that is: (i) a straight line if they are in phase (a); (ii) an ellipse if they are out-of-phase (b); (iii) in all other cases (c) of intermediate phase the ellipse has rotated axis (Figure 4.3). The parametric resonance applies to a harmonic oscillator with vibrating support (Figure 4.4), such as the small oscillations of a pendulum whose suspension point oscillates vertically (Figure 4.6). The parametric resonance (Figure 4.5): (i) occurs for excitation frequency in a narrow range around the frequency of the first harmonic and its submultiples (a) with decreasing amplitude; (ii/iii) in the presence of damping the range of excitation frequencies is narrower (b) and there is an increasing threshold amplitude of excitation (c). The parametric resonance (Table 4.2) is distinct from ordinary resonance either linear (Table 4.3) or non-linear (Table 4.4).

A non-linear restoring force (Figure 4.7) associated with a potential (a), in the absence of damping and forcing, leads (b) either to closed limit cycles or open paths. For example, (Figure 4.8a) the hard spring (b) has limit cycles only (c), whereas the soft spring (d) has limit cycles (open paths) between (outside) the (e) separatrices. The soft or hard spring (Figure 4.8) corresponds to a cubic term in the non-linear restoring force, and to a quartic term in the potential, preserving the equilibrium at the origin (Figure 4.9b); a quadratic term in the non-linear resting force corresponds to a cubic term in the potential (Figure 4.9a) and the inflexion at the origin shifts the average position towards the lower potential. For a damped (Figure 4.10) non-linear oscillator forced close to the natural frequency or half of it, there is non-linear resonance at the natural frequency with amplitude increasing towards higher frequencies as the amplitude of forcing increases (a–c); if the forcing amplitude exceeds a threshold there are amplitude jumps and a hysteresis loop (d). In the case of forcing at the first (second) harmonic [Figure 4.14 (4.15)] the non-linear resonance at the natural frequency may have amplitude jumps (thresholds) as well as frequency ranges of suppression of oscillations. The limit cycle (Figure 4.10d) described in opposite directions either extracts (or provides) oscillation. An example of a limit cycle with growing oscillation is the "flutter" of a control surface of an aircraft (Figure 4.11a) subject to non-linear aerodynamic forces. The oscillation grows until the control surface hits the bump stops with loss of control, structural damage, or both. The flutter may be avoided by reducing the aerodynamic forces by flying at lower speed, thereby reducing (dotted line) the flight envelope (solid line) of the aircraft in terms of accessible combinations of velocity v and altitude z. The "breaking wave" sequence (Figure 4.10a–d) of non-linear resonance is analogous to water waves (Figure 4.12): (a) in the deep ocean the waves are linear because their amplitude is small compared with the depth; (b) when approaching a sloping beach the amplitude is no longer small compared with depth, and the waves steepen and break. Another example of non-linear wave is (Figure 4.13) a sound wave of large amplitude in a compressible medium: (a) the wave speed is larger (smaller) in the compression (rarefactions) phase; (b) thus the compression front advances relative to the lagging rarefaction tail until a shock wave forms.

The self-excited dynamo (Figure 4.16) in co(counter)rotation has a diverging (decaying) electric current, whereas a (Figure 4.17) self-excited electromechanical dynamo has a velocity tending to a finite non-zero (zero) value asymptotically for long time. The finite asymptotic solution or non-zero steady state occurs in the presence of: (i) linear amplification that prevents decay for small amplitude; or (ii) quadratic damping that prevents growth for large amplitude. The result is a steady state v_+ at an intermediate amplitude (Figure 4.17). In the case of linear and quadratic damping, decay to zero v_- is inevitable. The asymptotic non-zero steady state v_+ (Figure 4.17) of an electromechanical dynamo (Figure 4.16) may be analogous to magnetic field generation in planets and stars due to the combination of rotation with ionized fluids containing positive and negative electric charges. An example is the magnetic fields of the earth and sun and their interaction (Figure 4.18) in the solar-terrestrial system. A different example of non-linear damping (Figure 4.20) concerns the quadratic damping (a) of a harmonic oscillator (b). A case of quadratic damping and large amplitude motion (Figure 4.21) is the circulating pendulum (Figure 4.19d), passing several times through the highest (Figure 4.19c) and lowest (Figure 4.19b) positions, until it oscillates (Figure 4.19a) and finally comes to rest at the lowest position of stable equilibrium. In the absence of damping there are three possible cases of motion (Figure 4.19) of a circular pendulum (Table 4.5): (i) oscillation (a) around the lowest point (b) if it cannot reach highest point (c); (ii) circulation or revolution (d) if it passes through the highest point (c) with non-zero velocity; or (iii) the intermediate case between (i) and (ii), which is stoppage at the highest point (c). All the preceding linear and non-linear oscillation, stability or instability problems involve the solution of differential equations (Table 4.6), for example by exact or approximate methods (Tables 4.7 to 4.10).

The preceding examples of non-linear mechanical and electrical illustrate some properties of dynamical systems with bifurcations (Figure 4.22). The theorem on the stability boundaries (Figure 4.23) shows that (a) upward (downward) facing segments are stable (unstable), indicated by solid (dashed) lines (b); four examples of the application of the theorem on stability boundaries are given (Figure 4.24), including suppression of oscillations (a, b), or amplitude jumps with an hysteresis loop (c, d). The super(sub)critical Hopf bifurcation (Figure 4.25) involves [a(b)] a stable (unstable) spiral changing to an unstable (stable) limit cycle through an unstable (stable) center, with the sequence taken in opposite direction. A mixed non-linear or van der Pol oscillator with linear damping depending non-linearly on the displacement has a subcritical Hopf (fold or cusp) bifurcation [Figure 4.25a (4.26)] for weak (strong) damping. The hysteresis loop (Figures 4.10d, 4.24d and 4.26) also appears (Figure 4.28) in the cusp bifurcation (a), corresponding to a surface in three dimensions with a double fold (b); the fold lines are projected on the plane as two curves converging at a cusp (c). The fold (cusp) bifurcation [Figure 4.28 (4.27)] corresponds to a quadratic (cubic) term in the non-linear restoring force, and a cubic (quartic) term in the potential. The cusp bifurcation or catastrophe (Figure 4.27) is illustrated by the potential (a), location of equilibrium points (b), their stability (c) and value of the potential there (d).

The iterated map (Figure 4.29) consists of the intersections of the integral curves by the directrix, that is a straight line passing through an isolated singularity. It can indicate (a) deterministic motion (b) or chaos (c). The scenarios of transition of a laminar to turbulent flow may involve an infinite sequence of bifurcations, or just three or possibly two or one; in the case of three Hopf (Figure 4.25a, b) bifurcations (Figure 4.30) at successive times (t_1, t_2, t_3) each adds a frequency $(\omega_1, \omega_2, \omega_3)$ and they may be: (i) all three visible (a); (ii) just two visible (b); or (iii) only one visible (c). The flow separation over the wing of an airplane (Figure 4.31); that is, the transition from (a) laminar to (b) turbulent flow, may lead to (c) departure from controlled flight and (d) entry into a hazardous spinning flight condition. The problem of competing populations (Figure 4.32), leads to a limit-cycle oscillation (a) with time delay (b). The index (Figure 4.33) of a singularity (a) corresponds to the winding number measuring the number of times the tangent vector to the integral curves goes around the singularity (b); the indices of isolated singularities can be added (c), including the singularity at infinity (d). The examples of the index of singularities and equilibrium points include (Figure 4.34), besides the ordinary point (h), viz. the quadrupole (a) [dipole (b)] [center $\equiv$ vortex (d)], tangent (e) [inflection (f)] point, and spiral (g) [saddle (i)] points. The qualitative theory of differential equations is a complement (Table 4.6) to exact and approximate analytical solutions and to numerical methods. The numerical methods are based on discretization (Figure 4.35) of derivatives of first (a), second (b), and higher orders. The numerical approximation of exact analytical solutions (Table 4.9) is more accurate than direct numerical methods (Table 4.7); the accuracy of the latter improves (Table 4.8) for smaller step size.

Bibliography

The bibliography of *"Non-linear Differential Equations and Dynamical Systems"*, that is the second book of volume IV *"Ordinary Differential Equations with Applications to Trajectories and Oscillations"*, and sixth book of the series *"Mathematics and Physics Applied to Science and Technology"*, adds the subject of "Non-linear Differential and Integral Equations". The books in the bibliography that have influenced the present volume the most are marked with one, two, or three asterisks.

Non-Linear Differential and Integral Equations

*Andronov, A., Vitt, A. and Khaikin, S.E. *Theory of oscillators*. Dover 1966, New York.

Boglioubov, N. and Mitropolski, I. *Les methodes asymptotiques en théorie des oscillations non-lineaires*. Gauthier-Villars 1962, Paris.

Cordeanu, C. *Differential and integral equations*. 1969, 2nd edition 1971, reprinted Chelsea 1977, New York.

*Davis, H. T. *Non-linear differential and integral equations*. United States Atomic Energy Commission 1960, reprinted Dover 1962, New York.

*Frank, P. and von Mises, R. *Die Differential- und integralgleichungen*. Friedrichs Vieweg 1930, reprinted Dover 1961, 2 vols., New York.

Galaktionov, V.A. *Geometric Sturmian theory of non-linear parabolic equations and applications*. Chapman & Hall 2004, London.

*Gilmore, R. *Catastrophe theory for scientists and engineers*. Wiley 1981, New York, reprinted Dover 1993, New York.

Gorenflo, R. and Vessala, S. *Abel integral equations*. Springer 1991, Heidelberg.

Hiorns, R.W. and Cooke, D. *The mathematical theory of the dynamics of biological population*. Academic Press 1981, New York.

*Hirsch, M. and Smale, S. *Differential equations, dynamical systems, and linear algebra*. Academic Press, 1974, New York.

Hochstadt, H. *Integral equations*. Wiley 1973, New York.

Kanwall, R.P. *Linear integral equations:theory and technique*. Academic Press 1971, New York.

Krasnov, M. L., Kisselev, A. I. and Makarenko, G. I. *Equations integrales*. Nauka 1976, Editions Mir 1977, Moscow.

Lefschetz S. L. *Differential equations: geometric theory*. Dover, 1963, New York.

Lovit, W. V. *Linear integral equations*. McGraw-Hill 1924, New York, reprinted Dover 1950.

Meiss, J. D. and Mackay, R. S. *Hamiltonean systems*. Adam Hilger 1957, New York.

Polynin, A.D. & Manzhinov, A.V. *Handbook of integral equations*. CRC Press 1998, Boca Raton, Florida.

*Polyamin, A.D. and Zaitsev, V.F. *Handbook of non-linear partial differential equations.* Chapman & Hall 2004, London.

Reissig, R., Sansone, G. and Conti, R. *Non-linear differential equations of higher order.* Noordhoff 1969, Dordrecht.

Richards, P. I. *Manual of mathematical physics.* Pergamon 1959, Oxford.

Saaty, T. L. and Bram, J. *Non-linear mathematics.* McGraw-Hill 1964, New York.

Volterra, V. *Theory of functions and integral and integro-differential equations.* Blackie, London, reprinted Dover 1959, New York.

References

1695   Bernoulli, J. *Acta Erudita* **59–67**, 537–557.

1697   Bernoulli, J. "Problema pure geometricum Eruditis propofisum: De conoitibus spheroidibus Quadam. *Acta Evuditorium* 113–124.

1724   Ricatti, J.F. "Animadversiones in equationes diferentiales secundi gradus". *Acta Erudita Supplementa* **8**, 66–75.

1763   D'Alembert, J.R. *Histoire de l'Academie de Berlin*, 242.

1873   Mathieu, E. *Cours de Physique Mathematique.* Gautiers-Villars, Paris.

1881–6 Poincaré, H. Mémoire sur les courbes défenies par une equation differentielle. *Journal de Mathematiques* **7**, 375–422; **8**, 251–296; **1**, 167–244; **2**, 151–217.

1883   Floquet, G. *Annales Scientifiques de l'École Normale Superieure* **12**, 47–88.

1883   Floquet, G. Sur les équations differentielles à coefficients periodiques. *Annales Scientifiques de l'École Normale Supérieure* **12**, 14–46.

1922   Pol, B. van der On oscillation hyrterisis in a triode generator with two degrees of freedom. *Philosophical Magazine* **43**, 700–719.

1931   Voltena, V. *Leçons sur la theorie mathematique de la lute pour la vie.* Gauthier-Villars, Paris.

1943   Hopf, E. Abzweigung einer periodischen Lösung von einer stationären Lösung eines Differentialsystem. *Akademie der Wissenschaten Leipzig, Mathematische – Naturwissenschaften Klasse* **95**, 3–22.

1971   Ruelle, D. & Takens, F. On the nature of turbulence. *Communications in Mathematical Physics* **20**, 167–192.

Index